MW01517689

AGING: DECISIONS AT THE END OF LIFE

INTERNATIONAL LIBRARY OF ETHICS, LAW, AND THE NEW MEDICINE

Editors

VOLUME 12

The titles published in this series are listed at the end of this volume.

AGING: DECISIONS
AT THE END OF LIFE

Edited by

DAVID N. WEISSTUB

Philippe Pinel Professor of Legal Psychiatry and Biomedical Ethics,
Faculté de Médicine, Université de Montréal, Canada

DAVID C. THOMASMA

Professor and Fr. English Chair of Medical Ethics,
Neiswanger Institute for Bioethics and Health Policy,
Loyola University of Chicago Medical Center, U.S.A.

SERGE GAUTHIER

Professor of Neurology, Neurosurgery and Psychiatry, Centre for Studies on Aging,
Faculty of Medicine,
McGill University, Canada

and

GEORGE F. TOMOSSY

Ross Waite Parsons Scholar,
Faculty of Law,
University of Sydney, Australia

KLUWER ACADEMIC PUBLISHERS
DORDRECHT / BOSTON / LONDON

A C.I.P. Catalogue record for this book is available from the Library of Congress.

ISBN 1-4020-0182-7
ISBN 1-4020-0183-5(Set)

Published by Kluwer Academic Publishers,
P.O. Box 17, 3300 AA Dordrecht, The Netherlands.

Sold and distributed in North, Central and South America
by Kluwer Academic Publishers,
101 Philip Drive, Norwell, MA 02061, U.S.A.

In all other countries, sold and distributed
by Kluwer Academic Publishers,
P.O. Box 322, 3300 AH Dordrecht, The Netherlands.

Printed on acid-free paper

TABLE OF CONTENTS

SUICIDE

RESEARCH ETHICS

PREFACE

Decisions at the End of Life is the last volume in this trilogy on *Aging* conceived for the *International Library of Ethics, Law, and the New Medicine.* In the first book, *Culture, Health, and Social Change,* our authors contested some of the historical and modern paradigms on aging, demonstrating the heterogeneity of the elderly population and corresponding impact on social health policy formulation. This theme continued in the second volume, *Caring for Our Elders,* which canvassed a range of issues concerning ethical and socio-legal structures influencing elder care. Some of the most emotive topics in the study of aging are addressed in this final set of reflections: quality of life, palliative care, euthanasia, suicide, and research ethics.

QUALITY OF LIFE

The artificiality and varied success of attempts to measure quality of life once again underscores the highly subjective nature of the human experience of aging. Nevertheless, "quality of life" is a concept that lies at the heart of our ethical vocabulary when debating issues affecting individuals in the final stage of life, and serves as an anchor for social policy decisions and judicial determinations.

Peter J. Whitehouse, Jesse Ballenger, and Sid Katz provide us with a unique perspective: "Quality of life can be measured objectively by others based upon mental state, income, housing, and social relationships, but it also includes an internal subjective sense of well-being that has wisdom associated with it." They refer not only to the cognitive aspect that underlies wisdom as an integrative process, but also to its relation to social and ethical commitments. Enhancing individual and collective wisdom is a critical goal for individual enhancement of quality of life, as well as the health of communities and the planet as a whole.

Fernando Lolas identifies "quality of life" as a concept that has several dimensions: subjective, multidimensional, complex, and dynamic. Taken together, these factors demonstrate why bioethics discourse on aging "is neither uniform nor unitary." Indeed, as Prof. Lolas describes, solutions to problems in this area are particularly elusive due to the character of *change* inherent in the aging process. Social and individual definitions rarely coincide, particularly when responding to end-of-life issues. This is not to imply that further discussion is without merit. Rather, he advocates a "proactive stance of a dialogic bioethics" that would serve as a "bridge between rationalities, disciplines, and people...between ages and generations."

PALLIATIVE CARE

"To live well, a person must know how to die well." So begins Linda L. Emanuel's reflection on palliative care. She describes how the quality of end-of-life care can yield benefits that go beyond the immediate experience of the patient. Regrettably, "a disorientation of professional values has allowed impoverished care for the dying." Prof. Emanuel explains how hospice and palliative care has emerged as a vital social component of the response to mortality, and that with their "movement into mainstream medicine and society, can be a source for our reorientation."

ASSISTED DEATH

Recent legislative developments in the Netherlands have again brought the issues of euthanasia and assisted suicide to the forefront of ethical, legal, and health policy debates. Gerrit K. Kimsma provides us with an overview of the Dutch position on physician-assisted death. He underlines that the practice was considered to be a crime *unless* performed by a physician complying with specific conditions established by law, which prescribe both substantive and procedural safeguards to prevent abuse. His exposé shows that the decisions that led to the current legal state of affairs were carefully considered, based upon extensive studies into the nature of medical decisions at the end of life. Prof. Kimsma addresses some major debates against the euthanasia position, such as the "slippery slope" argument, noting that the Netherlands is the only country in the world that has produced reliable data that refutes concerns about any suggestion of danger to the elderly from unwanted death. By no means does he advocate adopting a stance of complacency; rather, there is "a need for alertness." But for the moment, he argues that "the limits of what physicians should do have been reached and it seems to be time for a 'time out' and a period of evaluation of the present modes of ending lives."

Henk Jochemsen, on the other hand, argues that euthanasia in the Netherlands should *not* have been legalized, on both ethical and legal grounds. The author also provides a thorough background of the legal developments that have led up to the recent legislation changes. For example, he argues that the reporting procedure in the earlier system was flawed, and belied "any claim of effective regulation, scrutiny, and control." The recent amendments will not lead to a far higher rate of reporting, and in "precisely those cases in which the requirements have not been fulfilled." Prof. Jochemsen's medical and social/ethical arguments are rooted in "a concern for a health care and a society in which all human beings, independent of their capacities and condition are protected, respected, and cared for."

On a related theme, George P. Smith, II, examines what he describes as "the decisions and charades, or symbolic actions and pretensions, that the medical profession follows in death management – primarily to avoid legal consequences for negligence, malpractice, or homicide, but very often also because there are inappropriate guidelines for suitable treatment of terminal illnesses." He describes, for example, the use of resuscitative codes given by physicians to be taken when a patient suffers a cardiac arrest, leading to full resuscitation in inappropriate cases. Historically driven practices thus can sometimes overbear current clinical realities. Prof. Smith argues for the need to re-evaluate "the ethical validity, medical propriety, and legal correctness of terminal sedation as a normative standard of humane conduct and palliative treatment at the end of life."

Roger S. Magnusson draws upon interviews that he conducted with healthcare workers involved with informal HIV/AIDS-related euthanasia networks in Australia. Drawing upon their broader experiences, which extended to cancer and geriatric care, he explores what he refers to as the "aging and euthanasia policy cocktail." His discussion is divided into two areas: "bedside" issues, which "relate to the patient's subjective experience of illness and his or her reasons for wishing to die," and "systemic" issues concerning "the nature of the health care system within which suicide talk takes place, the broader cultural and legal environment, and the possible flow-on effects of a right to die." After a detailed treatment of these issues, Dr. Magnusson takes the cautious position of advocating legalization of euthanasia, but only within a heavily regulated statutory framework.

SUICIDE

Understanding suicide, from either a mental health or societal perspective, is a monumental and worthy endeavor that still requires extensive study. A question that one might ask, for example, is whether suicide among the elderly is more common in jurisdictions where euthanasia is unavailable. Lorraine Sherr and Fabrizio Starace emphasize the importance of considering "the total context of end of life and the gray area between suicide and euthanasia." The authors provide a detailed review of the literature that attempts to link terminal illness and end-of-life acts, including suicide and euthanasia. While the link seems clear in certain illnesses, this is not uniformly the case. A clearer understanding is needed in order to develop a successful societal response.

Naoki Watanabe, Manabu Taguchi, and Kazuo Hasegawa, on the other hand, relate the results of a specific study in which they contrast rates of suicide among elderly persons in different regional settings in Japan. They conclude that differences do exist between the individuals in two regions in terms of family composition and life style, quality of life and depression, and levels of social support.

In broad terms, the authors of both chapters illustrate the complexity of inter-related factors that can contribute to end-of-life acts.

RESEARCH ETHICS

The final three chapters in this volume are devoted to the topic of research ethics. While the question of involving elderly subjects in research is not strictly an "end-of-life" issue, the objective of medical research on conditions affecting the latter stage of an individual's life is invariably directed towards either improving quality of life or postponing death. In that sense, the motivations underlying the desire to conduct medical research draw upon concerns related to each of the topics canvassed thus far in this volume.

Moral and ethical discourse on the conduct of research with elderly subjects is complex and offers no clear-cut solutions, particularly in those cases where the ability of the individual to make autonomous decisions concerning participation in research is in question. Beginning with the classic work of Hans Jonas, Arthur Schafer provides a thorough analysis of the arguments in favor and against doing research with this population. He addresses problems of obtaining informed consent and the role of surrogate decision-makers who are faced with the dilemma of enrolling persons under their care in research protocols that may, ultimately, present unknown risks. As Prof. Schafer describes, charting a balanced path between societal and individual interests is not a simple task, but one that must be attempted.

Part of the solution to ensuring that this balance is achieved is proposed by David C. Thomasma, who advocates the vital role to be played by community consent in the decision-making process that determines whether or not vulnerable persons should be enrolled in research. Prof. Thomasma argues that surrogate decision-makers representing the community of caregivers "must be involved in the consent process, because they have an inherent interest in the outcome of the research."

Finally, George F. Tomossy, David N. Weisstub, and Serge Gauthier, argue for the enhancement of regulatory controls to ensure that only *ethical* research is conducted in a given jurisdiction, irrespective of whether the research is funded by private or public sources. The authors recommend that effective regulation must include independent oversight and enshrine fundamental principles within a statutory framework.

The conclusions that can be drawn from these three contributions include recognition of the importance of research on conditions or illnesses affecting elderly persons, but whose rights cannot be blindly sacrificed in the name of societal objectives. Special efforts need to be taken in order to ensure that adequate safeguards are in place, including alternative structures for providing consent on behalf of persons with impaired decision-making ability, and the implementation of appropriate regulatory mechanisms.

David N. Weisstub, David C. Thomasma, Serge Gauthier, and George F. Tomossy.

Montreal, Chicago, and Sydney, September 2001.

ACKNOWLEDGMENTS

We wish to thank each of our esteemed authors who contributed their collective wisdom from across many disciplines to this volume. We also extend our deepest gratitude to those who generously participated in the peer review process, and thereby provided authors with the benefit of their insights. We are deeply indebted to Robbin Hiller and Diane Kondratowicz, both of the Neiswanger Institute for Bioethics and Health Policy at Loyola University Chicago Medical Center, for their assistance in editing and preparing this book for publication. Finally, we thank our publishing editor at Kluwer Academic Press, Anne Ultee, and her staff, Helen van der Stelt and Nellie Harrewijn, for their constant encouragement. Financial support for this project was graciously provided by Aventis Canada, Novartis Canada, Bayer Canada, and Janssen-Cilag Canada, under the auspices of the Centre for Studies on Aging, McGill University, and the Chaire de psychiatrie légale et d'éthique biomédicale Philippe Pinel, Faculté de médecine, Université de Montréal.

CONTRIBUTORS

JESSE BALLENGER
Instructor, Dept. of History, Case Western Reserve University, U.S.A.

LINDA L. EMANUEL
Buehler Professor of Medicine and Director, Buehler Center on Aging,
Northwestern University, U.S.A.

SERGE GAUTHIER
Professor of Neurology, Neurosurgery and Psychiatry, Centre for Studies on Aging,
Faculty of Medicine, McGill University, Canada.

KAZUO HASEGAWA
Professor Emeritus, School of Medicine, St. Marianna University Tokyo, Japan.

HENK JOCHEMSEN
Lindeboom Chair of Medical Ethics, Free University of Amsterdam, and Director,
Prof. dr. G.A. Lindeboom Institute, Ede, The Netherlands.

SID KATZ
Emeritus Professor of Geriatric Medicine, Columbia University, New York, U.S.A.

GERRIT K. KIMSMA
Associate Professor, Department of Philosophy and Medical Ethics, Free University
Medical Center, Amsterdam, The Netherlands.

FERNANDO LOLAS
Professor of Psychiatry, University of Chile, Santiago, Chile.

ROGER S. MAGNUSSON
Senior Lecturer, Faculty of Law, University of Sydney, Australia.

ARTHUR SCHAFER
Professor of Philosophy and Director, Centre for Professional and Applied Ethics,
University of Manitoba, Canada.

LORRAINE SHERR
Reader in Clinical and Health Psychology, Royal Free & University College
Medical School, London, United Kingdom.

GEORGE P. SMITH, II
Professor of Law, The Catholic University of America, U.S.A.

FABRIZIO STARACE
Director, Consultation Psychiatry & Behavioral Epidemiology Service, Cotugno Hospital, Naples, Italy.

MANABU TAGUCHI
Research Scientist, School of Medicine, St. Marianna University Tokyo, Japan.

DAVID C. THOMASMA
Professor and Fr. English Chair of Medical Ethics, Neiswanger Institute for Bioethics and Health Policy, Loyola University of Chicago Medical Center, U.S.A.

GEORGE F. TOMOSSY
Ross Waite Parsons Scholar, Faculty of Law, University of Sydney, Australia.

NAOKI WATANABE
Associate Professor, School of Medicine, St. Marianna University Tokyo, Japan.

DAVID N. WEISSTUB
Philippe Pinel Professor of Legal Psychiatry and Biomedical Ethics, Faculté de médecine, Université de Montréal, Canada.

PETER J. WHITEHOUSE
Professor, School of Medicine, Case Western Reserve University, U.S.A.

CHAPTER ONE

PETER J. WHITEHOUSE, JESSE BALLENGER & SID KATZ

HOW WE THINK (DEEPLY BUT WITH LIMITS) ABOUT QUALITY OF LIFE

The Necessity of Wisdom for Aging

All human beings age, and it seems intuitive to claim that they wish to do so wisely and with a high quality of life. Yet, wisdom and quality of life are problematic concepts. Despite attempts to develop some objective, quantifiable measures, both concepts have an ineluctably subjective element. But however difficult it is to deal with these concepts with scientific precision, they are nonetheless essential to human beings of all ages and, over the past decade, there have been renewed attempts in various fields to bring them into scientific and scholarly discourse. Quality of life has emerged as a major topic in gerontology, only recently, though integrating the concept into clinical practice and public policy has been challenging. Wisdom has been in decline in academics for decades, though works have appeared over the last decade in psychology, (Sternberg 1990) philosophy, (Lehrer 1996) and management (Srivastva and Cooperrider 1998) calling for a renewal of wisdom as a central concept. This paper examines the complex relationship between aging, wisdom, and quality of life, offering a framework for an integrative model. Suffering and loss, which can threaten quality of life in old age, can also be essential for the development of wisdom. Wisdom in turn may be the foundation for experiencing a renewed quality of life as we age.

We claim in this essay that the *concepts* of aging, wisdom, and quality of life "evolve" historically and culturally. We can examine through time the change in pattern of word usage. Our claim, for example, is that the concept wisdom has been out of favor in academic disciplines. There is an intimate relationship between word and concept. The neurolexical entries of the words "wisdom" and "quality of life" include vast networks of connections to other parts of the mind/brain that involve various memories and sensory modalities. Such abstract words can be more or less idiosyncratic or shared with others. Rich words have vast power to promote health

D.N. Weisstub, D.C. Thomasma, S. Gauthier & G.F. Tomossy (eds.), Aging: Decisions at the End of Life,
1-19.
© *2001 Kluwer Academic Publishers. Printed in the Netherlands.*

(or illness) of human organisms. Let us start our inquiry about aging, wisdom, quality of life, and their interactions with this caveat about limits

Aging, wisdom, and quality of life are to a limited extent about limits. Academic discourse on any, but especially these topics, should begin by acknowledging these limits, and defining the boundaries of inquiry. Knowing what is omitted – knowingly or unknowingly – by one's self or ourselves (or especially what might be unknowable), is paradoxical knowledge. Yet, such information used wisely is essential for survival of life.

What are the limits as we see them of this paper? First, we set some specific limits on content. Much of our paper may be characterized as the shared intuitions emerging from a collaborative writing project. For example, religion and spirituality are critically associated with wisdom and quality of life but we can only offer some sketchy thoughts about this content area. We are also limited in that we deal with abstract constructs. Practical wisdom and quality of life, as concepts, are best learned from stories of human beings living in nature.

Second, a discussion of limits should include aging, quality of life, and wisdom themselves, as each has a sense of limits included in the concepts. Aging is limited by birth and death, at least biologically; quality of life is limited by the resources of the planet and the needs of our fellow human beings; and finally, wisdom is limited by the claim that wisdom recognizes itself as an ever evolving, perhaps ultimately, unachievable state. An academic discourse must be about balance as well as about limits. Here we led with wisdom – our ability to appreciate vitality. The life-course of the essence (or spirit) of each of us is on a temporal trajectory that we call aging. Aging is something we all do, not just the old. Elders serve to guide the following generations. Intergenerational learning is critical.

Our lives have diverse QUALYS associated with them. But it is our integrated knowledge and values, i.e. wisdom that will be essential for our survival as a species. Without wisdom we cannot make difficult but necessary individual and social decisions to ensure health of the biosphere. Without wisdom we will be the wastrels of our own production.

A third set of limits has to do with our concept of an integrative model of aging, wisdom, and quality of life. Models in science take many forms from very detailed mechanistic-computational models to grand schemes linking general concepts. Our model (see Figure 1 at p. 16) is simple and should therefore be distrusted. It is also incomplete but, more importantly, prelusive and inceptive. In the spirit of limits, we are not claiming to have an integrative model, nor perhaps even that we know what a model is. Perhaps our metaphor of building a model includes a sense of the geographic context of the ideas and the manner by which a building scaffold can be raised. However, process must accompany structure. Change can occur through rewards and happiness or suffering and pain. Change can enhance or threaten quality of life in old age. Yet, change can be essential for the development of wisdom. Wisdom in turn may be the foundation for experience of renewed quality of life as we age. We will present the variables in relationships with each other in the model in several different ways, including using a diagram (Figure 1). None of these manners of expression is adequate to capture the complexities that we are trying to include in our modeling process. Nevertheless, we hope we can lead the reader with an intuition about what is emerging from our own evolving thought processes.

Finally, the particular perspectives of the authors limit this paper. All of us have as one of our lenses on life the care of older individuals. The three authors include interdisciplinary scholars trained in geriatrics, neurology, psychiatry, ethics, and the humanities. We have all worked to understand social attitudes toward aging and various medical conditions affecting the elderly, including Alzheimer disease.

As academics we celebrate intellectual activity – but what activity and to what end? Many academics are by label Doctors of Philosophy (i.e. lovers of wisdom) but in practice are narrow, socially marginalized experts. We hope that this essay inspires profligation in our universities about ideas to counteract what we have previously called asophia (Whitehouse et al. 2000). Perhaps, the more specific disease of academia is aphronesia – a lack of practical wisdom. Let us engage our universities in the enterprise of saving our planet.

Such an enterprise would require interdisciplinary work to the point of creation of transciplinary professors of the future (Potter 1971). We need to be the inceptors of new centers (and decenters) of creative energy. Information system technology will create collective thinking. Attending to relationships (e.g., mentoring) will make sure such knowledge is integrated with ethical values and caring. Universities have so much to do to become any kind of "uni" at all or for that matter, having any other than self-relevant "verse."

In our universities of the Future, work to create empirical data will be balanced with attempts to develop integrative models that help us all think wisely about the quality of life of our planet during our tenure as stewards.[1]

The intent of the paper was realized as the process began, i.e. to bring the authors together. The paper is only one temporary step in a process. While this writer-serving goal may serve to alienate a reader who considers that our main goal should be to provide him or her information, we make no apologies. This paper illustrates the complex relationships in the collaborative writing process that brought three individuals together with the reader in which we present our model. We must attempt to live the principles described in this paper and in this way the writing project evolves in parallel with our own efforts at personal development and social action. (We hope that in a small way our words assist our readers similarly.)

HISTORICAL BACKGROUND: PROMETHEUS AND THE TOWER OF BABEL

> Well I am certainly wiser than this man. It is only too likely that neither of us has any knowledge to boast of; but he thinks that he knows something, which he does not know, whereas I am quite conscious of my ignorance. At any rate, it seems that I am wiser than he is to this small extent that I do not think that I know what I do not know.
>
> Socrates, from Plato, *The Apology*

If it is true that claims of knowledge indicate folly rather than wisdom, attempting to sketch a history of wisdom and aging would seem foolish indeed. Even if the task were taken to be simply empirical, it would consume the lifetime of a devoted scholar, and spill out over many volumes. The burden of scholarship in the humanities and social sciences over the past two decades, however, suggests that the task of writing history is far from straightforwardly empirical. Language itself is unstable and shifting, constantly undermining our ability to say what we mean and mean what we say. Similarly, scholars have shown that historical accounts do not simply follow the facts wherever they may lead, but, even while scrupulously adhering to standards of scholarship, follow the plotlines of conventional fictional genres. Meanwhile, the study of cultural and historical context has extended to the foundations of all human knowledge, de-stabilizing even the grounds on which such studies can be made. Overarching historical narratives have been discredited as at best pitiful attempts of anxious elite to impose order on the world, and at worst insidious devices of oppression.

In such a situation, the very idea of writing history appears absurd. Yet, more than ever, it is an indispensable part of any responsible intellectual enterprise. History can no longer be regarded as an unproblematic chronicle of all that went before merely serving as backdrop for the real intellectual action. Rather, historical narrative is itself real intellectual action. It is the material from which foundational premises are constructed and an arena in which some of the most important conflicts must be confronted. Moreover, historical narratives, e.g. scientific progress or decline and fall, are embedded in all intellectual enterprises. Responsible scholarship in any field will thus seek to make explicit the historical narrative on which it is based as a way of articulating the commitments and assumptions that will inform the entire project, and acknowledge competing narratives and the frailty of knowledge in general. The *a priori* commitments and assumptions revealed in a discussion of the historical narrative on which a project is based may not wholly determine the results, but they surely help to define the sort of problem one is trying to solve, and a particular kind of problem often implies a particular kind of solution. In the remainder of this section, we briefly sketch out the competing historical narratives of wisdom and aging in the western world, and articulate our own view of the problem.

Two competing narratives shape most accounts of the history of wisdom and aging, and both hinge upon an assessment of the enlightenment ideal of rationality and the mastery of nature by science and technology as the road to human progress. One narrative describes the enlightenment ideal in heroic terms, and the rising authority of science and technological mastery of nature as genuine progress toward the fulfillment of human happiness. Reason is the titan Prometheus, courageously stealing the fire of scientific knowledge from the gods for the benefit of humanity. A second narrative sees the enlightenment ideal and the rising authority of science and technology as eroding the moral and cultural frameworks for human meaning. Science is the Tower of Babel, a monument to foolish pride, in which the drive for technological mastery results in the loss of a common language to express human meaning, and the potential for an organic human community is destroyed. Given what has been said in the previous paragraphs, neither of these narratives can be seen as authoritative on empirical grounds alone, though of course evidence remains an essential element in constructing persuasive historical narratives. Rather, their persuasiveness ultimately rests upon an *a priori* position regarding the enlightenment ideal itself. In general, the progress narrative continues to be articulated in the sciences, while the declension narrative is articulated in the humanities and some of the social sciences (anthropology, and to a lesser extent sociology). These narratives are at the heart of a series of acrimonious debates about science and society that have been characterized as the "science wars."

As described by Stephen Katz (1996, 27-8), the progress narrative appeared in the post-WW II geriatrics and gerontology literature as way of legitimizing the emerging field as a science:

> Characteristically positivist, these accounts separate gerontological knowledge from its social moorings and depict it as the direct outcome of a scientific quest to understand humanity more objectively and truthfully. Consequently, gerontological writers commonly map the history of gerontology along a seemingly unbroken path advancing from ancient philosophy (Hippocrates, Cicero, and Galen) and medieval theology to the foundations of modern medicine in the Enlightenment and finally the emergence of gerontology and geriatrics in the nineteenth and twentieth centuries.

As Katz's description suggests, such accounts are generally unsophisticated, viewing history as unproblematic – articulating the progress narrative without any evidence of an awareness of competing narratives, or that such an articulation implies commitment to an ideology. In claiming to be part of an unbroken quest for knowledge, these works blur distinctions and conflicts in the past, and tend to deny that scientific knowledge and wisdom are at odds. A notable exception is Raymond Tallis, a renowned British geriatrician who has, in several books, produced an extensive and thoughtful defense of the Enlightenment ideal against what he sees as the dangerous pessimism of modern and post-modern critics in the twentieth century. Tallis is well aware of counter-narratives and their power, and understands that the ideological implications of these narratives. He is unabashedly and unequivocally committed to the values and ideals of the Enlightenment. In his *Enemies of Hope,* he articulates the Enlightenment view that wisdom is an impediment to knowledge, and a barrier to the fulfillment of a humanitarian sensibility. In answer to those who lament the loss of wisdom at the hands of science and the colonization of old age by reductionistic biomedicine, he argues that "we should be cautious about advocating the return of a commodity [wisdom] that was in such rich supply when the only effective medicine was brutal surgery and bonesetting." Tallis claims that laments about the loss of wisdom that have accompanied the rise of modern medical science are generally accompanied by horror stories about the misuse of high-tech medical interventions and diagnostics, resulting in the harm and humiliation of patients. However, such horror stories do not imply the need for a return to traditional knowledge so much as a more rigorously scientific practice of medicine. "What these examples of stupidity call for are not some mysterious substance called 'wisdom'...possessed by a 'wise' physician, but less stupidity and less idleness. After all, everyone can see the stupidity of the error," he argues. "The next step is to see how it came about. Clever idiots make a more powerful case for non-idiotic cleverness than for something called 'wisdom'." In Tallis's view (1999, 171,175), "wisdom" is a dangerous anachronism:

> It traditionally comes in, or is concentrated in, pearls and it seems as if these pearls can be linked together in an amulet to ward off critical intelligence and to drive away the demons of actual expertise, common sense, genuine learning from experience, factual knowledge, and ordinary sympathy and kindness. For all of these latter threaten – along with accountability – to leave wisdom without a place to call its own....'Where is the wisdom we have lost in knowledge?' the poet asks. If wisdom is a form of unearned authority, a legitimacy from nowhere, our reply must be a cheerful: "Well lost."

Thus, Tallis does not dispute a narrative of the loss of wisdom, but challenges its meaning. The displacement of a broad concept of "wisdom" by science is not a tragedy to be mourned, but an achievement to be celebrated.

While Tallis's account is valuable as a corrective to the tendency of critics of modern science to romanticize pre-modern cultures – a danger that critics have often enough called attention to themselves, if less often actually managed to avoid – we are more persuaded by scholars who have seen the decline of wisdom in Western cultures as a significant loss. If the postmodern experience is characterized by fragmentation and the multiplication of competing and conflicting perspectives constituting a crisis of knowledge, as many scholars have suggested, then the decline of a multidimensional, integrative concept like wisdom might plausibly be thought of as both a symptom and a cause. More importantly, restoring some form of the concept of wisdom to scientific and scholarly discourse may invaluably enrich our ability to face challenges, such as human aging, that confront us.

A number of psychologists have recently issued a call for the rehabilitation of the concept of wisdom, and so it is worthwhile paying attention to how they have articulated the history of why their discipline cast wisdom aside. Daniel N. Robinson explains the abandonment of the study of wisdom by psychology in the nineteenth century as essentially the playing out of the implications of the Enlightenment. According to Robinson, from remote antiquity to the beginnings of modern philosophy in the 17th and 18th centuries, wisdom "was explicated in terms of providential gods, muses, astrological forces, a sixth sense, genetic bounty or accidents of nature" (Robinson 1990, 21). In the writings of a pantheon of enlightenment philosophers – Francis Bacon (1561-1626), René Descartes (1596-1650), John Locke (1632-1704), David Hume (1711-1776), Voltaire (1694-1778) and Jean Jacques Rousseau (1712-1778) to name but a handful – such explanations were laid open to devastating attack. Stern empiricism became the arbiter of all claims to knowledge, and any "wisdom" that rested upon metaphysical *a priori* commitments or an ineffable mystical or aesthetic sensibility was suspect. In this view, "wisdom" can be no more than "a scientific understanding of the laws governing matter in motion ... technical knowledge of how things work" (Robinson 1990, 21-2).

By the nineteenth century, philosophers led by Kant (1724-1804) and, more emphatically, Hegel (1770-1831) had developed an influential critique of empirical science as a philosophical system. Empirical science was limited to the phenomenal world. The great sages and artists of history had access to a richer store of non-empirical knowledge, transcendent knowledge in the *noumenal* world. This transcendent knowledge was available only through the practice of philosophy, religion, and art. Strands of this critique circulated throughout the west under diverse names such as idealism, romanticism, and transcendentalism; its major intellectual result was an intractable conflict between the humanities and the sciences. From the nineteenth century on, mainly poets and artists preserved the broad, pre-scientific concept of wisdom. Scientists eschewed such a fuzzy concept. "Accordingly, as psychology searched for scientific standing in the nineteenth century, its founders and leaders recognized that any number of ageless issues would have to be ignored, lest the new 'science' be corrupted by the older 'metaphysics.' *Wisdom* was but one of these issues, though time has come to question the wisdom of this decision" (Robinson 1990, 24).

Another psychologist, Aleida Assman, tells much the same story, but concludes more optimistically that wisdom may be making a return (Assman 1994). Assman traces the flourishing of different forms of wisdom from antiquity through the Renaissance, but argues that wisdom declined in the Age of Reason with the cultural construction of a dichotomy between science and wisdom. "Wisdom was now defined as the opposite of science...Knowledge had to be curtailed and disciplined, accommodated to abstract laws or universal principals before it could claim any dignity....Under these epistemological circumstances, wisdom became the shelter of debunked knowledge and forgotten lore." Scientific discourse meanwhile was impoverished, constituting itself by eliminating elements of human experience such as "religious revelation, mystical vision, knowledge of tradition, proverbial know-how, personal concerns, and subjective experience" (Assman 1994, 202-3). Assman sees greater continuity than Robinson, however, arguing that the work of Nietsche and Wittgenstein and more recent scholars grouped under the rubric postmodern have created possibilities for a renewal of wisdom:

> Wherever we find a growing recognition of the limits of knowledge, we see traces of wisdom. Modesty, as the primal virtue of the wise person, again appeals to the scientist whose skepticism was always methodological and never existential. Another aspect is the loss of an integrating perspective. Worldviews, modes of thought, and forms of life are no longer bound under an obligatory and universally accepted set of values. In the postmodern condition of existential pluralism, the neglected figure of wisdom is summoned back from her exile and is restored as a respectable figure (ibid., 203-4).

Cultural historians of aging have generally told a very similar story. Stephen Katz and Thomas Cole have separately argued that modern scientific discourse on aging has reductionistically focused upon the deterioration of the individual aging body, and thus eroded richer frameworks of communal and spiritual meaning. Prior to this, discourse on old age was concerned with the connections that the aged individual maintained with community and with God. In this pre-modern discourse, the physical and mental deterioration of old age was not ignored, but it did not necessarily generate intense anxiety. In a society ordered by principles of hierarchy and reciprocal obligation, disease and dependency were tolerated as a salutary illustration of humanity's ultimate dependence upon God.

Synthesizing the research of several historians, Stephen Katz characterizes representations of the aging body in Medieval Europe as "polysemic" – able to carry a diverse array of meanings connected to universal forces and moral principles. Although the aging body continued to be thought of in terms of Galen's four humors, more broadly it was a way for thinkers to "contemplate a richly metaphorical cosmos in which the human body was a spiritual extension of God's design...Through the kindred medical, philosophical, and astrological schemes of life, old age was positioned in a dialectical relationship between the physical and the spiritual, where the latter could overcome the limitations of the former." Katz formulates four axioms that characterize pre-modern aging: 1) old age combined the deterioration of the body with opportunities for renewal, redemption and transcendence in spiritual attainment; 2) pre-modern discourse on aging enveloped the aging body in a complex system of metaphorical codes, linking old age to a variety of material and spiritual forces; 3) the meaning of old age was not reducible to the aging body itself, but emanated from the universal order of which the body was but a single element; 4) death and dying at all ages was a communal, spiritual process, publicly enacted, connecting the worldly and eternal realms (Katz 1996, 30-33).

Thomas Cole describes the place of old age in colonial New England in similar terms. Representations of old age were rooted in religion, and the clergy remained the authoritative voice on old age. In their view, the losses of old age served as a new invitation for Christians to witness and participate in the balance and harmony of God's creation. While the physical and mental deterioration of old age usually entailed a loss of physical and social power for the old, it did not vitiate, even when the aged became demented, their fundamental role in the community – to acknowledge dependency and longing for the salvific grace of God (Cole 1992; Haber and Gratton 1994). As one minister put it, piety in the old hid the "imbecility of body and mind" that so often afflicted them; it gave meaning to the lives of old people "when they would otherwise be useless and burdensome." Long after the days of productive labor were over, the old could still serve God, their families, and communities through their piety, acting as "visible monuments of sovereign grace" (Cole 1992, 64-65).

Both Katz and Cole argue that this view of aging did not immediately change as Enlightenment ideals of rationality and human agency challenged Calvinist theology in the eighteenth century. Katz describes the persistence of the humoral framework that was at the heart of the medieval conception of old age, and a continued tendency to connect the aging to larger external forces – now thought of as the laws of nature. Cole argues that an appreciation of natural law and human agency led rational humanists of this period to view the loss and suffering in old age as inevitable and natural evils of existence that could be assuaged and ameliorated, but not eliminated. Ultimately, one simply had to accept deterioration and even death as part of the benevolent design of nature (Cole 1992, 103-4). Nonetheless, what Katz describes as the "blossoming optimism about human perfectibility" that characterized the Enlightenment tended to focus more attention on the issue of deterioration of the aging body. Many enlightenment treatises on aging were concerned with understanding the triumphs and tragedies of the aging body.

With the emergence of modern scientific medicine in the late-nineteenth century, this concern would lead to a reductionist focus upon the deterioration of the aging body as the essence of aging. Virtually all historians of aging have focused upon the rise of scientific medicine in the second half of the nineteenth century as a central development in the erosion of broader frameworks of meaning for old age in a way that parallels what scholars have claimed about the loss of a broad concept of wisdom. In Coles's (1992, 193-195) words:

> By the early twentieth century, aging had been largely cut loose from earlier religious, cosmological and iconographic moorings.... Laboratory scientists and research physicians attempted to cast off religious dogma and mystery surrounding natural processes. Rejecting transcendent norms and metaphysical explanations, they turned to biology in the hopes that nature itself contained authoritative ideals and explanations of old age...The formative literature of gerontology and geriatrics helped complete the long-term cultural shift from conceiving of aging primarily as a mystery or an existential problem to viewing it primarily as a scientific and technical problem.

Again, parallel with the history of wisdom sketched above, both Katz and Cole see the postmodern as a source for the creation of renewed frameworks of meaning for old age. Katz concludes with a call for the "undisciplining" of gerontology – an embrace of the indeterminacy and fragmentation of knowledge that would allow the creation of new models of aging free of the constraints and reductionist tendencies of the post-enlightenment medicalization of aging. Cole optimistically tries to envision such a model. "Despite the continued cultural dominance of the scientific management of aging, there are signs that the modern quest for a rational, healthy, and orderly life course have reached a limit – or at least a turning point," he writes. "A growing number of people have criticized the rigidity of traditional age norms and the segregation of lifetime into the three boxes of education, work, and leisure. We may be witnessing the emergence of a postmodern course of life, in which individual needs and abilities are no longer entirely subordinated to chronological boundaries." Though it is understandably difficult – and in principle impossible – to describe this postmodern life course with precision, Cole nonetheless suggests the basic form such models will take: "Vital postmodern ideals of aging will require a kind of cultural *bricolage*: taking valuable bits and pieces of our cultural inheritance and arranging them in new ways to meet current needs, circumstances, and particular communities" (Cole 1992, 239-40, 243). Contrary to the way Raymond Tallis would have it, for these authors, the view from postmodernity is decidedly optimistic.

Of course, these postmodern ideals of wisdom and aging may be easier to conceive of than to instantiate. For all of these authors, the way out of the post-Enlightenment loss of wisdom and devaluation of aging is through better ways of thinking about the problem. In that sense, these critics resemble the Enlightenment ideal they criticize. Though they disagree radically about the content of right-thinking, both the Enlightenment and its critics proceed on the assumption that better thinking – consisting of clear-headed rationality and commitment to scientific certainty on the one hand and an acceptance of doubt, uncertainty and contingency on the other – will lead to a better future. But it is not all clear that the impasse we face is simply a matter of right-thinking. In Cole's account, social structure plays an important role in forming, if not quite determining, cultural attitudes toward aging. For example, in discussing the continued tolerance for the inevitable suffering and losses of age in the Enlightenment, despite the heightened emphasis upon scientific mastery of the body, Cole suggests that the key is the persistence of hierarchical social relations in a Republican social order. "Acknowledgment of the intractable sorrows and infirmities of age remained culturally acceptable as long as men and women lived in families, churches, and communities regulated by principles of hierarchy, dependency and reciprocal obligation." In contrast, a liberal social order characterized by aggressive individualism and fluid social relations, made such tolerance impossible. "How could the ideology and psychology of self reliance be squared with the decay of the body? Only by denying its inevitability and labeling it as failure" (Cole 1992, 104). It is the intolerance for suffering and loss that characterizes a liberal social order, Cole argues, that drives the medicalization of aging in the modern world. But if attitudes toward aging are driven to a large extent by the social order, then it follows that we must change the social order before we can meaningfully change the way we think about wisdom and aging. As long as aggressive individualism and weak social relations characterize the social order, we may expect that – however highly we may claim to value them – neither wisdom nor old age will fare well.

This is not necessarily a pessimistic observation, nor does it render efforts to improve our ways of thinking meaningless. But it does suggest the enormity of the problem, and leads us to recall the wisdom of Simone De Beauvoir's oft-quoted concluding words of *The Coming of Age:* "Once we have understood what the state of the aged really is, we cannot satisfy ourselves with calling for a more generous 'old age policy,' higher pensions, decent housing and organized leisure. It is the whole system that is at issue and our claim cannot be otherwise than radical – change life itself" (Beauvoir 1972, 543). A conceptual model that intends to reclaim the dignity of wisdom and aging and link them to quality of life must find ways to integrate not only the post-Enlightenment scientific world view and its critics, but must address the social ground on which both of them stand.

Before the enlightenment wisdom was rooted in what we would now call organizations – tribes, guilds, churches, crime syndicates, etc. Society tries to support those that best meet social need. Organizational culture contributes to overall social culture. The dominance of business currently reflects short-term materialistic values. The key is to develop businesses or other organizations that realize short and long-term improvement in cultural values towards nature, for example, that are consistent with and in fact, demand a long-term materialistic view such as natural capitalism.

Science attacked such entrenched medieval wisdom. It threw out religion and with it lost spirituality. Now science has fragmented knowledge so much that it cannot be wise. What greater affront to wisdom is the narrow professionally selfish vaunted PhD degree (Doctor, i.e., teacher of love and wisdom)! Unless institutions of "higher" learning get a little lower so-to-speak, get out in the communities, and be more useful they will die as vital centers of learning.

Postmodern wisdom will mediate science as the dominant intellectual force because it respects science but puts it back into perspective. Social context and social construction are real.

CONCEIVING CHANGE: TOWARD THE INTEGRATION OF AGING, WISDOM, AND QUALITY OF LIFE

But, how can we change life itself? In the remainder of this paper we will discuss the concepts of aging, wisdom, and quality of life, and present a beginning for a model that integrates the three in a way that we hope suggests a valuable approach to the challenges of human aging, and perhaps to many other issues facing humanity as it enters the twenty-first century. Abstract models cannot bring about change, either great or small. Are they useless if they are limited to the world of ideas? Their value lies in translation into human action in the social world. We recognize that human action is complex, and cannot simply be directed the creation of abstract models. Nonetheless, we believe that subtle shifts in the meaning and use of key words, such as we suggest here, can both lead and follow broader cultural changes, and that cultural change will impact human behavior. Thus, we hope that our enterprise – tentative and playful as it is – will be a contribution to the changes necessary for continued survival of life on the planet.

Our model of aging, wisdom, and quality of life describes the dynamics of the interface between the subjective and objective worlds. Every individual has personal values that reflect his or her own wisdom and sense of quality of life. Every individual also contributes to a rule-governed world in which interactions among different subjective worldviews create a common sense of the external objective environment. The evolution of values is due to biological and cultural forces. Values (i.e. inclinations towards certain goals influenced by various emotions) have been viewed as providing classes of automatic behavioral responses to environmental perturbations.[2] Individual and collective wisdom are ways of thinking and living to fashion the unknowable future from the fleeting but tangible present based upon the always reinterpretable past. Wisdom must be seen in action-in-the real world of life, not just in the world of ideas.

Aging

Aging might be viewed as a ground for exploration of wisdom and quality of life. Aging can be viewed using geometric metaphors as a linear arrow through time, the flow of the universe in another dimension, or a circle of life. Everything is changing and thus all components of the universe age. Human aging itself is bounded by birth and death. Death is the endpoint of life for an individual human being as a biological entity, and may be viewed as a personal tragedy. Death from the point of view of an integrated ecosystem is an intrinsic and important part of the re-creation of biological diversity that permits response to changes in the environment. Death is essential to life.

What does the aging process of a human being contribute to the development of wisdom and/or the achievement of a good quality of life. Aging is accompanied by the accumulation of life experiences that may impact positively or negatively on intellectual and moral development, as well as aspects of quality of life such as financial well-being and social networks. Living to an advanced age may bring insights that come from looking back on a long series of experiences; it may also bring dementia. Long life may bring wealth, good health and an active social life; or it may bring financial deprivation, chronic illnesses, and social isolation.

From another viewpoint, quality of life and wisdom may seem related to aging in developmental terms. Moral and intellectual abilities apparently develop in a series of stages. Rather than simply being a neutral domain or dimension in which experiences have positive or negative cumulative impact, aging may be seen as a process in which there is a necessary order to the enfolding of wisdom and quality of life. In such a view, wisdom and quality of life can more likely be attained as human beings become older. Although stress, suffering, and loss are often viewed in unequivocally negative terms, they may be also crucial experiences in the development of wisdom. Wisdom in turn may be seen as an important aspect of enjoying a high quality of life. Drawing nearer and nearer to one's own death may create opportunities to develop wisdom. Wisdom, in turn, enhances the quality of life even at its close. In this view, sustaining community meaning, moral integrity, personal dignity, and self-worth are more important to creating wisdom and maintaining quality of life at an advanced age than eradicating disease and even suffering.

Aging occurs in the dimension of time. J.T. Fraser's Project entitled, "The Voices of Time" (Fraser 1923) was a cooperative survey of man's use of time as expressed by the sciences and by the humanities. Time is a rich concept not to be limited to the measurement of time as developed by the physical sciences. Our sense of time has a biological substrate (e.g. biological clocks associated with endocrine systems) but is also affected by the cultural environment in which children learn their temporal frameworks. Fraser calls for the development of a discipline such as chronosophy, as an interdisciplinary normative study of time. He suggests a need for a deeper study of time and its metaphors. Such understanding would enrich our notions of aging as individuals, as societies, and as life forms. Our conceptions of time, particularly how we envision our future selves and society, are critical to developing self-awareness and sharing our life stories.

Quality of Life

Quality of life is a multidimensional holistic concept. It includes an irreducible combination of these interdependent parts: elders (body, mind, and spirit), living and non-living environment, and life experiences in a spatial-temporal framework (Katz and Gurland 1991). The concept of quality of life has entered the medical literature only recently, but its use is increasingly rapidly.

Wisdom can viewed as knowledge of the total system in which life is embedded, including characteristics of the cybernetically organized components. Quality of life similarly can be portrayed as a web of spatially and temporally organized structures and processes. This view leads to a systems orientation to understand the determinant factors of living systems as irreducible wholes and the dynamic functions that are interwoven arrangements of all parts of the whole. (Holarchies as Koestler and Wilber point out offer a richer construct than hierarchies (Wilber 1995).

Quality of life emerges in a complex cultural and natural world. As measured in health studies, quality of life usually includes assessment of such factors as cognition, mental well-being, activities of daily living, social networks, financial security, housing, and environment. On the face of it, certain aspects of quality of life can be measured objectively by outside observers, for example, financial well-being and the ability to function. However, even these aspects remain elusive because a critical aspect of quality of life assessment is the subjective perception of one's own quality of life. Wisdom is at the core of this personal assessment of what constitutes a quality life (the yang in the yin of quality of life).

A crucial aspect of the experience of quality of life is the congruence of one's life goals with one's achievements. Individuals may set for themselves goals that are biologically or culturally difficult to achieve. However, if these goals are viewed by society as important to the individual then they raise the personal development stakes. Such a life adventure requires a greater degree of wisdom to achieve its goals. In general, however, if individuals set concrete goals for themselves such as a certain level of material well-being, then the experience of quality of life will be related more to the congruence of the expectation and the achievement rather than the absolute amount of material wealth. Money does not necessarily produce happiness but a positive self-directed life story does.

The ability to adapt to changes in the environment is at the heart of quality of life. Aging implies change and change implies the need for wisdom to find the balance between a conservative internal integrity and liberal response to biological and cultural evolutionary pressures. Biological evolution has created subjectivity as an adaptive mechanism. The interplay between subject and object is a cultural tool that may enhance survival. Integrating facts and values at the fuzzy interface between the subjective and objective is the task of wisdom in the service of a quality life.

Subjective and objective processes are complimentary as strategies to pursue a quality life. Objective processes operate according to fairly rigid and widely shared sets of rules. Subjective processes are relatively autonomous features of the whole person. Objective and subjective adaptive approaches are difficult to define but the processes that influence quality of life can be studied. Objective and subjective approaches have evolutionary and cultural history that serve to guide human beings in maintaining a balance between a stable status quo and different possible future states. Subjective processes and states are referred to by terms such as emotions, feelings, attitudes, and preferences. They provide a rich and diverse set of concepts and ideas that can potentially shift our perceptions of the objective world. The objective world is best assumed to exist, although philosophers have disagreed about how to prove this apparently obvious fact. However, saliency of the "real" world is less critical than the nature of the shared intersubjective perceptions that we have of the world and how it works. As Richard Rorty (Rorty 1989) and others have emphasized, the project for a post modern liberal democracy is to maintain the balance between rigid agreed-upon social rules and individually inspired variations in belief systems. Beliefs lead to behaviors necessary to respond to changes in the environment. Ultimately, subjective approaches are fundamentally spiritual in that the perception of freedom to act autonomously, even when combined with a rich understanding of the limitations of social context, gives life its meaning and purpose.

Society has evolved different ways in which interaction between individual subjective worlds and the intersubjectively created objective world can occur. Such interactions lead to changes in culture. Professionalism is one way societies have evolved to intensify the degree of objectification. Excessive objectification of knowledge can restrict the opportunity for cultural change, however. Excessive subjectification could lead to romantic chaos. Clearly, the cultural means for radical behavioral changes are deep but hard to fathom. The most important concept to guide change is what we value. As guides to behavior, values, and evolution of values are the essential forces of social change.

A dimension to quality of life not always considered by scientists is the degree to which religion and/or spirituality play a role. We will not attempt to differentiate spiritual and religious beliefs here, although spiritual tends to refer to individually held views of transcendent issues whereas religion relates more to socially shared patterns of behavior. Both such types of beliefs modify an individual's sense of wisdom. Some of the most commonly cited sources of wisdom are religious texts. Moreover, different religious belief systems present models for the ideal or good life. The evolution of spiritual belief systems is also a cultural and intellectual space where human beings can demonstrate the ability to adapt, yet preserve some fundamental beliefs.

The transformation of spirituality as a current cultural phenomenon in society cannot be discussed in detail. We recall our discussion of limits earlier here.

However, we will leave the reader with the thought that there is something intrinsically spiritual about wisely attending to the holistic systems in which our lives are embedded. A human wisdom directed toward preservation of life on the planet based upon a rich appreciation for the interrelationships between life and death may represent a new emergent belief system that returns to the roots of nature based religions. Such modified values may help protect the human race creating ecological havoc. How does the future of our environment and life on the planet relate to aging, wisdom, or quality of life?

In the beginning of this paper, we warned about the dangers of creating too-grand meta narratives. Hence, with some trepidation we describe the need to at least consider the potential need for a global bioethic. As originally proposed by Potter (1971) the world seems in need of different fundamental guideposts for creating our future. The survival of human and other species is fundamental worthwhile considering as an anchor in the process of creating values. The aging of our population and deterioration of our environment signals the need to develop new attitudes towards the ecosystem in which human life is evolving. Clearly our ability to heal or damage the biosphere is more connected to our cultural traditions than to our opportunities to evolve biologically in the short term. A global bioethic calls for personal responsibility for one's own behavior and commitment to collective efforts to improve the quality of life on the planet in general. This concept of a global bioethic takes us back to some of the original formulations of Van Rensselaer Potter and Aldo Leopold (Potter 1988) and away from the more modern conceptions of bioethics as response to medical technological innovations. The sources for this renewed sense of bioethics are many and include, for example, deep ecology. In this tradition, recognition of our fundamental and mystical connections to nature combine with an intellectual understanding, albeit limited, of the natural systems in which we participate. Our future will depend upon having the wisdom to understand complex systems and how they evolve in the future. As important, we also will need to make personal changes in our own behavior concerning, for example, resource consumption and social justice. Perhaps creating a global bioethic as an evolving concept to create discourse about the quality of life on this planet would be a wise course of action.

Wisdom

Like quality of life, wisdom is a multidimensional, integrative concept that is somewhat difficult to define. Clearly, one aspect of wisdom, considered particularly by psychologists, is intellectual ability that includes such cognitive aspects as the so-called executive functions: goals setting, planning, and monitoring one's own behavior and the behavior of others. However, wisdom also involves the integration of cognition with emotion, and the appraisal of human values. Wisdom involves balancing priorities using good judgment and considering many different features of a situation. In fact, it is an essential attribute of wisdom that it is applied pragmatically in real life situations, particularly ones own. The ability of the self to reflect critically upon itself is a part of wisdom. "Know thyself" as a recommendation for philosophical reflection is common to many ancient and modern wisdom traditions.

Another conception of wisdom portrays wisdom as an end in itself. That is to say, being wise also leads to living a good life, in both moral and other senses. A good life would have, in fact, considerable congruence with a high quality of life, raising the issues as to whether a life led wisely is usually or always a life of high quality. As mentioned earlier, Socrates, one of the archetypes of a wise person in Western culture, was said to have had his wisdom acclaimed by the Greek oracle. However, Socrates himself refused to accede to his own wisdom when told of the opinion of the oracle. It was in part recognizing his own ignorance that won Socrates a reputation for wisdom. Part of being wise is recognizing that wisdom is an ongoing process that recognizes limits. It was publicly recognizing the absence of wisdom in his fellow citizens that led to his death, however.

Figure 1: An Integrated Spacial Model of Aging, Wisdom, and Quality of Life

Human Experience through the Life Course

Adapting

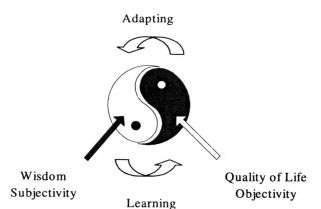

Wisdom   Quality of Life
Subjectivity   Objectivity
                    Learning

An integrated model of aging, wisdom, and quality of life

Creating an integrated model that incorporates aging, wisdom, and quality of life is not a small challenge. Models necessarily involve simplification, and whereas Whitehead enjoined us to seek simplicity, he also suggested that we remain suspicious of it. The concept of "integer" embedded in an integrated model implies wholeness and self-sufficiency. It is a complex job to think about systems, yet we must. However, we are helped enormously by our fellows in whom we trust.

Nonetheless, hate and fear threaten to dominate our spirit. A dominant value for integration is integrity in the sense of trustworthiness.

Our model is based upon the idea that wisdom and quality of life are integrated aspects of a unified human experience, which can thus best be represented by the oriental yin and yang concepts in apposition. Wisdom is the subjective aspect of quality of life, and quality of life is the objective aspect of wisdom. Wisdom is seen as a subjective concept intrinsic to the individual but affects objective quality of life. Quality of life can be measured objectively by others based upon mental state, income, housing, and social relationships, but it also includes an internal subjective sense of well-being that has wisdom associated with it. We intended there to be no difference between structure and process in the model in that the model is constantly in motion. The aging process energizes change. Learning is key for human beings and other life forms to gain cognitive knowledge from their environment. Such knowledge of knowledge permits the wisdom necessary to adapt to the changes in the environment. Such adaptive processes are key to quality of life. Those who age and fail to adapt become history before they die.

Thus, we visualize the model of quality of life and wisdom as a multiple dimensional circles, spheres and roundness'.[3] Since it is an integrated model, i.e., whole, both its more abstract and most concrete levels must parallel. In the intellectual ether and in real life, wisdom and quality of life can be integrated. How do we judge a person to be wise – by how they lead their life regardless of any objective characteristic, i.e., by the quality of their lives (and those around them). Such judgments cannot be reduced to simple formulae, however. The blind quadriplegic can be as wise and enjoy a higher quality of life, while a movie star, fabulously beautiful and wealthy may be miserable. Enjoying life quality is balancing self/society; family/career; action/strategy; and always learning. If expectations, goals, and successes match, then our lives can be happy at least for some precious period of time. Creating sacred time and space helps as well. Thus, enjoying a life of quality both builds and depends upon wisdom. Such individual wisdom would by necessity contribute to collective wisdom and improvement in the condition of life on the planet.

CONCLUSION

How can ideas about aging, quality of life, and wisdom be integrated into some general notion of human potentiality? Is this question answerable? Is it wise to pose such a question? One focus for integration is the very conceptions of life and death. Part of an assessment of quality of life in an older individual is how death fits into the picture. An ideal situation would be that quality of life would be maintained until the point that biological life ends. One can also ask what a wise individual might say about his or her own death, and about the end of life in general. Clearly, Socrates viewed his death in relationship to his society, as well as in the context of his own individual life. More than being congruent with personal and social conceptions of individual life, death needs to be seen as a part of normal ecological processes.

Individual death is a necessary part of the evolution of life and the appearance of future generations. It seems evident at this point in life's evolution on this planet that billions of people are threatening to extinguish not only their lives, but those of other species as well. Clearly, a greater wisdom is needed to understand these complex, long-term issues than we have evidenced so far. Many communities are growing increasingly old particularly in countries that have achieved some basic level of public health and industrialization. Thus, aging and environmental issues are the two most important long-term challenges that the human race faces. To find the wisdom to adapt our conceptions of quality of life to be congruent with the survival of our species and others is the principal challenge for the future.

Peter J. Whitehouse, Professor, School of Medicine, Case Western Reserve University, U.S.A.

Jesse Ballenger, Instructor, Dept. of History, Case Western Reserve University, U.S.A.

Sid Katz, Emeritus Professor of Geriatric Medicine, Columbia University, New York, U.S.A.

NOTES

[1] As mentioned, the individuals writing this paper not only represent different academic disciplines but are also different learners. Mentoring relationships link several of us, although these relationships are more collaborative than some conceptions of mentorship allow. Your authors are of different ages and at different stages of life. Most of all, we hold different viewpoints about effective ways for animating our reflections into action in the real world. We wrote this paper for ourselves. Our writing process reflected our ideas in action. Peter Whitehouse was asked by David Thomasma (editor) to contribute a chapter to this book. He first engaged his colleague Jesse Ballenger in the history department, who had worked with him on numerous collaborative writing projects, and then Sid Katz who, along with Barry Gurland, is writing a book on quality of life. We thank Barry for his insightful comments on early drafts and his commitment to these issues. We wrote both independently and collectively; we specifically did not try to reach consensus. Consequently, this document is dynamic and the ideas we hope will surface yet again in future writing projects.

[2] Barry Gurland, personal communication, 2000.

[3] Here we focus only on the core of the single yin and yang of wisdom. Arrows can be drawn in both directions connecting the two concepts of wisdom and quality of life. Perhaps they are dialectic but they seem too complex to be encompassed in a single dimension. Quality of life reflects the role of nature, culture, and society- the framework in which individuals live out their own lives. It is objective in contrast to wisdom that is more subjective. Wisdom is an attribute of the self that encompasses cognitive, ethical, and intuitive abilities. The interplay in the model can be characterized as the process of change. Different change words could be employed. For example, "learning" is a process of changing the self and potentially gaining wisdom through living one's life. Having wisdom allows one to "adapt," another change word, in ways to enhance individual and collective quality of life. The model expresses wisdom and quality of life as yin and yang in which quality of life represents a small circle within the area of wisdom that represents the objective aspects of wisdom. Similarly, wisdom is a component in the domain of quality of life that represents the subjective aspects of quality of life. Aging is itself the dimension along which adaptive learning occurs in a path that may lead the individual organism to a higher quality

of life and wisdom. The diagram requires another dimension (or more) to reflect temporal course. What would a fuller model permit in terms of action in real life? We envision a complex information processing system of human beings connected electronically with access to vast repositories of knowledge. Its role would be to create wisdom and enhance quality of life to the world. It would not be a big brother, nor even a big sister, but rather contribute to the democratization of knowledge. This could be a source of trust, as well as information.

REFERENCES

Assman, A. 1994. Wholesome knowledge: Concepts of wisdom in historical and cross-cultural perspective. In: *Life-span development and behavior,* eds. D.L. Featherman, R.M. Lerner, and M. Perlhutter, 187-224. Hillsdale, NJ: Lawrence Erlbaum, 187-224.

Cole, T. 1992. *The journey of life: A cultural history of aging.* Cambridge, UK: Cambridge University Press.

de Beauvoir, S. 1972. *The coming of age.* New York: Putnam.

Fraser, J.T. 1923. *The voices of time: A cooperative survey of man's views of time as expressed by the sciences and by the humanities.* New York: George Braziller.

Haber, C., and B. Gratton. 1994. *Old age and the search for security: An American social history.* Bloomington: Indiana University Press.

Katz, S. 1996. *Disciplining old age: The formation of gerontological knowledge.* University Press of Virginia.

Katz, S., and B. Gurland. 1991. Science of quality of life of elders: Challenges and opportunities. In: *The Concept and measurement of quality of life in the frail elderly,* eds. J.E. Birren, J.E. Lubben, J.C. Rowe, and D.E. Deutchman, 335-43. San Diego: Academic Press.

Lehrer, K. 1996. *Knowledge, teaching and wisdom.* Dordrecht: Kluwer.

Robinson, D.L. 1990. In *Wisdom: Its nature, origins and development,* ed. Robert Sternberg, 13-24. Cambridge: Cambridge University Press.

Rorty, R. 1989. *Contingency, irony, and solidarity.* Cambridge: Cambridge University Press.

Sternberg, R., ed. 1990. *Wisdom: Its nature, origins and development.* Cambridge: Cambridge University Press.

Srivastva, S., and D.L. Cooperrider, eds. 1998. *Organizational wisdom and executive courage.* San Francisco: New Lexington Press.

Tallis, R. 1999. *Enemies of hope: A critique of contemporary pessimism: Irrationalism, anti-humanism and the counter-Enlightenment,* 2d ed. New York: St. Martin's Press.

Van Potter, R. 1971. *Bioethics, bridge to the future.* Englewood Cliffs: Prentice-Hall.

———. 1988. *Global bioethics: Building on the Leopold legacy.* East Lansing: Michigan State University Press.

Whitehouse, P.J., J. Gute, and J. Ballenger. 2000. *Neuropsychology of wisdom.* First Virtual Congress of Neuropsychology on the Internet.

Wilber, K. 1995. *Sex, ecology, spirituality.* London: Shambhala.

CHAPTER TWO

FERNANDO LOLAS

ETHICS AND QUALITY OF LIFE IN THE ELDERLY

QUALITY OF LIFE: A BROAD CONSTRUCT IN A RELATIONAL CONTEXT

One advantage of the construct quality of life is its apparently neutral status. Life can be of good or bad quality and can change for the better or the worse. Personal satisfaction may shift from the positive to the negative and *vice versa*. What could be considered weaknesses, the polemic nature of the word quality, and the inexhaustible meaning of life may, in fact, be assets.

The construct can be used in different contexts. It may be employed *descriptively*. It can be used in an *evaluative* fashion. In some traditions, it is employed *normatively*, to dictate what can and should be the case. One of the less elaborated contexts will be pursued here, namely, as a *relational* construct. It will be argued that this use of the construct is the most appropriate to deal with bioethical issues.

Quality of life is employed as a criterion for decision-making in healthcare and other areas. Ethical tensions derive from a difference between what should be and what is the case in a given situation. The relational point of view underscores that there is always a difference between what people may want to be or to have and what they actually are or have. This applies to individuals and to groups. Without trying to apply a norm, and even reducing the evaluative component, assessments of quality of life can be used descriptively. Beyond description there is a relational stance, where description is turned into contrast and comparison, introducing the dynamic point of view so much needed when applying the construct to the life cycle.

The word "health" contains attributes reflected in the usual definitions, stressing a desirable state or condition or a quest for plenitude (Mordacci and Sobel 1998, 37). Health status, satisfaction with health care, environmental conditions, and bodily conditions are integral components of the concept of quality of life. It incorporates many domains: health-related and not health-related.

Assessment of quality of life is complex. There has been a broadening focus upon the measurement of health beyond traditional indicators such as mortality and

D.N. Weisstub, D.C. Thomasma, S. Gauthier & G.F. Tomossy (eds.), Aging: Decisions at the End of Life,
21-30.
© *2001 Kluwer Academic Publishers. Printed in the Netherlands.*

morbidity. The "positive side" of the scale is difficult to approach. Although it is recognized that health includes, but is not reduced to, absence of disease or ailments, there are more scales to measure anxiety and depression than to measure happiness or positive affects (Orley and Kuyken 1994). Well-being is an elusive concept. Behind it appears the notion of plenitude, perfection, and excellence.

A good reason to develop a comprehensive evaluation is the fact that the predominant model of medicine has changed. The impact of bioethical thinking has resulted in a "horizontalization" of the traditional vertical relationship between experts and lay people characteristic of modern professions. It is recognized that at least three incommensurable discourses express derangement, disorder, and disease. *Illness* identifies the subjective component, *disease* the technical labeling (the third person approach), and *sickness* the social evaluation of a disease state (another third person approach, a lay one). These are ways of expressing impairment or detriment from the perspective of the social practice of medicine. Individual limitations in the form of *impairment, disability,* and *handicap* reflect the consequences of disorder upon personal life. Evaluation of life quality ideally should be comprehensive enough to accommodate these different aspects.

A relational use of the construct quality of life implied in the bioethical context means to employ it as a *mediating link* between values, principles, norms, and actual behaviors. Both in terms of individual life and of social context (microbioethics and macrobioethics), assessments end up in comparisons between an ideal or a standard and an observed state of affairs. Whereas in the nonhuman sciences this comparison can be made against relatively fixed or constant standards, human life changes continuously and mediation between what is the case and the ideal is particularly difficult. The comparison is always biased by the position of the observer, his/her cultural background and the expectations of the society in which it is made.

FEATURES OF THE QUALITY OF LIFE CONSTRUCT

A common feature of most approaches is the realization that quality of life is a *subjective* evaluation, so personal in fact that any generalization runs the risk of being incomplete and partial. It is common experience that conditions, that according to common sense should be highly incapacitating, are not experienced in the same form by those afflicted by them. For an age-related evaluation this feature has particular significance since the standard discourse on aging consists mostly in accounts of frailties, incommodities, and limitations.

It is also recognized that quality of life is a *multidimensional* construct. Several dimensions should be considered, beyond the strictly biomedical paradigm. Levels of social functioning, spiritual concerns, environmental factors (objective and subjectively perceived), all must be included. Operationalization of domains has been performed in different forms, with categorical and dimensional approaches singly or in combination. Replication of procedures is difficult. Sometimes, questionnaires developed in one context cannot be used in others. This is relevant considering that challenges posed by a life-cycle approach are compounded by cultural differences in self perception and expectations associated with old age. The "disablement process" characteristic of old age in modern societies does not uniformly affect all spheres of life. Some manifestations of sexuality, for instance, are not expected from elderly people. The "social clock" prescribes and proscribes what is correct and what is not.

The construct quality of life is *complex*. In some domains the desired direction of change is positive, in others negative. It is known that life events of all sorts, positive ones included, induce stress. Too many events are undesirable, given the additive nature of components of the "life stress" complex. They can be as detrimental as total absence of events. It is suggested, however, that negative events cause more undesirable consequences than positive ones. If quality of life shifted toward the positive end, people would have less worries and would experience positive events. What qualifies as positive or negative may be difficult to ascertain if the full range of human experience is taken into account. Complexity challenges the development of appropriate quantitative or qualitative indicators. The influence of age is particularly pervasive as well as difficult to precise; some domains may be underrepresented in the aged population and sources of satisfaction may be reduced or different. As aging is a very individual process, what gives pleasure to a person may not be important for another.

Finally, the *dynamic* nature of quality of life concepts should be considered. Passage of time does not only mean that sources of satisfaction vary in quantity and quality. It also implies that the "set point" for each satisfactor may change, that is, the criterion by which events, circumstances, and conditions are evaluated. The cohort effect should also be included in this category of influences There is a difference between growing old in the 1960s or in the 1990s. Formative influences and historical events, even technological accessibility, are different. "Dynamic" means that sources of satisfaction may vary throughout life. There may be change in the reasons why they constitute sources of satisfaction and the extent to which they fulfill their role at different life stages. The challenge is complex. Development of appropriate indicators must meet the requirements of a moving target, be sensitive to cultural factors and be based upon some empirical notion of human life change.

In addition, a critical attribute of human life is meaning, and it stems from the spiritual make up of human subjects. Without it, life experience cannot be satisfactory or have a good quality. It may be argued that life of good quality is also *meaningful* life. People not only wish to be happy, they also want to have good reasons for being happy

These characteristics of quality of life: subjective, multidimensional, complex, dynamic, and meaningful not only pose problems for quantitative or qualitative evaluation of the construct. They also have implications for bioethical dilemmas confronting healthcare workers, policymakers, and society at large.

ASSESSING CHANGE: A CHALLENGE FOR LIFE-CYCLE STUDIES

The general historical tone in the discourse concerning aging and the elderly is one of hopelessness and helplessness. Dependency upon others, physical limitations, losses of all kinds (authority, income, relationships) characterize advanced age in most societies. Disablement, impairment, and vulnerability are common. Despair, resignation, adaptation seem to constitute appropriate responses to decay. Impairment, disability, and handicap are manifest. Several "models" are used to present conceptualizations of aging (Allert et al. 1994)

At the same time, the historical record contains expressions of satisfaction and well-being, of respect and appreciation, and of pleasure and enjoyment. It may be said that the discourse on aging, both objective and subjective, is neither uniform nor unitary.

What is evident in all accounts of aging is change. Throughout life, change and transformations are salient features of the human condition. Contrary to the traditional view of reality as an aspect of perceptions that does not change, human nature and reality are subject to change. Even traits, long-standing characteristics of persons, evaluated by conventional means, show changes with the passage of time.

For assessments, qualitative and quantitative approaches are used. The most challenging aspect of quantification refers to the measurement of change. For this, a theory of dynamic constructs is needed. There are essentially two types of approaches. One uses conventional, fixed, instruments. Changes are assessed by applying the corresponding instrument to the same subjects in different moments or to different populations. For many performances, when the same cohort is analyzed change is less evident than when two different groups are assessed. This is the case for memory functions and other cognitive processes. General conclusions, under these circumstances, are difficult to draw.

The dynamic nature of self-perception throughout life makes it necessary to evaluate according to the subjects' own set point. Thresholds for change are implicit in their spontaneous manifestations and, in many domains, not only in those singled out as relevant by instrument designers. A truly comprehensive evaluation, including all aspects relevant to a person, probably is beyond what is doable. As is the case in other areas, life is inexhaustible and what remains after all scientific and technical discourses have been applied is still relevant for evaluation and appraisal.

These considerations emphasize that change is an elusive concept. In assessing quality of life, it constitutes a particularly difficult problem to approach. All the more so when age is considered. This is probably one reason why so many metaphors are employed to tackle essential aspects of human life. This has been considered as a river, as a ladder, as a journey, as a road. By means of these graphical depictions insight is gained on some of the attributes related to change.

BIOETHICS: DIVERSITY AND DIALOGUE

Life is a continuum and social discourse singles out particular events, rites of passage, or circumstances for defining periods, transitions, and accomplishments. There are several discourses and their construction depends upon audience, emphasis, expectations, beliefs, and values. It is not the same to grow old in New York or in Lima. It is not the same to be German, Afro-American or Irish, man or woman, rich or poor. In all respects, diversity increases with age, not the reverse. This should be kept in mind when elaborating on the bioethics of quality of life in old age.

As stressed elsewhere, participation and dialogue constitute keystones in bioethical thinking, development, and application (Lolas 1999). By itself, participation improves well-being and helps in the developmental process of becoming adult and getting old. Dialogue is a generic term that covers informal and formal conversations, information, and knowledge exchange, but also a tool for living together, articulating differences and consensus, and reaching moral decisions. Dialogue is a particularly important element for addressing quality of life issues in old age and in the process of dying.

It should be remarked that aging is not only a biological process. It is essentially biographical change. Although the individual self is felt as ageless, both meaning and significance of human life qualify the process of changing along the axis of time. Meaning and significance replace physical appearance and sensory enjoyment as satisfactors in life as aging occurs. They refer to the subjective and the objective aspects of being relevant or important to others.

Bioethical analyses, at the beginning of the bioethics age, were devoted mainly to transitions and crises. Beginning of life and end of life afforded opportunities for reflection. Breakthroughs in science and technology were evaluated. Critical challenges were reflected upon. Population aging, increased longevity, demographic changes are outcomes of civilization not consistently dealt with by bioethical thinking. They produce problems and pose challenges of a very mundane nature. They lack the spectacular character of breakthroughs.

A bioethics for everyday life is needed to tackle these issues. Normal people are confronted daily with dilemmas of an ordinary nature, not with vital decisions. Aging and its associated changes are everyday phenomena, appear in all cultures, and need attention everywhere. Satisfaction, as beauty, presents itself in many different forms.

MEANING AND AGING

There are two aspects of the problem of meaning throughout life and especially in old age. One is personal meaning – what everyone believes his/her life is, has been, or should be. The other is societal meaning – what societies in word and deed manifest regarding a given individual.

These two aspects rarely coincide. The individual self, for instance, is ageless. Most people feel younger than they look or are looked upon. Functional age is not chronological age, for the better or the worse.

In most Western societies the prevailing model for meaning in old age is that of deficiency. To grow old is to accumulate deficits, to lose capabilities – in sociological terms, *disablement*. Aging not only means deterioration of mental and physical abilities and vulnerability to disease. It also entails reduction in responsibilities and the loss of the social role of economic producer, good, or service provider, and active member of the community. Other societies, especially in the East, emphasize the growth of wisdom and life experiences of the elderly, the role that aged people have in preserving traditions and custom. The shift away from an oral tradition to written records has been partly responsible for the diminished importance of old members of the community in preserving tradition. Wisdom is no longer personal wisdom but data stored in devices, non-human memory.

In psychodynamic approaches, the tension between individual expectations or realizations and evaluation by others is stressed along with the need to come to terms with personal life. As previously remarked, this has been variously expressed in metaphors such as a journey, a river, a series of stages, a ladder, or a fire. A pervasive metaphor, particularly relevant for tackling issues of aging, is that of a project, a vital plan that has to be formulated, consciously pursued, and thoroughly evaluated; meaning, in old age, that both individual and societal is associated with this notion. Stages in life have been variously associated with cognitive development, adoption of social roles, moral judgment, challenges to face, tasks to accomplish, beliefs to have, and ways of articulating the world around. From a logic of rational certitudes most people in mature years derive toward a logic of conviction. Religious beliefs take prominence and autonomy of judgment is achieved. "Age conscience," as a variable, is associated not only with chronological age, but also with capacity for introspection and life events. Its importance for a full appreciation of quality cannot be overestimated (Kohlberg 2000).

The contrast between personal and social meaning is important for the development of appropriate frameworks. Linking this notion with that of change, it may be said that the pace of change differs for the individual and for the community. These two points of view rarely coincide and reflections suitable for the individual may not necessarily be useful for the group. Many frustrations for elderly people derive from different outlooks, especially diverse conceptions about who is useful, why, and how.

In the search for meaning, it is important to distinguish between the personal and the societal level. The tension between individual and group appears in every discussion of bioethical principles. Beneficence, for instance, always implies setting priorities and limits within a community. Even autonomy must be qualified and reformulated. The legitimate exercise of autonomy is in itself a strong component of life quality for any individual. To be in control of one's own life is felt to be essential for good life. Autonomy may be severely restricted in old age for some individuals. Dependency on others is felt as painful and degrading, particularly when it is not accompanied by the feeling that help is provided, not in exchange for money, but as an expression of a caring attitude.

In final analysis, the mediating role of the construct quality of life consists in harmonizing individual demands with societal interests. This harmonization constitutes an ethical imperative insofar as it involves choices and options. Values can be considered universals of meaning about what is good and what is bad. Values imply options, preferences, and behaviors translated into formal *prima facie* principles that can be filled with contents through specific rules of conduct. The ways in which principles are understood or realized in individual lives and in the life of society may differ. Every member is *homo duplex*, at the same time individual and part of a collective. The shaping of expectations, satisfactions, and anxieties occurs at the interface between individual systems of meaning and the culture of the community informed by values expressed in traditions throughout the generations.

Contrary to pedagogy, that increases similarity between persons, *gerogogy* (education for old age) is by definition an exercise in diversity. Individual characteristics should be respected in the context of each culture. This is one challenge to bioethical considerations about quality of life in old age. A common myth about old age is to assume that everybody ages the same way. One of the most significant changes needed in stereotypes concerning the elderly is the appreciation of diversity. Contrary to most formulations and declarations, that tend to speak of the elderly as a homogeneous group, discourses should be shaped to accommodate the varieties of experience aging entails. Adaptations to society and rights that can be demanded from society are different for different individuals. Respect for persons means also respect for their individual codes of conduct within the society.

Diversity is not by and in itself a negative factor − however complicated the picture it creates might be. Resources for successful aging are available in almost every society. The creative use of moral reflection consists basically in finding ways of making them usable for different persons. Also, in shaping productive beliefs about the tasks ahead and beliefs conceived of as unifying constructions that help in providing meaning to life.

END-OF-LIFE DECISIONS

One of the greatest moral paradoxes in postmodern societies is that longevity is sometimes accompanied by a lack of resources and assistance in the final years of life. Adding years to life does not always mean adding life to years. This is important when talking about quality of life in the elderly. Much of the question regarding meaning and dignity revolves around this issue: to live longer for what, with what purpose, and at what price (Hanson 1994). The very idea of becoming a burden for families or institutions or having to go through the ordeal of poverty and infirmity lead to pessimistic thoughts in many. It encourages thinking about assisted suicide, euthanasia, and orders not to resuscitate.

The two domains of bioethical reflection are the individual and the societal. Microbioethically, each person has to come to terms with his/her own life, develop a sense of appropriateness for the way he has treated others in the past and receives treatment as an elder person. Macrobioethically, societies take options on generational solidarity, justice, beneficence, dignity, and other formal principles translated into concrete rules of conduct in different contexts and cultures. In both domains, the dialogical principle prevails. Persons establish dialogues with their consciences and with their cultural past as available in books, traditions, and beliefs. Societies, and groups within societies, establish dialogues and develop forms of participation in which values are confronted, options selected, and decisions taken.

The relational quality of the construct quality of life should not be construed as relativism. However, the dialogical principle involved in health care should be consistent with behavior at the end of life, be it the personal own life or that of others. In this stage, acts can be divided into those that are *self-regarding* and those that are *other-regarding* in nature. Decisions about when to die and how to die are among the ones that occupy much time in old age. The able-bodied status of the healthy youth and adulthood is replaced by infirmity, malady, and dependency. An irreversible handicap Confronting death in the era of technology poses serious challenges, in fact the risk of being condemned to life under unpleasant and painful circumstances (Cowart & Burt 1998). Subjectively felt good quality depends heavily upon the ability to voluntarily accept or refuse treatments, indicating a degree of control over one's own body and future. As a matter of fact, health is the experience of life as a promise of good, and the need to be in charge is a strongly felt one in all stages of life, dying included. For the competent individual who is about to die there must be a time for reassuring that his/her wishes will be considered and respected. The traditional paternalistic stance of the health professions clashes with an apparent desire on the part of many elderly to be left alone or to be relieved from pain. As Boyd (1999, 159) suggests, many fears and limitations associated with old age and the proximity of death do not stem from the elderly but from middle-aged people, who project onto their parents their own anxieties regarding death and dying. A strong case can be made that myths associated with frailty and vulnerability in old age reproduce conventional knowledge shared by society at large and do not stem from actual experiences. Again, sharing meaning and anticipating old age ("*gerogogy*") is a moral imperative to deal with frustrations and anxieties near the end of life.

Intergenerational dialogue, in keeping with the dialogical foundations of bioethics, is a key issue in policymaking for achieving good quality of life for the aged population. It is expected that the recognition of diversity due to age will be incorporated to the other, more established diversities, and be integral part of decision-making bodies entrusted with legislation and regulation of social policies.

CONCLUSIONS AND SUGGESTIONS

Perhaps the social practice most intimately connected to aging in Western societies is medicine. In it the prevailing metaphor is that of a deficiency state in need of help and support. Therapy as ministration becomes the way in which society responds to aging, thus converting it into an anomaly or a disease state. The hegemony of the deficiency model permeates even the scientific discourse on aging, reproducing conventional knowledge and societal prejudices even where it seems most objective and detached.

From the moral point of view, this leads to unrealistic expectations about what to expect from social practices in general, and from medicine in particular. Realistically, medicine may contribute more to the quality of life showing its limitations, not by continuing to foster the notion that it can make a reduction in mortality or an increase in life expectancy under good life conditions. This message should reach people in all walks of life and stimulate self-gerogogy, self-help, and respectful dialogue. As Callahan (1994, 41) puts it, medicine should resist that "most powerful of all medical temptations, the vision of a direct and invariable correlation between medical progress and human happiness, that conflates the quest for meaning and the quest for health. They do not go hand in hand."

Development and maintenance of a sense of moral solidarity between the generations depends upon a public dialogue on the significance of old age in the common life of society and on a firm understanding that values are shared meanings at the individual and at the societal level. Micro and macrobioethics are needed for a full realization of expectations regarding quality of life in old age and near the end of life. Workable forms of relationship between the generations should be developed, ones based upon realistic assumptions about human nature and not on ideal obligations or universal duties. Problems, such as rationing of health care services, support, and solidarity, demand the continuous effort of dialogical construction, deconstruction, and reconstruction for demands and needs, and expectations change. The ethics needed should not be reactive, but proactive: it should anticipate challenges, not simply react to them (Callahan 1987; Lolas 2000).

The proactive stance of a dialogical bioethics should also include anticipation of challenges and preparation for old age at all stages of life, establishing meaningful bonds with those who established the traditions, with those who are contemporaries, and with those who will be in the future. Bioethics is a bridge between rationalities, disciplines, and people. It is also a bridge between ages and generations.

Fernando Lolas, Professor of Psychiatry, University of Chile, Santiago, Chile.

REFERENCES

Allert, G., G. Sponholz, and Helmut Baitsch. 1997. Chronic disease and the meaning of old age. *Hastings Center Report* 24(5): 11-3.

Boyd, K. 1999. Old age: Something to look forward to ? In *The goals of medicine. The forgotten issue in health care reform*, eds. M.J. Hanson, and D. Callahan, 152-61. Washington D.C.: Georgetown University Press.

Callahan, D. 1987. *Setting limits*. New York: Simon and Schuster.

———. 1994. Aging and the goals of medicine. *Hastings Center Report* 24(5): 39-41.

Cowart, D., and R. Burt. 1998. Confronting death: Who chooses, who controls? *Hastings Center Report* 28(1): 14-24.

Hanson, M. 1994. How we treat the elderly. *Hastings Center Report* 24(5): 4-6.

Kohlberg, L. 2000. *Die Psychologie der Lebensspanne*. Frankfurt/Main: Suhrkamp.

Lolas, F. 1999. *Bioethics*. Santiago de Chile: PAHO Regional Program on Bioethics.

————. 2000. *Bioética y antropología médica*. Santiago de Chile: Mediterráneo.

Mordacci, R., and R. Sobel. 1998. Health: A comprehensive concept. *Hastings Center Report* 28(1): 34-7.

Orley, J., and W. Kuyken. 1994. *Quality of life assessment: International perspectives*. Berlin, Heidelberg, New York: Springer Verlag.

CHAPTER THREE

LINDA L. EMANUEL

PALLIATIVE CARE

A Weak Link in the Chain of Civilized Life

To live well, a person must know how to die well. For every death well died, a few live on greatly enriched and many live on bettered by that gift. This chapter seeks to explore some of the less tangible but powerful benefits of quality end-of-life care that have implications beyond the patient's experience. It first explores the possibility that hospice and palliative care approaches may be an important component of a robust response to mortality, not only for individuals but also for societies. It goes on to explore the notion that professionalism properly understood is well exemplified by hospice and palliative care philosophy and by its history. It further suggests that disorientation with regard to professional values has allowed impoverished care for the dying. Finally, it explores the possibility that hospice and palliative care can reduce damage and assist survival despite devastating losses for individuals and society.

HOSPICE AND PALLIATIVE CARE AS PROFESSIONALISM IN ACTION

The hospice movement began as a counter-current fringe movement. Similarly, the palliative care discipline remained a field of little stature in medicine until recently. In many ways, these movements reinstituted professionalism where it was much needed and where society was little focused. Perhaps the blend of human need and social oblivion created fertile soil. The movement was not self-conscious about professionalism, but it was self-conscious about caring for human need, and this was sufficient to generate a microcosm of professionalism in action. In the last few years, this movement has penetrated mainstream medicine and social awareness, and the welcome it is receiving provides support for the view that it can return a needed source of care and professionalism.

D.N. Weisstub, D.C. Thomasma, S. Gauthier & G.F. Tomossy (eds.), Aging: Decisions at the End of Life,
31-47.
© *2001 Kluwer Academic Publishers. Printed in the Netherlands.*

Whole-Person Care

While medicine was taken up with the enthusiasm of the scientific era and the sense of the technical imperative, the hospice movement was enacting the larger goals of care for the ill that embrace care as well as cure (Saunders 1998). From the Hippocratic Oath to the Physician's Prayer to the many modern iterations of medical codes of ethics (AMA 2000), all medical codes of ethics have embraced these dual goals of cure and care.[1] Hospice initially focused on the care aspect of this dual goal, mobilizing all goals of care and restricting its population to those for whom cure was impossible. More recently the palliative care movement has joined forces with the hospice movement, seeking to bring the care philosophy to those for whom curative intent is still relevant as well (Doyle 1998). This sets the stage for a full rendition of professional medical care that pursues both cure and care.

The hospice and palliative care movements are founded upon the precept that care must include all the major dimensions of an ill person's experience. The modern movement's founder, Dame Cecily Saunders (1998) identified these dimensions as: physical, mental, spiritual, and social. In the physical domain, the aim is to relieve pain and other forms of suffering. This is so both for the sake of relief *per se* and also to allow attention to the other dimensions of life. Physical suffering, it was observed, commands attention and prevents the personal and interpersonal work of dying that can allow a peaceful departure. The consequences to a person's psyche of dealing with major and often disfiguring bodily changes, losses in physical capacity, and diminished expectations for the future, are profound. This dimension is acknowledged as valid and in need of attention and care as well. Similarly, the importance of a person's spiritual dimension as he or she faces the end of life was given high priority in the days of ancient hospice, when the prime intent of hospice was to provide the best possible preparation for a person's journey to heaven. In the modern, ecumenical hospice movement, the spiritual dimension remains important and accommodations are made for a full range of spiritually meaningful practices that people bring (Speck 1998). Further, the social context of a person's life is supported. While not specifically articulated as such, hospice builds upon the reality that a patient who is fragmenting as a person can retain a sense of wholeness if his or her family, friends, and community are supportive and if the different parties can adjust to the necessary role changes.

Family Care

The role of the family caregiver, with all its burdens, and the importance of the patient as a person who is part of a family context is endorsed (Monroe 1998; Emanuel et al. 2000). A growing body of literature supports the view that families take responsibility for a large proportion – about one-fifth – of the health care budget even in the USA (Emanuel et al. 1999; Arno, Levine, and Memmott 1999; HCFA 1998). Evidence is also accumulating that support for the caregivers can help with patient well-being; one randomized controlled trial, for instance, found that institutionalization was delayed by about one year in the group that received family counseling and support groups (Mittleman et al. 1996). To provide support to the family as well as the patient, hospice began to define the work of dying. It seems to involve settling relationships, passing on roles, leaving good memories, coming to peace in existential matters, and leaving practical matters in shape for the survivors. This work takes more than just the patient. It involves the people who have shared their lives with him or her (Parkes 1998). Hospice pioneers began to identify patterns of mourning. The importance of helping the patient to prepare loved ones for life without him or her, of anticipatory mourning by families and of mourning by patients for each loss of function or role tends to be identified. The exhaustion of caregivers was noticed and hospices respond by offering respite care. After death, hospices remain involved, identifying family members at risk of complex grief, distinguishing normal from pathological grief, and providing periodic bereavement care for long periods of time. The result of this relationship between hospice and the family is often described in mystical terms referring to the privilege of giving and receiving care and of engaging in this work. Many individuals and family groups feel transformed. Studies, while not yet numerous or advanced, also document tangible gains. For instance, shorter bereavement periods are observed for those who have received counseling than those who have not (Bass, Bowman, and Noelker 1991).

Charity Sources of Service Yield a Source of Support

There is also a practical consequence to supporting families. Hospice care can, in turn, rely upon families. The power of the community to be greater than the sum of its parts can be realized. Since the hospice movement relies greatly upon charity and family care, this is important from a logistical and financial point of view. But the potential for this synergistic creation of family and professional empowerment may not have been fully understood even today. Neither, perhaps, was it appreciated that this is another element of professionalism that was under-realized in society but well implemented by the hospice movement. This kind of care was mobilizing communities; it was engaging society in the enactment of the professionally protected value: care.

THE PROFESSIONS

To see hospice and palliative care as an enactment of professionalism begs the question: What is professionalism? Why does it matter that professionalism is well enacted?

Professionals as Protectors of Values

In modern and ancient history and across geographic and national divides, the professions have had key roles in society (Wynia et al. 1999). Most if not all civilizations worthy of the term have found it necessary to have professions (Emanuel 2001). Many civilizations on the brink of crisis have had major problems in their professions (Alexander 1949). Similarly, efforts to expand power without the civilizing effects of committed professions have been the subject of many entries in the annals of notoriety (Grodin and Annas 1996).

Why might this be so? Professionalism involves a contract between the professions and society of expertise in exchange for certain privileges (Friedson 1970), but this is not the professions' essential reason for being (Wynia et al. 1999). Professions exist in order to protect vulnerable values in society. This is enacted mostly by caring for people in potentially or actually vulnerable situations (Pellegrino 1990). History repeatedly shows that certain values, including care for the ill, but also justice for all, fairness for all, and education and respect for religions, all yield to other motivations such as power from time to time. Professional protection is not perfect, but it seems to help, and in doing so it preserves important elements of civilization. The major professions – medicine, law, religion, and education – are bound over by society to protect their designated values (Krause 1996; Brint 1994).

Imbalances in Medicine's Values

Modern United States society may not be on the brink of crisis now, but it is arguably somewhat disoriented, with its professions being caught up with self-interested or other imbalanced motivations (Johnston et al.; Emanuel and Bienen 2001). Five contributing elements to this imbalance can describe the perspective. These and other elements together provide a reasonable interpretation of how the profession failed to sufficiently preserve and promote the values of care for the ill so that hospice and palliative care was unwelcome, and the revitalizing sources that it could bring were not realized.

One element has been the growth of an autonomy-promoting consumerist mentality in society. As with the other elements, this has had important benefits in society as well as unwanted contribution to disorientations. Benefits include the reality that abrogation of human rights is perhaps harder in a society that emphasizes autonomy. Consumerism seems to foster flourishing economies without which warfare and epidemics are more likely. Incentives to listen to people can be helpful – in medicine as in general – since much wisdom exists in the voices of those who need and have experienced the services (Millenson 1999). Abolition of this or any of the other elements would be a mistake. Nonetheless, these more business-minded and politically correct Western approaches have provided some of the backdrop in which biotechnological approaches to self-enhancement have diverted medicine's needed focus on care of the ill and protection of public health (Emanuel and Emanuel 1996). Many parties promote the business – accountable and politically – accountable approach to medicine. The pharmaceutical industry drives medical decision-making by influencing physicians and consumers alike with effective social marketing techniques, and influences policy with well-funded lobbyists on Capitol

Hill. The insurance industry seeks to wield its influence in parallel ways. Large employers are similarly active as they try to manage the large portion of their expenditures that go to employees' health and retirement benefits (Emanuel 1997). None of these parties have failed to notice that much expenditure in medicine goes to intervention at the end of life.

A second imbalance relates to the value of cure versus that of care. As so often, one source of this disorientation in medicine has been the success it has enjoyed. The twentieth century scientific era has seen the medical profession reach a zenith of social respect and power. With the biomedical sciences, medicine was able for the first time to deliver results and command authoritative explanations. The goals of medicine were to save lives and eradicate disease. As laudable as these goals were and are, the price was a relative diminishment of the values of care in a broader sense of the term. Patients complained of feeling treated as if they were objects, organ symptoms, or diseases. The focus of clinical care was more on the failing organ than on the human being (Starr 1982).

In a related development, the goal of medicine and aspects of society more largely became that of overcoming death itself. People with incurable illnesses became unwelcome evidence of failure. Death became a taboo in medicine as well as society (Aries 1974). The dying and the bereaved began to feel undignified and abandoned. One underpinning of medicine, as well as that of other professions, has always been a supposition that individuals can or must reveal relevant facts regardless of social taboos and personal reasons to otherwise guard ones' confidences. When the profession adopts a taboo, then candor and truth-based therapeutic approaches become problematic. When the profession adopted and promoted the society-wide fantasy that medicine could prevail over death itself, and death became a failure and a taboo, caring approaches to inevitable dying and healing after bereavement posed a problem to medical professionals (Field and Cassel 1997; Emanuel, von Gunten, and Ferris 2000). Not surprisingly perhaps, curricular contents and collegial mentoring relationships dropped these subjects or delegated them to junior members of the team often without offering instructive guidance. The skills were lost to the discipline except insofar as individuals discovered and nurtured them in their own selves and experience-guided processes of learning (Rabow et al. 2000).

A fourth element has been the standard of wealth to which physicians and others in health care have become accustomed. In the era of fee-for-service medicine, few questioned the legitimacy of the system. Oblivious to the inherent conflict of interests in the system, and with it the incentive to provide expensive and marginally necessary care, the medical industry boomed, the national economy became dependent upon the medical industry, and the primary objectives of patient care and public health received rather less attention (Moreim 1995; Rodwin 1993). Profitable procedures predominated over less profitable forms of therapy. Articles noting the high costs of care at the end-of-life seemed to bring this wealth-oriented trend into opposition with quality care for the dying.

Similarly, the selfish element of United States society has allowed physicians to look past the abysmal state of health care among the very poor and the working poor. Professionalism should not permit its members such blindness; professionals are bound over to seek out and solve such problems. Great nations may have become great by their use of sweat labor, but their standing as true civilizations is diminished by records of failure to care for sectors in the population to the point that the failure affects morbidity and mortality. True civilizations have professionals who prevent or at least earnestly seek to correct that. While groups within the medical profession have become active in these matters, they are small and enjoy too little support from the profession's major institutions.[2]

A fifth feature may well be the reductive definition of professionalism that omits the central moral element, as noted above. The historical origins of this eviscerated understanding can probably also be traced to the ascendancy of science as the central explanatory framework of medicine. As specialization began, specialty groups broke away from general medical groups who kept moral codes at the center of their identity (Stevens 1998). As specialty groups gained power, they returned to the need for moral codes (Latham and Emanuel 1999). But in the long transition when specialty groups were powerful and disinclined to any clear ethics, the most operative understanding of professionalism was the social science rendition that described a social contract and omitted the virtues of care and the moral imperatives that arise from the vulnerabilities of illness and mortality. This may have both permitted and exacerbated the amnesic rendition of medicine that suffered from an under-emphasis on care and service. Hospice and palliative medicine did not seem to fit in this cultural context.

Whether these five are the pertinent elements or some different blend, the reality that current enacted professionalism has been missing the mark has been increasingly clear to a number of commentators (Kassirer 1995). The ill are arguably less well voiced than the largely well, and the dying may be the most voiceless of all. When the medical profession is not fully devoted to advocating for the ill, the enforced silence of illness is perpetuated and the wisdoms of the human conditions that result from witnessing or learning about the experience of illness and death are under-realized. In the search for reorientation, hospice and palliative care became more interesting to society. Educational and service delivery programs began to emerge in the mainstream of medicine and society.[3]

SURVIVING BENEFITS TO INDIVIDUAL AND SOCIETY

In the meantime, the care provided by hospice was showing the benefits it provides to people beyond the dying patient who received the central focus of care. Disseminated benefit to society more broadly is characteristic of the activities of true professionalism. It is in this penumbra of direct professional action that professionalism connects with and provides a civilizing element for its society. To see this in regard to hospice care, it is relevant to further examine the above-mentioned benefits to immediate survivors after the patient dies.

Intimate relationships

One of the devastating features of bereavement is the loss of the person with whom the meaning of life was routinely shared. "I-thou" relationships, as Buber called them, are the context of meaning for people (Buber 1965). Loosing such relationships can plunge a person into meaninglessness unless there is some possibility of meaning in other socially and personally acceptable relationships. By providing company and sharing both in anticipation of loss and after the loss, hospice workers and volunteers can provide the family members with experiences of shared meaning that supplement the family unit (Toseland and Rossiter 1989; Blake-Mortimer et al. 1999). Possibilities for making meaningful connections elsewhere may no longer be unthinkable or inexperienced.

Social Roles, Vested Roles, Missions

Loss of a family member usually entails a change in personal identity with respect to social role. Depending upon who has died, the husband becomes a widower, the mother a childless woman, the child takes on adult roles, the uncle takes over the care of additional children, and so on. Like other role transitions in life, these can go well or not well (Richardson and Kilty 1991; Richardson, Bermans, and Piworwarski 1983).

Part of the work of dying, as noted already, is to prepare the survivors for this change in role. The manager reaches the point when she can no longer work and she passes her role to the assistant manager. The dying mother gives her children blessings that speak to their future role. Children quarrel over who takes the decision-making role when the parent is sick, and somehow thereby prepare for their roles when the parent is gone. The grandfather shows his grandchildren how to remember his ways as they take on additional responsibilities to care for their aunt. The terminally ill son comforts his parents by explaining that their role and his are already fulfilled (Kushner 1981). More subtle yet profound roles, such as roles of dependency and support, can also be passed on. The dying father passes to his son the role of supporting mother in emotional as well as practical ways. The ambivalent spouses or competing siblings settle what they can of their issues, giving legitimacy to the inherited roles and perhaps preempting disabling guilt in the survivor.[4] Each of the transitions can be profoundly disturbing and accomplishing them can demand large amounts of personal effort.

In each transition, vesting survivors with a new role can give a source of pride, a reason to live on that honors the person who died, that allows the person who died to "live on within the living," that gives the survivor a place in society that is honorable. Even in the face of wholesale social decimation, this can be true (Kroeber 1973).

Bereavement Care, Memories and Legacies

The memories of how a loved one died can affect a survivor for the rest of his or her life. While many hope for a peaceful transition that includes warm memories and a tranquil continuing presence in the bereaveds, others need their loved ones to fight until the end. The needs of the bereaveds may not be the same as those of the person who dies. Whatever memory and intangible legacy a person leaves behind, dying and caring for dying offers the opportunity to complete whatever they need to. If this work can provide a transitional "emotional survival package," the bereaved may be protected from the decimation of loss.

For instance, a child who lost her mother at 11 years old reflected upon the fashion she found of making her mother's memory and legacy live for her; in all she did, she would ask herself if it would have made her mother proud. This coping mechanism was partly offered by her mother. As the mother prepared for dying, she told her daughter that she would always live on inside her. The mother was notable in that she did this without prompting; many need assistance in figuring out what it is that they can give to their survivors to help them live on. Partly the daughter created the coping mechanism by focusing the acknowledged presence of her mother in her mind to provide a kind of daily compass for sizing up situations and decisions. Many bereaveds could be afforded a slower, gentler introduction to coping if others helped them come to these kinds of realizations and internal orientations.

Many hospices provided both professional counseling and volunteer support groups to which caregivers and bereaveds could belong, so providing experience not only with hospice people but with other members of the community. Importantly, these affiliations are born around the work of bereavement, the experience that otherwise might be difficult to share. Profound experiences that cannot be shared place the person at high risk for becoming isolated in that dimension. So these relationships became a kind of preventive care, preempting some of the paralyzing forms of grief that can afflict the bereaved (Parks 1998).

Frameworks of Meaning, Belonging, Bridging, and Preserving Values

Legitimization of these transitions allows for an important affirmation of a person's origins. Whether it is a young child who must go forward without parents, or parents who must live out their lives after their child has died, origins are an anchor to the present and the future. Affirmed in where one comes from, it may be easier to move forward. Knowledge of the past helps to bridge to the future. Often, survivors remark that the hard part is remembering. They may note a continuing sense of loosing one's origins, of loosing the realities they once lived by and for (Parrish 1972). The work of bereavement can articulate or otherwise affirm these owned identities and origins, and the witnessing of them will have made them more secure. The supported survivor can own and belong to their origins better, even when these origins are past and gone.

Most, if not all, of us live within interpretive assumptions that allow us to discern meaning in events (Parsons 1954). Many live by interpretive frameworks that encompass dying long before they face death themselves. For these people the work of dying and bereavement need not create new frameworks, but rather discover their existing resources. Others need to create frameworks that can accommodate the challenges of mortality. Whether existing or newly created or modified, the work of bereavement makes these interpretive frameworks emerge. When the work of dying and bereavement is done with people beyond the immediate family, a community can also emerge that endorses and lives by – in part or in full – a common framework of meaning (Kleinman and Kleinman 1991; Kleinman 1986; Berdes 2001).

One of the challenges of being bereaved is to preserve the values espoused by and embodied by the deceased. All of these above activities and states lend affirmation to the values that the deceased person espoused in their relationship to the person who is now bereaved. Sometimes, the values of the deceased person were unhealthy, and additional personal work may be needed to find something that is comforting to hold on to. Either way, the question is how survivors can preserve the values that they wish to and need in order to provide continuity for their lives and the lives of their family and community. One of the features of having provided the experience of care by hospice and its volunteer team is that care can help the good values to live on and the difficult aspects to settle down.

Survivors of traumatic experiences that fall well beyond the range of experience of those they return to or live with after the experience suffer a paralyzing isolation that seems to block healing (Glicksman and Van Haitsma 2001). In cases of epidemics, such as the AIDS epidemic, an opportunity often exists to bring survivors together to share experiences. For a survivor, it can be critically important to find someone or a community of people with the ability to not only empathize but to resonate and validate experience based upon parallel experience. Partnerships based upon this alone can be profound and lasting (Strumpf et al. 2001).

One bereaved husband remarked, one year into his efforts to raise an infant baby girl, that "every day it gets harder, not easier." His wife had died in a car crash, and he had not had the benefit of preparation for dying. Because they had recently settled in a new country, he had not found a community to support his mourning. There is growing awareness that of all deaths to suffer, a death alone is perhaps the worst. But, there is less awareness that of all bereavements to suffer, to mourn alone is perhaps the hardest.

Sources of suffering and of healing for survivors impact individuals and the larger community. Each source of suffering can spiral down into pathological states for individuals and communities or they can move forward positively (Parkes 1998). When professions are so disoriented and society is so little infused with its values that the dying and the bereaved suffer alone or receive care that diminishes the importance of what they face, the chances then fall that an individual can die well or recover in a healthy fashion. Strong and long lasting civilizations provide extensively for the bereaved, with rituals that insist on remembering the life cycle that includes death and that insist on making a community supportive and on providing public recognition of the dead. In these rituals are the seeds of new communities; communities that confirm the importance of the bereaved's origins with the deceased and that can move on with a shared past and a shared future in a newly coalesced community that is made authentic by connecting the past with the future.

Preserving the Root Civilization

As protectors of essential but vulnerable values, professionalism is particularly important in times of social crisis. The values of care, justice, education, and others are values that can help to bring about healing from social crises. When these values are little understood, little articulated, and little lived by, recovery can be especially difficult. Occasionally people face tragedies so large that not only their personal world risks annihilation with the loss of a loved one but an entire community or nation risks annihilation. Sometimes just a few remain and must face the challenge of rebuilding a lost community. Sometimes those who remain are too old or too young or too sick or too hurt to readily accomplish this task.

Hospice workers have so often seen the seemingly impossible renewal in the face of tragedy that it seems to them that miracles are expectable, even normal. Something about treating the dying as whole people with a whole family and community context, something about emphasizing anticipatory mourning and bereavement support, something about emphasizing care when cure is impossible, has provided the survivors with the ability to go on. The survival kit that hospice has seen them use has included: a strong identity with their origins with the now deceased; a secure, authentic passage of role; a few people or a community to share a framework of meaning; and a reconstituted community with whom to recreate the lost community.

Fitting Within Financial Constraints

All of these activities may seem expansive and cost intensive. Quality end-of-life care is not cost free. After all, it takes time, and peoples' time tends to be an expensive commodity. However, hospice uses rather little highly paid time. Rather it creates an interdisciplinary team of professionals, most of whom command less burdensome salaries than physicians do, and non-professionals. It mobilizes volunteers and maximizes family support. This is a model used most effectively by societies in crisis, and is possibly the most cost-effective model available. One study examined a single element of hospice and palliative care when it conducted a randomized trial of support groups for family caregivers in a Veterans' Administration population. It found that the intervention cost approximately $700 per person as compared to $7,500 per person in the control group for whom additional care was eventually necessary (Peak, Toseland and Banks 1995).

In sum, it is a reasonable hypothesis or interpretation to state that the professionalism enacted by hospice and palliative care may be demonstrating how it can contribute to the stability of civilization even in the face of significant challenge. The next section will consider possible tests of this idea by asking if hospice and palliative care can mitigate the damage of some current challenges to society.

CURRENT NEEDS IN SOCIETY: CAN QUALITY END-OF-LIFE CARE HELP?

End of life issues have been intertwined with two major challenges in United States society, being both part of the answer to urgent concerns and the beneficiary of the creative energies that the challenges engaged. These are the physician-assisted suicide and euthanasia movement and the AIDS epidemic.

Physician-Assisted Suicide and Euthanasia Movements in the U.S.

Debate will continue as to whether the demand for legal assistance in the voluntary ending of a terminally-ill patient's life is moral and defensible and whether it benefited the end-of-life care movement (Emanuel 1998b). The issue here, however, is only to indicate that quality care at the end of life could provide a large part, and perhaps almost all, of the answer to the root causes of suffering that were making people interested in physician-assisted suicide and euthanasia.[4]

Empirical studies have repeatedly confirmed that people considering the option were troubled by mental suffering such as depression, fear of abandonment, fear of indignity, and fear of (more than actual) pain. Similarly, they were troubled by social suffering, especially fear of being a burden and of spending family financial resources that were intended for others (Van der Maas and Emanuel 1998). The aging are a particularly vulnerable group in all these spheres since they may be widowed and alone or their spouses may be sick; they have more chronic disabilities including cognitive losses. Frailty is likely to be a predisposing factor to all the fears that correlate with interest in physician-assisted suicide or euthanasia.

These sources of suffering may be especially acute because of the culturally promoted death denial that prevents anticipation of and experiences of dying well. Hospice and palliative care addresses each of these needs. A practical and non-judgmental approach to the moral controversy allowed hospice and palliative care workers to get on with caring for the sources of suffering, treating depression, providing dignity in care and attention, and addressing the burden on the caregiver and family (Emanuel 1998a). Anecdotal evidence suggests that this approach allows the entire question of physician-assisted suicide or euthanasia to melt away in the face of a common pursuit of quality palliative care, both for the patient and for those he or she cares about.

Respect for the person and the connected community and care for both were the ingredients that had been missing. The drive to physician-assisted suicide and euthanasia was a manifestation of the suffering that occurs when it is missing. Hospice and palliative care could provide the missing care. Commentators have begun to argue that the physician-assisted suicide and euthanasia movement, like the individual requests, is receding. A return to professionalism in the form of hospice and palliative care seems to have provided an important part of the most effective and respectful response to the movement.

Empowering AIDS-Affected Communities

When the gay community in the United States became affected by the AIDS epidemic, the world began to witness a remarkable renewal in that community. The strength, compassion, and resourcefulness of the response was responsible for the renewal of this community in the face of loss. It was also responsible for a number of other startling changes. It generated a profound reexamination of the way in which scientific research is done and the way in which drugs are made available. Perhaps most important for this thesis is the fact that it demanded explosion of taboos that were preventing open support of the needs of the dying and bereaved. And emerging from it all was the power of care itself; it provided a great boost to the affected community and to the hospice and palliative care movements.

The international community is waking up to the fact that AIDS has all but wiped out an entire generation of adults in some African countries, and China, India, Russia, and Eastern European regions are following similar trends. With the aged and children being left alone to head households, often with a number of infants who are already HIV infected, and some of the older children too, the prospects for survival as a civilization with roots and a future are dire.

Part of the task again involves explosion of taboos that prevent people from doing the work of dying and the work of bereavement. While the social structures may be strong, and while some of the societies may be healthy and full in their approach to death and use of ritual, in many it remains nearly impossible to discuss AIDS and its routes of transmission. Medical workers in these societies should consider it their duty, as professionals, to address these issues despite taboos.

Isolation of AIDS sufferers prevents the work of dying. Empowering them seems a better approach. Otherwise, bereaved children are left without the sources of renewal that might enable them to go on. As appalling and impossible as it may seem that teenagers must head households and bear the burden of sustaining income generating or agriculturally productive work, the reality is that some youngsters can manage this if they are supported, even with remarkably little. The history of modern America's origins includes many stories of orphaned children who crowded into household already desperate from their own losses (Polk 2000). Since the choice for social policy is between assisting the youngsters and turning away in desperation, the social and moral calculus is simple. As a matter of human decency, and as a matter of stability for civilization, assistance is warranted and a top priority. Beyond the argument of obligation is the possibility of a better chance of survival as an identifiable tribes and cultures rather than annihilation for many groups. Africa is ahead of India, China, and Russia in its virtual loss of a generation. However, these continents could be not far behind.

Hospice and palliative care is primarily about care for the dying. While people may resist this type of care because of fears of dying, the type of care itself is not a taboo. It provides an opportunity for professionals to address what is a taboo without violating norms in a destructive fashion. What hospice and palliative care can also bring is a vehicle for doing the work of preparation for dying and the work of bereavement despite the taboos around AIDS. When the trust of families has been established education about preventive health measures can also be introduced, making another bridge back to sustainable recovery. This in turn can be linked to other AIDS/HIV prevention programs (Pfizer Foundation 2001). A major element in any effective strategy for averting further catastrophe in all countries should probably be the effective implementation of hospice and palliative care.

Violent Conflict

It is even possible that the hospice and palliative care approach could assist in providing experiences of humane care that could help prevent and limit local violent conflict. Consider teen violence. Recently, hospice programs in the United States have begun to involve teenage volunteers (Lo 2001). Accounts of the maturing experiences that this offers lends hope to the idea that youth may glean the same grounding effects on older adults (Romer 2001; Mahood and Romer 2001). At a time in life and society when youth violence is high, it is possible that these experiences of care and quality dying can provide an alternative avenue for otherwise destructive tendencies. It is also possible that the hospice and palliative care approach could help recovery from destructive conflicts that have already occurred. This may be true in local communities, where children who are not supported as they face losses are more likely to seek the company of gangs and other unstable social groupings.

It may also be true in wartime that hospice and palliative care approaches could be helpful. Possibly this is too optimistic. But it is also possible that revealing common elements of humanity, in the ways that occur with witnessing death and caring for those involved, could make the cruelty of war more apparent. Empathic experiences of the depth of suffering that occurs around death, and some community knowledge of the different degrees of suffering when death is violent or sudden, can help in this regard. Understanding the types of creative approach to crisis can make the disposition of a culture positive rather than negative. The counterpart to understanding cruelty is to have an alternative. Experiences of creativity in caring can provide assistance in seeking creative alternatives.

Similarly perhaps, recovery from the losses of war can be assisted by hospice and palliative care activities. Needed sources of survival may match those needed to recover after the losses wrought by major epidemics. The stresses of being involved in the dissolution of life into earth or the explosion of life into dismemberment and then earth have parallels. Each may cause a trauma and a resulting stress disorder that can perhaps be mitigated by finding company in the experience and a community in that company. Perhaps that community can recover better (Parrish 1972; Emanuel 1995).

Practical Implementation of and Limits to the Model

Naturally, there are limits to what hospice and palliative care can do. While a strong argument can be made for including this approach in preventative and in recovery programs, it has not been systematically tested. Even if it were to be demonstrated successful, it is not likely to be even close to a panacea. It can foster and synergize with a wide range of other recovery mechanisms but it cannot be all of them.

Currently, plans to implement or augment training in hospice and palliative care are under development in a number of Eastern European and African countries. Each pursues its own design and implementation. In many cases the initial motive in providing this training and care was simply to meet the needs of the dying. But equally, there is understanding that this kind of care assists the family and community and a related goal or happy side effect of the initiatives may be to disseminate care practices into the daily life of people who are trying to pull through in the face of or after crisis.

CONCLUSION

This chapter has examined some of the beneficial effects of quality care at the end of life. Articulating a view of professionalism, it has suggested that modern western society has been disoriented in its professionalism and that, as a result, end-of-life care has suffered. It has suggested that the absence of attention to dying and care has allowed lapses in our civilization. Explaining the elements of hospice and palliative care that exemplify professionalism, the chapter has further argued that its growth on the fringes of society and then its movement into mainstream medicine and society can be a source for our reorientation. Four crises have been considered in this chapter. First, the physicians-assisted suicide and euthanasia movement seems to have responded well to the caring approach of hospice and palliative medicine. Second, the AIDS crisis in the United States has galvanized a set of resources characteristic of what hospice and palliative medicine can provide. Third, the AIDS crisis in Africa and beyond, it was postulated, could benefit significantly, even perhaps providing a component of survival for affected groups. The approach and its implementation could provide a source of improved stability for the world more generally. Fourth and perhaps related, violent conflict was considered and the possibility explored that hospice and palliative care could be an element of preemptive approaches and an element in reaching resolution and survival after conflict has occurred. Some proposals to implement and proposals implementing this kind of care were described.

Linda L. Emanuel, Buehler Professor of Medicine and Director, Buehler Center on Aging, Northwestern University, U.S.A.

NOTES

[1] See: Oath of Hippocrates. In *Ethics in medicine: Historical perspectives and contemporary concerns,* eds. S.J. Reiser, A.J. Dyke, and W.J. Curran, 5. Cambridge, Mass: MIT Press, 1977.

[2] See, for example, Physicians for National Health Policy (www.pnhp.org).

[3] See: American Association of Colleges of Nursing, *The ELNEC curriculum,* Available at: URL: http://www.aacn.nche.edu/elnec/curriculum.htm (updated April 24, 2001); EPEC.net (Education for Physicians on End-of-Life Care), *The EPEC curriculum.* The EPEC Project Northwestern University Medical School, Chicago IL 2000. Available at: URL: http://www.epec.net; and Millbank Memorial Fund 2000. *Pioneer programs in palliative care: Nine case studies.* New York: Millbank Memorial Fund.

[4] See AMA Amicus Brief for *Washington* v. *Glucksberg,* 521 U.S. 702 (1997), Chicago, IL: American Medical Association.

REFERENCES

Alexander L. 1949. Medical science under dictatorship. *New England Journal of Medicine* 241: 39-47.

American Medical Association (AMA). 2000. *Code of ethics.* Chicago, IL: AMA.

Ariès, P. 1974. *Western attitudes toward death: From the Middle Ages to the present.* Trans. by P.M. Ranum. (Orig. title *Essais sur l'histoire de la mort en Occident*) Baltimore: John Hopkins University Press.

Arno, P.S., C. Levine, and M.M. Memmott. 1999. The economic value of informal caregiving. *Health Affairs* 18: 182-8.

Bass, D.M., K. Bowman, and L.S. Noelker. 1991. The Influence of caregiving and bereavement support on adjusting to an older relative's death. *Gerontologist* 31: 32-42.

Berdes, C. 2001. *Sense of community in residential facilities for the elderly.* Ph.D. diss., Northwestern University.

Blake-Mortimer, J., C. Gore-Felton, R. Kimerling, J.M. Turner-Cobb, and D. Spiegel. 1999. Improving the quality and quantity of life among patients with cancer: A review of the effectiveness of group psychotherapy. *European Journal of Cancer* 35: 1581-6,

Brint, S. 1994. *In an age of experts: The changing role of professionals in politics and public life.* Princeton, New Jersey: Princeton University Press.

Buber, M. 1965. *Between man and man.* New York, NY: Macmillan.

Emanuel, L.L. 1995. The privilege and the pain. *Annals of Internal Medicine* 122: 797-8.

———. 1997. Professional standards in health care: Calling all parties to account. *Health Affairs* 16: 52-4.

———. 1998a. Facing requests for physician-assisted suicide: toward a practical and principled clinical skill set. *JAMA* 280: 643-7.

———, ed. 1998b. *Regulating how we die.* Cambridge, MA: Harvard University Press.

———. 2001. Ethics and health care services. In *Essential issues for leaders: Emerging challenges in health care,* 37-64. Oakbrook Terrace, IL: Joint Commission Resources.

Emanuel, L.L., and L. Bienen. 2001. Physician Participation in Executions: time to eliminate anonymity provisions and protest the practice. *Annals of Internal Medicine* [in press].

Emanuel, E.J., and L.L. Emanuel. 1996. What is accountability? *Annals of Internal Medicine* 124: 229-39.

Emanuel, E.J., D.L. Fairclough, J. Slutsman, H. Alpert, D. Baldwin, and L.L. Emanuel. 1999. Assistance from family members, friends, paid care givers, and volunteers in the care of terminally ill patients. *New England Journal of Medicine* 341: 956-63.

Emanuel, E.J., D.L. Fairclough, J. Slutsman, and L.L. Emanuel. 2000. Understanding economic and other burdens of terminal illness: The experience of patients and their caregivers. *Annals of Internal Medicine* 132: 451-9.

Emanuel, L.L., C.F. von Gunten, and F.D. 2000. EPEC Series: Gaps in end-of-life care. *Archives of Family Medicine.* 9: 1176-80.

Field, M.J., and C.K. Cassel, eds. 1997. *Approaching death: improving care at the end of life. Committee on Care at the End of Life, Institute of Medicine Report.* Washington, DC: National Academy Press.

Friedson, E. 1970. *Professional dominance: The social structure of medical care.* Chicago: Aldine.

Glicksman, A., and K. Van Haitsma. 2001. Distant and recent trauma – The experience of Soviet holocaust survivors. *Journal of Clinical Geropsychology.* [Forthcoming]

Grodin, M., and G. Annas. 1996. Legacies of Nuremberg: Medical ethics and human rights. *JAMA* 127: 307-8.

Health Care Financing Administration (HCFA), Office of the Actuary, National Health Statistics Group, Oct. 29, 1998. Cited in *Always on call: When illness turns families into caregivers,* ed. C. Levine, 6. New York: United Hospital Fund of New York, 2000.

Johnston, S., S.R. Cruess, R.L. Cruess, and L.L. Emanuel. *The inherent conflict of roles in medicine's professional associations: An exposition and management strategies.* [Submitted manuscript].

Kassirer, J.P. 1995. Managed care and the morality of the marketplace. *New England Journal of Medicine* 333(1): 50-2.

Kleinman, A. 1986. Understanding the impact of illness on the individual and the family. *Journal of Psychosocial Nursing and Mental Health Services* 24: 33-6.

Kleinman, A., and J. Kleinman. 1991. Suffering and its professional transformation: toward an ethnography of interpersonal experience. *Culture, Medicine & Psychiatry* 15: 275-301.

Krause, E.A. 1996. *Death of the guilds: Professions, states and the advance of capitalism: 1930 to the present.* New Haven, CT: Yale University Press.

Kroeber, T. 1973. *Ishi: Last of his tribe. The true story of a boy from a lost civilization.* New York: Bantam Books.

Kushner, H. 1981. *When bad things happen to good people.* New York: Little, Brown & Co.

Latham, S., and L.L. Emanuel. 1999. Professionalism challenged. In *Medical ethics and professionalism,* ed. R. Baker, A. Caplan, S. Latham, and L. Emanuel. Baltimore: Johns Hopkins University Press.

Lo, J. 2001. On being a teen volunteer. *Journal of Palliative Medicine.* 4: 129-31.

Mahood, S., and A. Romer. 2001. The hospice teen volunteer program at the hospice of the Florida Suncoast. *Journal Palliative Medicine* 4: 117-28.

Millenson, M. 1999. *Demanding medical excellence: Doctors and accountability in the information age.* Chicago, IL: University of Chicago Press.

Mittleman, M.S., S.H. Ferris, E. Shulman, G. Steinberg, and B. Levin. 1996. Family intervention to delay nursing home admission: A randomized controlled trial. *JAMA* 276: 1725-31.

Monroe, B. 1998. Social work in palliative care. In *Oxford textbook of palliative medicine,* eds. D. Doyle, G.W.C. Hanks, and N. MacDonald, 867-80. Oxford: Oxford Medical Publications.

Moreim, E. 1995. Balancing act: The new medical ethics of medicine's new economics. Washington, DC: Georgetown University Press.

Parrish, J. 1972. *12,20 and 5: A doctor's year in Vietnam.* New York, E.P. Dutton.

Parkes, C.M. 1998. Bereavement. In *Oxford textbook of palliative medicine,* eds. D. Doyle, G.W.C. Hanks, and N. MacDonald, 995-1010. Oxford: Oxford Medical Publications.

Parsons, T. 1954. *Essays in sociological theory.* London: Free Press.

Peak, T., R.W. Toseland, and S.M. Banks. 1995. The impact of a spouse-caregiver support group on care recipient health care costs. *Journal of Aging and Health* 7: 427-49.

Pellegrino, E. The medical profession as a moral community. *Bulletin of the New York Academy of Medicine* 66: 221-32.

Pfizer Foundation. 2001. Major study in Uganda to identify best community-based approaches for HIV/AIDS prevention. Press release (June 25). Available at:
http://www.pfizer.com/pfizerinc/about/press/ugandastudy.html

Polk, W.R. 2000. *Polk's folly. An American family history.* New York: Doubleday.

Rabow, M., G. Hardie, J. Fair, and S. McPhee. 2000. End-of-life care content in 50 textbooks from multiple specialties. *JAMA* 283: 771-8.

Richardson, V., S. Bermans, and M. Piwowarski. 1983. Projective assessment of the relationships between salience of death, religion, and age among adults in America. *Journal of General Psychology* 109: 149-56.

Richardson, V., and K.M. Kilty. 1991. Adjustment to retirement: continuity and discontinuity. *International Journal of Aging and Human Development* 33(2): 151-69.

Rodwin, M.A. 1993. *Medicine, money and morals: Physicians' conflicts of interest.* Oxford: Oxford University Press.

Romer, A. 2001. Mutual developmental benefits for teen volunteers and persons at the end of life. *Journal of Palliative Medicine* 4: 113-6.

Saunders, C. 1998. Forward to *Oxford textbook of palliative medicine,* eds. D. Doyle, G.W.C. Hanks, and N. MacDonald. Oxford: Oxford Medical Publications.

Speck, P. 1998. Spiritual issues in palliative care. In *Oxford textbook of palliative medicine,* eds. D. Doyle, G.W.C. Hanks, and N. MacDonald, 805-14. Oxford: Oxford Medical Publications.

Starr, P. 1982. *The social transformation of American medicine: The rise of a sovereign profession and the making of a vast industry.* New York: Basic Books.

Stevens, R. 1998. *American medicine and the public interest: A history of specialization.* Berkley, CA: University of California Press.

Strumpf, N., A. Glicksman, R. Goldberg-Glen, R.C. Fox, and E.H. Logue. 2001. Caregiver and elder experiences of Cambodian, Vietnamese, Soviet Jewish, and Ukranian Refugees. *International Journal of Aging and Human Development.* [In press]

Toseland, R.W., and C.M. Rossiter. 1989. Group interventions to support family caregivers: a review and analysis. *Gerontologist* 29(4): 438-48.

Van der Maas, P., and L.L. Emanuel. 1998. Factual Findings. In *Regulating how we die: The ethical medical and legal issues surrounding physician-assisted suicide.* Cambridge, MA: Harvard University Press.

Wynia, M., S. Latham, A. Kao, J. Berg, and L.L. Emanuel. 1999. *Physician professionalism in society.* New England Journal of Medicine 341: 1612-5.

CHAPTER FOUR

GERRIT K. KIMSMA

ASSISTED DEATH IN THE NETHERLANDS AND ITS RELATIONSHIP WITH AGE

AGE AND ASSISTED DEATH

In a volume on aging, with an expose on the issue of euthanasia and assisted suicide in the Netherlands, at least two questions must be addressed. The first one is a general description of the Dutch position on and practice of physician-assisted death, be it either euthanasia or assisted suicide. The second question concerns the relationship between age and euthanasia. This relationship for some or many, depending upon the point of view, is or is perceived to be problematic. The most commonly seen version of that problem is a claim that: if euthanasia were a socially acceptable option, then older people must be aware that they might be a population at risk for unwanted "euthanasia."

I venture some preliminary remarks and allow myself some digressions in this contribution, for the sake of liveliness. The rigidity of my paper may suffer from it, but I also hope that the understanding of what we are doing in the Netherlands is enhanced positively by this liberty.

I place the above word "euthanasia" in quotation marks because in the Netherlands, and I hope in the rest of the world as well, we would call the unwanted ending of a life murder. In many countries to assist someone in ending a life, even if desired, is a crime. Where assistance is not considered a crime, as in Switzerland, a different approach has developed between the official medical position and what lay people, organized in "Exit," are allowed to do. Assistance in dying might be allowed, if and when in the view of the attending physicians it would be a considered and inescapable option to alleviate the suffering of a patient with whom communication is impossible. But then the gray area between unacceptable shortening of life and the doctrine of double-effect has been crossed, and in the latter case a majority of physicians might find this approach acceptable. The Dutch position, however, is linked to voluntary euthanasia as an option for a suffering yet fully conscious patient who has no more alternatives to improve his medical

D.N. Weisstub, D.C. Thomasma, S. Gauthier & G.F. Tomossy (eds.), Aging: Decisions at the End of Life,
49-65.
© *2001 Kluwer Academic Publishers. Printed in the Netherlands.*

condition, with only a short stretch of life to live that is, in the opinion of a patient, miserable and beyond human dignity. It is *this* option of choice of a human being to decide the time of his own death that seems to be the troublesome issue for many. This conviction of mingling with what should be left alone, at least not be left to choose for an individual human being, runs like a deep current through the history of Western thought. Contrary to a commonly held opinion that this idea has its roots in Christianity, similar ideas can already be found in early Greek thought, even though the opposite positions, that one can choose to die at a whim, can also be observed, as with the Stoa. In Plato for example, objections against suicide are formulated to the effect that life is like a pre given period in time, that cannot be dealt with at will: one must maintain his station like a soldier or stay the presence as a prisoner should stay his term of conviction. In one form or another, the idea is that life should be lived as long as the final breath, as if there is something like a "natural" life span that extinguishes like a candle. It is not equal to a possession that can be handled as one chooses. Notions like these return in most objections against assisted death, be it physician-assisted or lay-assisted.

More outspoken legalistic versions of this idea can be observed in the Middle Ages, where suicide carried a penalty, sometimes even a death penalty, because one's body and one's capacity to earn belonged to a nobleman, who should not be robbed of worldly possessions. In this period, Church oriented and state oriented convictions together had the effect that people who committed suicide were not allowed to be buried in a normal fashion, in the normal places, enlarging the omen of evil for them.

The modern argument involving the elderly is one particular form of this general reservation towards notions of (assisted) suicide.

PHYSICIAN ASSISTED DEATH IN THE NETHERLANDS: A GENERAL PICTURE.

I shall draw some lines of argument and hope to take away some misperceptions in this area about the Dutch practice of euthanasia.

In the first place, the history of euthanasia in the Netherlands *as a public issue* is a short one, of no more than 25 years. There is, unlike in Germany, Great Britain, and the USA, neither extensive long-term history nor short periods of intense public debate. When this debate initiated in these countries around 1870 and around the turn of the century, there was no Dutch participation (Meerman 1991). Neither was there a long history in the area of organizations to advance the cause of voluntary euthanasia: the Dutch Society for Voluntary Euthanasia was established in 1973, after the "cause celèbre" of the case that made physician-assisted death an option without criminal prosecution (Kimsma and Van Leeuwen 1998). In the USA and in Great Britain, euthanasia societies were established in the thirties of the previous century, while proposals to legalize physician-assisted death date back even earlier. Both in the states of Ohio and Iowa laws to allow euthanasia in cases of incurable diseases or accidents were introduced and defeated already in 1906 (Persells 1993; Emanuel 1998).

In the second place, *euthanasia* and *physician-assisted suicide*, jointly captured under the term *physician-assisted death*, have become legally accepted practices in the Netherlands since the acceptance of a bill in the First Chamber of Dutch Parliament in April 2000. This bill actually codifies what has been the practice in this country. For people who tend to think that something is either legal or illegal, the until recent Dutch position is rather confusing, but that has to do with an approach to certain practices with a difference between the formal and the factual status. Euthanasia fell under this category of tolerance, in Dutch: "gedogen." It boiled down to the simultaneous existence of both a formal prohibition and a factual tolerance. If I may allow myself a comparison, then this seemingly contradiction comes closest to the Roman Catholic approach to certain morally debatable practices: a strict approach in the pulpit, a more lenient approach in practice, more or less the difference between public and private.

This actually confirms the "status aparte" of euthanasia, both medically and legally. Medically speaking *euthanasia is not normal medicine*: there is no medical standard. A patient may ask for an active end to life, but a physician can refuse on grounds of conscientious objections. Legally speaking, there is a status aparte, because this concerns a crime, where the perpetrator/physician has to report his own transgression, in clear violation of a principle of Roman law known as "nemo tenetur" (Woretshofer 1994).

In the present law, euthanasia and assisted suicide are considered crimes unless they are performed by a physician who is required to comply with certain conditions of a careful medical practice. Since jurisprudence in 1974, the message has been relayed that shortening the life of a suffering patient would not necessarily lead to criminal prosecutions if conditions were met. That is one important aspect any student of the Dutch situation must understand quite clearly. It is the legal establishment and not primarily the medical establishment that made physician-assisted death an acceptable act. Only after "the law" allowed this option, the Royal Dutch Medical Society as well as the official Dutch organization changed its official position and started working together with the government to develop safeguards. This development, of course, could neither take place without a popular grounding nor without adequate support of both legal and medical professional peers: a majority nowadays of more than 90% of the Dutch support the option of euthanasia as does the profession with almost similar figures. In a certain way, one can see this development as a response by these professions to the changes of the time. The seventies are the period of individualism, liberties and self-determination, as opposed to group identities, paternalism, hierarchy and authority, especially religious authority (Griffiths, Bood, and Weyers 1998).

The *safeguards* against abuse are to be found in the conditions that will not lead to a prosecution in the case of an unnatural death that is being reported. These conditions are 4 in number, two material conditions, and two more procedural.

The two *material conditions* are:

1. presence of voluntary request of a competent patient, and
2. presence of a state of suffering that cannot be redressed because there is no cure available.

The two *procedural conditions* are:

1. consultation by a medical colleague, to establish whether the material conditions have been fulfilled, and
2. extensive and detailed reporting for outside evaluation and checking.

The wording of these conditions has been developed in the court cases after 1974 through 1994. Two cases have reached the highest Dutch Court, in 1984 and 1994, and have resulted in the continuance of physician-assisted death. Since 1994 there is a procedure that falls short of legalization of assisted death, but nevertheless functions as such. It is a procedure that is part of the Burial Law, with the obligation for physicians to report any case of unnatural death, and with a wording on the conditions that need to be fulfilled in order to prevent prosecution.

The Dutch Government introduced a bill in 1999 to decriminalize physician-assisted death and the bill passed in 2000. This bill contains prudential safeguards, similar to the safeguards that are in place while the criminal nature of these acts is still on the books. Parliament had to decide whether to turn the proposal into a law. Given the combination of the present Government parties, the Dutch now have the first nationally valid law on physician-assisted death in the world.[1] One of the safeguards in this bill is that physicians must report any case of unnatural death to the legal authorities, as they must nowadays. Their report is relayed by a local community physician to one of five Regional Euthanasia Evaluation Committees that evaluates the facts of the case. It consists of a physician, an ethicist and a lawyer, acting as extended representatives of "the public" as opposed to a limiting of this evaluation to the medical profession.

In this committee the acceptance of cases of euthanasia is decided and the evaluation is based upon the submitted reports and is marginal in nature and in principle. Only when questions arise with respect to the conditions, physicians are invited to provide further information, including furnishing details in writing, and sometime appear in person before the commission. Then the marginal evaluation becomes complicated and sometimes leads to a recommendation for further investigation by the "medical police," the state organization of medical inspection (Van Leeuwen and Kimsma 2000).

To avoid a political stalemate between the opposing Christian Democrats and the advocating Socialists, who formed Dutch government, it was decided in 1988 to initiate research into the extent and nature of medical decisions at the end of life. This research was extremely successful and was repeated in 1995. A third, similar, project is intended to be carried out for 2001, with a probable report in 2002. The presently known figures are considered reliable because of the legal immunity for the interviewed physicians (no or less socially desirable or "safe" answers) and the confirmation between retrospective and prospective parts of the study.

I shall select some figures out of the multitude and concentrate on the "medical decisions at the end-of-life," including physician-assisted death (Van der Maas, Van Delden and Pijnenborg 1991; Van der Wal and Van der Maas 1996). The extent of the involvement of the medical profession in 1990 was in 38% of all deaths, and in 1995 in 42%. The medical decisions as a group are subdivided in: physician-assisted death (1995:3,4%, in absolute figures 4,500 people), alleviation of pain and symptoms with the possible effect of a shortening of life (1995:18,5%) and death in the course of withholding or withdrawing life-prolonging treatment (1995:20%).

Of these figures there are a few that were disquieting and were the reason for reflection on further policy developments. One concerns the category of patients whose life was shortened without their explicit request and the other figures concern the number of reported cases.

The subcategories of *physician-assisted death* are: euthanasia, physician-assisted suicide and, the "discovery" of a group of patients whose life had been shortened in the course of their care, while they were *incommunicado: the ending of life without a patient's explicit request (LAWER)*. There has been a debate between proponents and opponents about this particular way of grouping, with the argument that the group of patients whose life was or might have been shortened by the use of pain medication should be counted also among the physician-assisted death group. In countries other than in the Netherlands, this action falls under the heading of "passive euthanasia," but in the Netherlands the term passive euthanasia has been abandoned. The justification for this change in terminology is that the adjective "passive" does not do justice to or even denies the active involvement of physicians in the decision-making. Physicians act, even though patients die a slower death because of these actions, both in the area of alleviation of pain and suffering, and the area where treatments are withheld or withdrawn. The background of this change in terminology is a growing consensus based upon the limitation of the Dutch definition of euthanasia. Since about 1980 we tend to speak of euthanasia only when there is an explicit request of a patient to have a life ended. This means that the medical subject, *the patient, has been replaced by the legal subject, a competent person,* who can make a treatment contract with a physician. Behind this semantic shift is the presupposition that this replacement of subjects by changing the definition automatically would solve the medical problem, where physicians sometimes feel obligated to end life by treating the suffering of a patient, even when one cannot communicate with that patient. This difference of opinion, between what is morally speaking good medicine (the mercy argument) and what is legally speaking good medicine (the autonomy argument), still needs to be resolved. It is for some physicians difficult to maintain that these conditions, the request and the suffering, are additional: both must be present in order to prevent prosecution. When the suffering of a patient is hard to bear for the observers and caretakers, the "rationality" of compassion or mercy may induce a physician to shorten life, but this act strictly speaking, does not fall under the category or physician-assisted death. One of the answers to these problems lies in early discussions between doctor and patient about the patient's preferences for medical treatment at the end-of-life when no cure is possible. But there is anything but a consensus on this point. It is an unacceptable consequence for many, even in the Netherlands. On the other hand, for Dutch physicians in general, the upheaval about the existence of the LAWER group was hard to understand, especially because of the dominant approach elsewhere of "terminal sedation." For Dutch physicians, and not only for them (Orentlicher 1998), terminal sedation boils down to the active ending of a life that is similar to LAWER, even in those cases where the patients" preferences of an easy self chosen death could have been sought (Quill and Kimsma 1997). Apparently, there are mechanisms of judgment at work that tend to cloud a clear view on what really happens and there is resistance to self criticism of present procedures, both in the Netherlands and outside.

There is a second set of figures that is disquieting for the Dutch: it concerns the number of cases of physician-assisted death that are being reported. Of the 1990 number of cases of PAD of 3700 only 18% of the cases were reported, of the 1995 cases of 4500 only 41% were made known through the agreed upon channels.

The reasons for non-reporting have been investigated also and the following reasons were given Van der Wal and Van der Maas 1996, 90-2):

- desire to save oneself and the family from
  the burden of judicial inquiry   55%
- desire to save the surviving relatives judicial inquiry: 30%
- fear of prosecution   36%
- desire of relatives to be saved from judicial inquiry   31%
- not all the requirements were met   30%
- this is an exclusive between physician and patient   12%
- there had not been an explicit request   5%
- fear for reactions of relatives   5%
- other   16%

Between the reported and the non-reported cases there is no difference in patient characteristics. The difference lies in the attitude of the physicians and perceptions of mainly the legal inquiry. Some physicians experience the act of reporting as a social obligation that they integrate into their interventions to end life, some even welcome the possibility to justify the act, almost akin to confession; other physicians mainly object to the burden of a judicial evaluation. Those who do not report tend to be older physicians with more experience in physician-assisted death, even before lower legal barriers and a higher risk of punishment existed. This gap between actual and reported cases points to a deep cleft of perception between the law and medicine where physician-assisted death is concerned. For the legal profession, any case of euthanasia is potentially a crime until shown otherwise and physicians are criminals until they have fulfilled their duties under the law. For physicians, assisted-death is one of the most difficult forms of health care with much personal and emotional involvement. This involvement is not being done justice in the most literal form within the legal procedure. One of the major points of irritation used to be the length of time "Justice" took to send a final decision of non-prosecution to the individual physician. Sometimes this legal procedure took more than two years and created anxieties for the involved physicians.

To answer the problem of underreporting the Departments of both Justice and Health Care in the Netherlands have developed plans with the Royal Dutch Medical Society for regional Euthanasia Evaluation Committees that went into effect in November 1998. The intention is to lessen the bureaucracy of assisted death by underlining the decriminalization, by "putting the law at more distance" and the expectation is that physicians will report in increased numbers. These regional committees have been functioning for more than two years now and the first report on their activities has appeared, the second one is to appear in may 2001.[2] One of the prudent conclusions is that the medical profession apparently has adopted a wait and see approach: the number of reported cases in absolute terms has not increased, with the additional remark that we have no reliable data on the right numbers of physician-assisted death. It may even be that the number of actual cases of PAD has decreased. The projected research will provide the answers to these conjectures.

There are deficiencies observed in the reporting of the physicians who performed euthanasia or assisted suicide. These deficiencies may well reflect the ambivalent position of the reporting physicians: one needs to describe the suffering of a patient and at the same time one needs to justify his actions in the light of this suffering. What becomes visible in these descriptions is the stressing of physical symptoms, such as pain, asphyxiation and nausea, possibly in an attempt to justify oneself and, next to providing information, also build up a defense for legal inquiries.

Another policy to increase reporting has been the development of professionally supported and educated networks of consulting physicians. This project is State supported through grants of the Ministry of Health, Education and Welfare. It started in 1996 in the city of Amsterdam as Steun Consultatie Euthanasie Amsterdam, meaning Support Consultation Euthanasia Amsterdam, known by its acronym *SCEA*. The project consists of a loosely organized group of about twenty physicians who are available for consultation around the clock, both for the formal consultation of the law but also for advice and support in case of problems or questions. The program's success has lead to a subsidized follow-up on a national scale, called Steun Consultatie Euthanasie Nederland, translated as Support Consultation Euthanasia Netherlands, shortened as *SCEN*. Already, the positive difference between the consultations of the SCEN consultants and other physicians has been noticed over the past year. The SCEN program will be continued another three years until each region in the Netherlands has been organized and educated. It is being evaluated for its effects and problematic areas in need of improvement (Onwuteaka-Philipsen 1999).

A third development that must have material effects in the area of life care concerns objections to the Dutch practice of physician-assisted death. The objection is that requests for euthanasia or assisted suicide may originate on deficient knowledge of palliative medicine. The core argument may have come from the "grand old lady of palliative care" Dame Cicely Saunders, who early on has argued that adequate palliative care would prevent the emergence of requests for assisted death (Saunders 1969). This suggestion is partly based upon the observation of foreign visitors that the Netherlands has only a small number of hospices, places where the terminal patients can die in an agreeable setting. This lack of hospices is often mentioned in one breath with a lack of palliative care, both as a body of knowledge and as a practice of care for terminal patients. Even though there is no comparative research in this area between countries, not even representative data in this area on national scales, the Dutch have concluded that palliative care should receive more attention. Through the founding of five national centers for research and education, the goal is to increase the quality of knowledge of palliative care by doing research on what happens, develop standards of care, and teach these standards at medical schools and through postgraduate programs. The net effect of these developments might also become instrumental in the policies to increase the quality of terminal care and the assurance of physicians that they have applied whatever was possible for their patients in their terminal care.

EUTHANASIA AS A RISK FOR THE ELDERLY

The moral position or claim that euthanasia might be a risk, especially for the elderly, stems from opponents of assisted death. It is one argument out of a cluster of arguments that have one common denominator: the fear of a slippery slope of causing undesired death for the vulnerable in society once euthanasia is an option. This cluster of arguments has a long history. In the systematic public debate it has a history of more than 130 years, but its history in the Netherlands is rather short; a little more than 25 years, in fact, as I have said, that is practically all of the Dutch history of euthanasia.

In the area of old age and fear of euthanasia, the first professor of nursing home medicine in the Netherlands, Michels, published interviews with elderly inhabitants of "his" nursing home for terminal patient, "Kalorama," in the Catholic area around the city of Nijmegen in 1973. Nursing Home medicine in the Netherlands has been developed into a medical specialty, with specific expert knowledge and skills that are taught at the university level. Nursing homes, in a nutshell, are state funded and not-for-profit institutions. Inhabitants are admitted on the basis of their medical needs and do not pay for their stay. Michels (1973) describes the results of 20 interviews with mainly older terminal patients:

> all these discussions were with severely ill patients, who were familiar with their situation, two weeks or several months before their death, (with the exception of patient 19 and 20, who both are still alive.)(at that time of course, GKK) These discussions have corroborated that those involved do not want euthanasia, as they do not regard it as a possibility to help them. On the contrary, for most of them a discussion about euthanasia is a threat. They do not feel safe in our hands anymore.

In a direct commentary Dutch sociologist Van Heek, in one of the first well-balanced books on euthanasia in the Netherlands, calls these conclusions scientifically invalid. The methodology of interviewing is deficient. In the first place, these patients had not requested euthanasia and, in the second place, these patients were very much aware of the fact that the interviewer was opposed to euthanasia. Van Heek adds that it is logical that not much information on the frequency of requests for euthanasia is available as patients, obviously, will not express a desire that might imply an intervention that the law forbids and many medical colleagues will disapprove (van Heek 1977).

The period I am writing about is the seventies, when the discussion on euthanasia in the Netherlands really took off and turned into a public issue, but hardly reached the intensity or ferocity of the abortion debate.

The question on age, euthanasia, and fear is also expressed along different lines, with different arguments, that in their combination I have called a "cluster." Both in the past but also in recent debates it is claimed that the patient's awareness of a physician's positive inclination to allow euthanasia might, in effect, weaken the patient's resistance to fight the disease. Also, the possibility that a physician initiating discussion about euthanasia effectively may cause patients to limit their contacts to their own disadvantage, is another argument. An even more subtle version of this slippery slope argument is to be found in the claim that a physician's positive attitude towards euthanasia may subconsciously induce a terminal patient to pick up a message that asking for an end to life is the socially desirable thing to do. The common denominator of this type of argument is the characterization of the "weak patient" and the "powerful physician," the latter being able to influence a dependent patient, even without speaking, by using body language and practicing silence where speech might be desired. It is interesting to notice that defenders of the right to assisted death in some of their arguments have similar opinions of their doctors. They tend to distrust them also, but use a different objection: in this case physicians are supposed to be prone to deny patients a right to assisted death, against the patients" wishes. There is no research to substantiate these negative claims of influencing patients, so we do not know their validity. One of the reasons for a second opinion by another physician in the Dutch procedure is meant to check this possibility.

The same question, should older people be afraid of their physicians, is repeated years later than Michels did by Canadian journalist Ann Mullens, when she visited the Netherlands in 1995. Her question raised eyebrows and disbelief, even of opponents, since it was considered a preposterous one for the moral sensitivity of the average Dutchman with whom she spoke (Mullens 1996). One opponent of euthanasia in these interviews even makes it quite clear that he does not wish to change family physicians, even though the physician is known to hasten the death of his terminal patients with a request to die. There is trust between doctors and patients in the Netherlands and this is no fairy tale. There are long-term relationships between families and physicians, and these relationships matter when terminal patients wish to die. In general, there is neither evidence nor fear that euthanasia is carried out against the wishes of patients.

Yet, this position is challenged in the international literature by opponents who are invited to observe and comment on the Dutch development. Sometimes this claim is even fuelled by Dutch opponents who, unfortunately, rarely voice their opinions in the Dutch journals. This is one of the reasons, but certainly not the only one, that there still is a strong current in the international euthanasia debate on the reality of expected danger of unwanted euthanasia, once the spirit is let out of the bottle, especially in Germany, but also in other countries. Similar positions have been taken up all along the course of the euthanasia debates. And this danger does not only concern the aged, but also the handicapped, the mentally sick, and the feeble minded. Over the past fifteen years handicapped people in Germany have, on several occasions, made public discussions impossible by disrupting scientific meetings, even some meetings where end-of-life care was not even on the agenda.

Fears of a slippery slope of unwanted euthanasia for these groups, once assisted death would be allowed for the self-conscious terminal patients with a request, can be observed already in the first international debate on euthanasia in Great Britain around 1870. "Accepting euthanasia will change the attitude towards the suffering of others. The standard of "unnecessary suffering" will become lower and lower."[3]

What can be observed also is that, for some, euthanasia is not only for the terminally ill, but also an option for society to clean itself from undesirables. A call for elimination of undesirable social elements is a strong characteristic of the German debates, both around 1910 and again in the twenties.

The Nazi T4 Program in the late thirties has shown how far a totalitarian regime can go in the extermination of unwanted human beings. The Nazi regime turned ideas into a grim reality. But, it is a mistake to think that Germany was an exception in all aspects and only totalitarian regimes are a sufficient explanation for these horrible developments. There was a much larger and more general support for limiting social rights of the lesser capable groups in society, on a European scale but also in the USA and Canada. Forced sterilization programs, the opening moves of the eventual extermination, for example, expressing the idea of improving the human gene pool by eliminating the participation of individuals with undesirable characteristics or behavior, can be observed all over the Western Hemisphere (Schneider 1990; Carol 1995; Broberg and Roll-Hansen 1996; Hasian 1996; Mazumdar 1992; Soloway 1995). What these sterilization programs had in common was an ideology of health care and eugenics, expressing the ideology of social Darwinism. It was a combined view on the social obligation of physicians to participate in the programs to improve the race by ridding the gene pool of undesired participation; in Germany, the upbeat to the elimination or murder of the participants (Friedlander 1995; Klee 1983).

These fears of a slippery slope are not only historical.

Over the past years physician-assisted death, on more than one occasion, has been branded as potentially conducive to a slippery slope from euthanasia on request to ending the life of suffering patients incapable of expressing their opinion justified by arguments of mercy. One example of this position is the statement of the German Society of Psychiatry, Psychotherapy and Neurology, in a meeting in Hannover on April 15, 1997. Aldenhoff, its representative, shows this apprehension very clearly: "Let us look at the historical dimension. The German psychiatric community has carried out the concept of "Killing of life unworthy of living" in a special way. It appeared before and to a large extent independent of National Socialism and during the Third Reich with active assistance of leading German psychiatrists was brought to a horrible highpoint. Today we know that those psychiatrists who actively and with intent organized the killing of theirs and others' male and female patients, in the judgment of their immediate circles, often appeared to be empathic, engaged and humane. In considering the historical details one can observe that, even in those circles independent of National Socialism, who concentrated in the first place on euthanasia to alleviate suffering, mix the killing on request with the killing against their wishes. We cannot simply find a common ground for this mixture in a rational way; but we can draw the lesson from this historical reality that the demarcation between these two quite clearly different motives – alleviation or murder – in reality apparently is not stable" (Aldenhoff 1998). It is obvious that the psychiatrist concludes that clear lines between mercy and murder are difficult to draw and even more difficult to maintain.

If we look at these past and present developments and positions, is there a lesson to learn? Do these facts and fancies increase our understanding of both the opponents and defenders of assisted death? How do these developments relate to the question of the elderly and actual euthanasia practice in the Netherlands? Historically speaking, unlike in Germany, there has never been a hint of a significant relation between the eugenics movement and euthanasia in the Netherlands (Noordman 1988). And speaking about the present, in a recent publication we have investigated the relationship between age and euthanasia and found no evidence of "geriatricide" (Muller, Kimsma, and Van der Wal 1998).

It must be stated before all else that the Netherlands is the only country that nationwide has produced reliable data. All other data on assisted death are merely local, provincial, or regional and are difficult to compare because of differences in survey questions, formats, and definition of terms. These Dutch data show convincingly that most elderly people do not die through assisted death. In five Dutch studies it is clear that assisted death occurs most frequently in the age groups between 65 through 79 years, followed by the age brackets of 50 to 64 years. In the age groups 0 to 49 years and over 80 years old, assisted death is least often performed (Van der Maas, Van der Delden, and Van Pijnenborg 1992; Van der Wal et al. 1991; Muller et al. 1995; Onwuteaka-Philipsen, Muller and Van der Wal 1997; Van der Maas et al. 1996). The reason of this particular age distribution is the predominance of assisted death in cancer patients. Eighty percent of the people who ask for euthanasia and assisted suicide suffer from cancer, making the age distribution of assisted death similar to the age distribution of cancer. There is no indication whatsoever that there is a predominance or overrepresentation of elderly in the data on assisted death. On the contrary, one might even defend it as the numbers of assisted death versus natural deaths in nursing homes are far lower than in hospitals (1 in 800 deaths versus 1 in 75 deaths). And, again, the numbers in hospitals are far lower than the numbers in the homes of patients (1 in 75 versus 1 in 25 patients who die at home). As we stated with respect to the last known data, "In 1995, 55 697 people aged 80 years and older died in the Netherlands, 756 through assisted death. In particular, the thought that in nursing homes (where mainly elderly people are admitted) assisted death is administered to the very elderly was refuted by the results reported by Van der Wal et al. (1996). This study was conducted among nursing home physicians to determine the characteristics of somatic patients who received assisted death. It was found that, of all patients with a serious somatic disease who died during the investigation period (4,5 years;n=44 100), only 77 deaths involved assisted death. As was the case in other studies, the majority of the patients who died as a result of assisted death were in the age category of 65 to 79 years. Moreover, two nationwide studies showed that in 1990 and 1995, in absolute numbers, no more than 20 to 70 nursing home patients died as a result of assisted death (Van der Wal et al. 1992; Van der Maas, Van Delden, and Pijnenborg 1992; Van der Wal et al. 1994; Van der Wal and Van der Maas 1996).

The conclusion seems fair that any suggestion of a danger for the elderly of unwanted death is not warranted by the available data, keeping in mind that the Dutch data are the only reliable and representative data we have. That might be the end of the discussion on the dangers that were perceived at the beginning of this chapter. Yet, there is more to it. This conclusion is valid but not necessarily the end of the question on perceived dangers.

Rationally speaking, the danger of a slippery slope only is a firm and empirically proven conclusion if one compares at least two periods in time. In this case, we do not have these data, other than the two Dutch studies, of 1990 and 1995. These studies do not allow any other conclusion than the one I defended above; neither proof of a slippery slope nor of an additional danger for the elderly in the area of assisted death. Thus there are no data to defend the slippery slope.

But that is spoken rationally. Essentially the slippery argument is not a rational argument, but a psychological one. Behind the slippery slope, anxiety shows. This anxiety is expressed clearly by the representative of the German Society of Neurologists, Psychiatrists and Psychotherapists that I quoted above. Aldenhoff expressed the anxious conviction that the moral fabric of German society would not be able to hold a line of undesirable developments, exactly because those who one cannot trust to defend that line cannot be singled out. That is one side of the slippery slope and for some, at least in Germany, it is the reason for their staunch defense against any leniency in the area of assisted death; the murderers are just as much among us as the compassionate assistants and they look alike. That is a message that I would not support for the Dutch present situation, but I can understand this opinion from someone of a nearby country, with such a horrible past. Given the fact that the Dutch population supports the option of euthanasia and the institutions of both law and medicine underwrite its reality, hardly another conclusion is possible that there is a deep difference in trust in the moral fabric between Dutch and German and other societies.

Yet, there is no reason for complacency, but a constant need for alertness. It is not inconceivable that in the Netherlands the lines that have been drawn will be extended to include the presently excluded Alzheimer patients, the psychiatric patients, the neonates, and the comatose, the group of the legally incompetents. The

Royal Dutch Medical Society (KNMG) in 1998 has published the papers of a Committee on the Active Termination of Life (CAL), the results of more than ten years of deliberation, to stimulate a public discussion as opposed to the inner circle of physicians, lawyers and theologians. The explicit intention of this book is to discover the limits of life ending interventions in cases of unbearable, irremediable suffering of patients who have been competent but are no more (Alzheimer, psychiatric patients, the comatose adults) or patients who never were competent (the neonates). There are at least two elements in this area that are basic.

The first is an intention that was expressed already in the first Dutch Court case of Leeuwarden in 1974, where *the terminality of a patient's condition* was not turned into a basic condition that would prevent prosecution from euthanasia or assisted death. The judge's argument was that, for some patients with a chronic disease with suffering, the demand of terminality would in effect be a discrimination against chronic patients.

This argument of discrimination essentially is a legal argument, but is supported by quite a number of individuals and institutions. The legal position on "terminality" is that it is not a pragmatic instrument; one cannot use the term to distinguish between those that qualify and those that do not.

The second element is pertinent for the present. As I stated above, the Dutch Government in 1999 has passed a bill to decriminalize physician-assisted death. This bill is considered problematic even for some proponents because at least two issues. One issue is the option of euthanasia for minors between 16 and 18 years old who, for Dutch law can operate at this age as adults in medical matters. The other issue, more in the focus of my line of thought, concerns the status of advance directives in the case Alzheimer's patients. In this law it is made possible for Alzheimer's patients to receive assisted death if and when they have an advance directive. There are some Catch-22 elements in this proposal.

Essential for the present procedures is the condition of a second opinion of an independent doctor to assess the "material" conditions: the nature of the request and the aspect of suffering. The request must, first of all, be voluntary. It must also be persistent, as opposed to being an expression of panic and fear. As specifications of suffering, the continuous nature, being unbearable, serious and without hope for improvement are elements that must be checked.

When these conditions are to be checked by the consulting physician, the legal description of this function implies a check of *the here and now qualification* of these conditions. A patient must repeat the desire before the consulting physician to have a life ended. In the case of an Alzheimer's patient who has lost the capacity to judge and orient himself as to the consequences of decisions and actions, there is no legally valid communication possible. The Alzheimer's patient has to answer questions that he does not comprehend, cannot answer and, therefore, in this line of thought, an Alzheimer's patient cannot qualify for assisted death.

The new element in the proposed law on assisted death implies that in the case of an advance directive, specifying the desire to have life ended in case of Alzheimer's disease, a physician who will follow through on that desire and actually end the life of that particular patient will also not face criminal charges.

This position has the elements to surpass the paradox that, for proponents, has been a stumbling block and a reason for bewilderment; advance directives serve the purpose to direct medical interventions for circumstances when one is incapable of participation in medical decision-making. They serve as guidelines for physicians and families when patients are too ill to decide themselves. So, popular thinking has it that, in the case of Alzheimer's disease, an advance directive is the perfect way to make clear that one does not wish to continue living once the capacity to decide has been eroded as a consequence of the disease. Ergo, one has made an advance directive for a particular situation, once this situation has arrived, even when this means that one cannot decide anymore, than guidelines of the advance directive should be carried out. That presently is the official position of the Dutch Voluntary Euthanasia Society. In the parliamentary debate of the law the secretaries of both Health Care and Justice maintained that the advance directives are intended to prevent a physician from being accused of murder because the condition of an actual request cannot be fulfilled, *Alzheimer's not being the sole and only medical condition of a patient.* In effect, this will mean a difference of opinion between a legal conception of advance directives, in the case of a decision to end a life, and a medical conception. For some proponents of the law, the wording of this law does allow euthanasia with Alzheimer as the sole condition.

Opponents of this line of thinking, among whom are the Dutch Society of Nursing Home Physicians, hold a clear-cut position. This position is the rule that assisted death is not normal medicine; there is no professional standard that has to be followed. Even though patients may have written down their wishes about a time of incoherence and loss of personality, that does not imply that physicians should follow these desires. From a moral point of view, most physicians consider ending the life of a happy Alzheimer's patient incompatible with the intentions of helping suffering patients die. From a lay perspective, not-following these deeply felt wishes is akin to lying to patients and being disrespectful of persons. With the issue of suffering of Alzheimer's patients a new chapter in the book of suffering is written. The nursing home physicians maintain that there are the "happy" Alzheimer's patients who seem to enjoy the little that life still has to offer. There is much weight given to the legal aspects in the debate, besides the procedural condition that a consultant must check the nature of the request here and now. Legally speaking, the nature of advance directives that carry weight are directed at refusal of medical care, not at wanted or desired "medical care." They are in legal terms negative declarations and, as such, carry legal weight, as in many other European countries and in the USA. But in this case of Alzheimer's patients, the advance directive is a "positive" declaration of intent. And for a positive declaration of intent there must be a medical justification, according to a medical standard, that in this particular case is not part of the professional standardized care.

At the time of writing this contribution, the outcome of a Parliament debate has become clear. Opponents of physician-assisted death see this development as a clear sign of the reality of a slippery slope. Proponents of assisted death consider the development as a consequence that somehow must be dealt with to prevent discrimination of suffering patients, including the incompetents such as Alzheimer's patients. The position of those physicians who should carry out this task is clear also: they do not see themselves in the position of carrying out this intervention. What they do see as options is an intervention of withdrawing medical treatment that has no further medical goal or the acceptance of a shortening of life as part of the alleviation of suffering.

After all of this has been said, it is difficult to foresee the possibility that the elderly really have a reason to fear that their life will be shortened, against their will, by physicians in the Netherlands. Yet, we have to remain alert and cautious, because the limits of what physicians should do have been reached and it seems to be time for a "time out" and a period for evaluation of the present modes of ending lives.

Gerritt K. Kimsma, Associate Professor, Department of Philosophy and Medical Ethics, Free University Medical Center, Amsterdam, The Netherlands.

NOTES

[1] The Australian Northern Territories were the first State with legalized euthanasia, but the law has been repealed by the Central Government in Canberra.

[2] *Regionale ToetsingsCommissies Euthanasie*, Jaarverslag 1998/99.

[3] Mr. Tollemache on the Right to Die, *The Spectator*, 15-2-1873, 206-207.

REFERENCES

Aldenhoff, J. 1998. *Stellungnahme aus Sicht der Deutschen Gesellschaft für Psychiatrie, Psychotherapie und Nervenheilkunde (DGPPN), (Position of the German Society for Psychiatry, Psychotherapy and Neurology (DGPPN),* In *Beihilfe zum Suicid, (Assisted Suicide),* ed. G. Ritzel, 45-6. Regensburg: S.Roderer Verlag.

Broberg, G., and N. Roll-Hansen. 1996.*Eugenics and the welfare state. Sterilization policy in Denmark, Sweden, Norway and Finland.* East Lansing, Michigan: Michigan State University Press.

Carol, A. 1995. *Histoire de L'Eugenisme and France. Les Medecines et la Procreation XIXe-XX siecle.* Paris: Seuil.

Emanuel, E.J. 1998. Why Now? In *Regulating how we die. The ethical, medical ,and legal issues surrounding physician-assisted suicide,* eds. L.L. Emanuel, 175-20. Cambridge: Harvard University Press.

Friedlander, H. 1995. *The origins of Nazi genocide. from euthanasia to the final solution.* Chapel Hill: The University of North Carolina Press.

Griffiths, J., A. Bood, and H. Weyers. 1998. *Euthanasia and law in the Netherlands.* Amsterdam: University Press.

Hasian, Jr., M.A. 1996. *The rhetoric of eugenics in Anglo-American thought.* Athens: University of Georgia Press.

Kimsma G.K., and E. Van Leeuwen. 1998. Euthanasia and Assisted Death in the Netherlands. In D.C. Thomasma, T. Kimbrough-Kushner, G.K. Kimsma, and C. Ciesielskli-Carlucci, 35-71. *Asking to die. Inside the Dutch euthanasia debate.* Dordrecht: Kluwer Academic Press.

Klee, E. 1983. *Euthanasie im NS-Staat. Die "Vernichtung lebensunwertes Lebens." (Euthanasia in the NS State. The destruction of life unworthy of living).* Frankfurt am Main: Fischer.

Mazumdar, P.M.H. 1992. *Eugenics, human genetics and human failings. The eugenics society, its sources and its critics in Britain.* London: Routledge.

Meerman D. 1991. *Goed Doen door Dood te Maken (Doing well by killing).* Kampen: Kok.

Michels, J.J.M. 1973. Euthanasie en werkelijkheid. (Euthanaisia and reality) In *Menswaardig sterven (Dying with Dignity),* eds. J. Immerzeel, et al., 313-25. Nijmegen: Faculty of Medicine.

Mullens, A. 1996. *A timely death: Considering our last rights.* Toronto: Alfred A.Knopf Canada. 107-156.

Muller, M.T., G.K. Kimsma, and G. Van der Wal. 1998. Euthanasia and asssisted suicide: Facts, figures and fancies with special regard to old age. *Drugs and Aging* 13(3): 185-91.

Muller, M.T., G. van der Wal, and J.Th.M. van Eijk, et al. 1995. Active euthanasia and physician-assisted suicide in Dutch nursing homes: Patient characteristics. *Age and Ageing* 24(5): 429-33.

Noordman, J. 1988. Om de zorg voor het nageslacht (Caring for posterity). Nijmegen: Sun.

Onwuteaka-Philipsen, B. 1999. Consultation of another physician in cases of euthanasia and physician-assisted suicide. Ph.D. diss., Free University of Amsterdam..

Onwuteaka-Philipsen, B., M.T. Muller, and G. van der Wal. 1997. Euthanasia and the elderly. *Age and Ageing* 26(6): 487-92.

Orentlicher, D. 1998. The Supreme Court end Terminal Sedation. An Ethically Inferior Alternative to Physician-Assisted Suicide, in: *Physician Assisted Suicide. Expanding the debate,* eds. M.P. Battin, R. Rhodes, and A. Silvers, 301-12. New York/London: Routledge.

Persels J. 1993. Forcing the issue of physician assisted suicide: Impact of the Kevorkian case on the euthanasia debate. *Journal of Legal Medicine* 14: 93-124.

Quill, T., and G.K. Kimsma. 1996. End-of-life care in the Netherlands and the United States: A comparison of values, justifications and practices. *Cambridge Quarterly of Health Care Ethics* 6 (2): 189-205.

Sanders, C. 1969. The moment of truth: Care of the dying person. In *Death and dying,* ed. L. Pearson, 49.79. *Current issues in the treatment of the dying person.* Cleveland: Case Western Reserve University Press.

Schneider, W.H. 1990. *Quality and Quantity. The Quest for Biological Regeneration in Twentieth-Century France,* Cambridge: Cambridge University Press.

Soloway, R.A. 1995. *Democracy and degeneration. eugenics and the declining birthrate in twentieth century Britain.* Chapel Hill: The University of North Carolina Press.

Van der Maas, P.J., J.J.M. van Delden, and L. Pijnenborg. 1991. *Medical decisions concerning the end of life.* The Hague: SdU Uitgevers. (Dutch)

———. 1992. Euthanasia and other medical decisions concerning the end of life. *Health Policy* 22: 1-262.

Van der Maas, P.J., G. van der Wal, I. Haverkate, et al. 1996. Euthanasia, physician-assisted suicide, and other medical practices involving the end of life in the Netherlands 1990-1995. *New England Journal of Medicine* 335(22): 1699-705.

Van der Wal, G., J.Th.M. van Eijk, H.J.J. van Leenen, and C. Spreeuwenberg. 1991. Euthanasia and assisted suicide in the home situation: I. Diagnoses, age and sex of the patients. (Dutch) *Ned Tijdschr Geneeskd* 135(35): 1593-9.

Van der Wal, G., M.T. Muller, L.M. Christ, M.W. Ribbel, and J. Th.M. van Eijk. 1994. Voluntary active euthanasia and physicians asissted suicide in Dutch nursing homes: Requests and administration. *Journal of the American Geriatric Society* 42(6): 620-3.

Van der Wal, G., J.Th.M. van der Eijk, H.J.J. Leenen, and C. Spreeuwenberg. 1992. Euthanasia and assisted suicide: How often is it practiced by family doctors in the Netherlands? *Family Practice* 9(2): 130-4.

Van der Wal, G., and P.J. van der Maas. 1996. *Euthanasia and other medical decisions concerning the end of life: Practice and notification procedures.* The Hague: SdU Uitgevers. (Dutch)

Van Heek, F. 1977. Euthanasie als sociologisch vraagstuk (Euthanasia as a sociological issue). In *Euthanasie (Euthanasia),* eds. P. Muntendam, et al., 48-70. Leiden: Stafleu.

Van Leeuwen, E., and G.K. Kimsma. 2000. Probleme in Zusammenhang mit der ethischen Rechtfertigung zur medizinischen Sterbehilfe: Auseinandersetzung mit der Euthanasie und ärtzlichen Suizidbeihilfe. In *Medizinethik und Kultur, Grenzen medizinischen Handelns in Deutschland und den Niederlanden,* eds. B. Gordijn, and H. Ten Have, 433-57. Fromann-Holzboog: Stuttgart-Bad Canstatt.

Woretshofer, J. 1994. *De Meldingsprocedure levensbeeindiging strafrechtelijk bezien.* The procedure to report in the light of criminal law. *Tsch Gezondheidsrecht* 7: 410-21.

CHAPTER FIVE

HENK JOCHEMSEN

WHY EUTHANASIA SHOULD NOT BE LEGALIZED

A Reflection on the Dutch Experiment

The experience of the Netherlands continues to be cited as illustrative of the euthanasia debate that is going on in many countries. The parliamentary debates on the legalization of euthanasia (November 2000 in the Second Chamber and April 2001 First Chamber) have drawn a lot of international attention. But, before this legalization in the Penal Code this country had adopted a legal regulation of euthanasia and, before and after that, extensive surveys into the practice of euthanasia had been carried out. The fact that the Dutch example is cited both by those who favor the legislation of euthanasia and those who reject it demonstrates that empirical data in themselves do not settle an ethical or juridical issue.

In this chapter I will argue against legalization of euthanasia, largely on the basis of the Dutch experience. I will follow two approaches: a juridical and an ethical. The juridical argument is based upon the assumption that any form of legal regulation of euthanasia is only acceptable if it would guarantee that the state, i.e. legal authorities, would in principle be able to assess each case.[1] The crucial question is whether the Dutch have succeeded in establishing a regulation of euthanasia that permits its application in individual cases under certain conditions, while at the same time maintaining a strict control of the practice as a whole. This paper will analyze the available data on Dutch euthanasia practice in the light of that question.

The ethical approach will discuss a number of medical ethical and social ethical problems that are raised by a legal regulation of euthanasia.

This chapter is divided into the following sections. Section two contains the clarification of some concepts that play a crucial role in the euthanasia debate. Section three presents information on the regulation and the practice of euthanasia and considers whether the requirements are followed. Section four discusses new developments after the publication of the second main survey (1996), considering the question whether they involve an improvement of the situation. In section five the ethical approach against (legalization of) euthanasia is presented. Section six provides the conclusion of the chapter.

D.N. Weisstub, D.C. Thomasma, S. Gauthier & G.F. Tomossy (eds.), Aging: Decisions at the End of Life, 67-90.
© *2001 Kluwer Academic Publishers. Printed in the Netherlands.*

CLARIFICATION OF CONCEPTS

In the Netherlands euthanasia is narrowly defined as "the intentional shortening of a patient's life at the patient's explicit request." In other words "euthanasia" is defined to mean only "active, voluntary euthanasia" and does not include intentional life-shortening by omission ("so-called passive euthanasia") or euthanasia without the patient's request (whether "non voluntary" if the patient is incompetent or "involuntary" if the patient is competent). For ease of exposition, I will follow the Dutch definition. Apart from the request of the patient this definition of euthanasia, as well as most others, contains two crucial elements. First, that the shortening of life, in other words the death of the patient, is intended, that is wanted, if not as an end in itself at least as a means to an end (ordinarily the ending of suffering). Second, that this intention leads the person to an action that he would not otherwise have performed and that, indeed, aims at the shortening of the patient's life. Hence, in my opinion euthanasia is not solely defined by the particular intention to shorten life. For this would mean that medical care informed by the acceptance that the patient will die and that is provided by a physician who personally feels that for the patient it would be desirable to die soon, could hardly be distinguished from euthanasia, whereas it could be normal medical treatment. Nor is euthanasia defined only by a shortening effect on the life of the patient. That would bring any course of action that would not maximally extend the life of the patient under the definition of euthanasia. This would either lead to an inhuman medicine that would essentially aim at the prolongation of the physical life of patients, or broaden the meaning of the term euthanasia to the extent of becoming virtually meaningless.

The important elements of the definition of euthanasia can be clarified further by discussing the differences between euthanasia and other actions of physicians that may seem to be euthanasia but are not. It has been agreed in the Netherlands that the following three categories of actions should not be considered as euthanasia (Leenen 1989, 520):

- stopping or not beginning a treatment at the request of the patient;
- withholding a treatment that is medically useless;
- pain and symptom treatment with the possible side-effect of shortening life.

These courses of action will briefly be discussed.

a) On the basis of the rule of informed consent, the patient is entitled to refuse treatment or to withdraw consent. When the physician respects the refusal of treatment of a fully competent patient and the patient dies soon after the withdrawal of life-supporting treatment, this is not to be classified as euthanasia. This does not mean that such situations cannot become at least ambiguous. In the first place because decisions of patients are often very much influenced by the information that the physician provides on the burden of the possible treatment, its prognosis etc., and by *the way* in which this information is provided. Hence, the performance of the physician can provoke the refusal of a medically useful treatment when in the opinion of the physician it would be better for the patient if he were "allowed to die." Secondly, because the patient can be suicidal, e.g., because of a depression, and refuse a proportional medical treatment in order to die. If the physician thinks this would be better for the patient, he could go along too easily with the patient's decision without assessing whether the decision of the patient is really a voluntary and free choice. This is not to deny the legal right of a competent patient to refuse any treatment, but to point out that a non-treatment decision by a physician *can* be morally ambivalent if not highly problematic.

b) Treatment that is medically useless should not be provided and can or should be withdrawn when it is provided. Medically useful treatment fulfills the following criteria:

- the treatment is effective and proportional, which means that the expected benefits of the treatment outweigh its expected or possible risks and burdens for the patient;
- the evaluation of benefits, risks and burdens for the patient should only take into consideration medical criteria and should not be an evaluation of the value of the life of the patient;

Good medical practice has always included stopping disproportional medical treatment. Calling this, in general, "passive euthanasia" is confusing. If the term passive euthanasia can be applied usefully at all it is to indicate situations in which a physician does not offer or provide a medically proportional treatment to a competent or an incompetent patient respectively, with the intention that the patient will die. In such cases the intention to shorten life has informed the action (in this case an omission) of the physician that resulted in a sooner death than otherwise (probably) would have been the case. On the other hand, the withdrawal of disproportional medical treatment does not (necessarily) intend the death of the patient, but intends not to cause discomfort that is not outweighed by benefits for the patient.

These considerations from the caregiver perspective do not deny the fact that a patient or his proxy has to give consent for any treatment. But (a) the requirement of consent does not rule out the professional responsibility of the physician, (b) the physician is not obliged to provide treatment that by the profession is clearly considered futile, and (c) the decision of the patient/proxy is often highly influenced by the physician.

The fact that in practice there is a gray area between clearly proportional and clearly disproportional treatment, in which patients and physicians would decide differently, does not rule out the importance of this principle for medical practice.

c) Pain treatment aims at the alleviation of the patient's suffering. It may and, in some cases, almost certainly will shorten the life of the patient. However, actions must be defined according to their aims and not according to their side-effects. So, the shortening of life as an unavoidable side-effect of proportional pain and symptom treatment is morally acceptable on the basis of the principle of double-effect and should not be classified as euthanasia.

Here again the conceptual distinction between this course of action and euthanasia does not rule out that in practice the distinction may seem vague. It is obvious that, in practice, pain treatment can shift to euthanasia by increasing the medication beyond doses needed to effectively treat the pain and symptoms with the intention to shorten the patient's life. Yet this distinction is crucial in the evaluation of medical care at the end of life. I will return to some of these points below.

THE REGULATION AND THE PRACTICE OF EUTHANASIA

The 1994 Regulation of Euthanasia and Physician Assisted Suicide

In order to interpret the data on the fulfillment of the requirements for euthanasia, it is necessary to know the legal regulation of life terminating actions that form the framework for the requirements.

The basis of the legal regulation of euthanasia that became effective in 1994 is the decision of the Supreme Court (in 1984) that a physician who has committed euthanasia can, in cases of an objectively established "conflict of duties," appeal to a defense of "necessity" (Penal Code art. 40). This conflict concerns, on the one hand, the duty to obey the law that forbids euthanasia and assisted suicide (Penal Code art. 293, 294) and, on the other hand, to alleviate suffering. The government approved the Supreme Court's decision, thereby accepting euthanasia under certain circumstances.[2] The conditions establishing a "conflict of duties" were essentially: a free, well-considered request, unacceptable suffering with no other reasonable possibilities to alleviate the suffering, and consultation of the physician by a colleague (Leenen 1989; Kastelein 1995).

In 1990 the government presented a new legal regulation on the basis of the courts' decisions. The prohibition of euthanasia and assisted suicide was maintained in the Penal Code. At the same time the procedure by which physicians *report* death in cases of euthanasia, assisted suicide, and life-terminating actions without an explicit request, was given a statutory basis by amending the law on the Disposal of the Dead.[3] In terms of this procedure, a physician who has terminated a patient's life informs the local medical examiner (municipal pathologist), who inspects the body externally and takes from the attending physician a statement that contains the relevant data (the patient's history, request, possible alternatives, consultation with a second physician, intervention). This report, together with an evaluation by the local medical examiner, is checked by the Public Prosecutor who considers if the termination of the patient's life was contrary to the Penal Code as interpreted by the courts. So, to the conditions mentioned above was added the requirement of reporting each case of euthanasia, assisted suicide, and life terminating action without an explicit request.

The Data on Euthanasia Practice

The results of two important surveys on end-of-life decision-making by Dutch doctors in the year 1990 and 1995 were published by Van der Maas et al. (1991) and

Van der Wal et al. (1996), respectively.[4] The latter survey sought particularly to ascertain the incidence of intentional life-shortening by doctors; the extent to which they complied with their duty to report such cases and the quality of their reporting. The most important quantitative data are summarized in table 1. (See also the chapter written by Kimsma in this volume). Critical discussions by this author and others have been published elsewhere (Jochemsen 1994; Jochemsen and Keown 1999; Keown 1995). Here I will only use these and some additional data in order to answer the question whether euthanasia practice is well under control of the legal authorities. This will be done by examining to what extent the requirements have been fulfilled.

Fulfillment of the Requirements

The criteria that should be fulfilled by a physician who performed euthanasia in order to successfully appeal to necessity are: the request must be voluntary, well-considered, and persistent; there must be unacceptable suffering that cannot be alleviated by reasonable medical means; and consultation and reporting is required.

Table 1. Main quantitative data on end of life actions by physicians in The Netherlands

		1995	1990
1.	Death cases in the Netherlands	135.500 (100%)	129.000 (100%)
2.	Requests for euthanasia	9.700 (7,1)	8.900 (7)
3.	Euthanasia applied	3.200 (2,4)	2.300 (1,8)
4.	Assisted suicide	400 (0,3)	400 (0,3)
5.	Life terminating action without specific request	900 (0,7)	1.000 (0,8)
6.	Intensification pain and symptom treatment:	20.000 (14,8)	22.500 (17,5)
	a. explicitly intended to shorten life	2.000 (1,5)	1.350 (1)
	b. partly to shorten life	2.850 (2,1)	6.750 (5,2)
	c. taking into account a probable shortening of life	15.150 (11,1)	14.400 (11,3)
7.	Withdrawing or withholding treatment (including tube feeding):	27.300 (20,1)	22.500 (17.5)
	a. at the explicit request of the patient	5.200 (3,8)	5.800 (4,5)
	b. without explicit request of the patient		
	b1. Explicitly to shorten life	14.200 (10,5)	2.670 (2,1)
	b2. Partly to shorten life		3.170 (2,5)
	b3. Taking into account a probable shortening of life	7.900 (5,8)	10.850 (8,4)
8.	Intentional termination of life of newborn babies		
	a. without withdrawing/withholding treatment	10	
	b. withdrawing or withholding treatment plus administration of medication explicitly to shorten life.	80	
9.	Assisted suicide of psychiatric patients	2 – 5	

Voluntary, well-considered and durable request
It is clear that in a substantial number of cases doctors intentionally terminate or at least shorten the lives of patients. There are 900 cases of terminating the patient's

life without an explicit request. But, in addition, doctors increase pain medication explicitly with the intention to shorten life in about 2000 cases; in a few hundreds of cases without an explicit request.[5] Furthermore, in about 14,000 cases treatment was withheld or withdrawn with the explicit intention of shortening life. It is very well possible, though we do not know for sure, that in a considerable percentage of these cases the treatment had become disproportional. So, withholding or withdrawing it would have been ethically justified. Yet, taking into account a statement of L. Pijnenborg, a co-author of Van der Maas in the 1991 report, those data should be taken seriously. Pijnenborg (1995, 18) writes: "If a physician administers a drug, withdraws a treatment or withholds one with the explicit purpose of hastening the end of life, then the intended outcome of that action is the end of the patient's life." According to this statement the death of patients should be considered the result of a physician's action in 14.200 + 2000 cases (categories 6a and 7b1 of Table 1), in addition to the 4.500 cases (categories 2+3+4 of Table 1). In many of these cases there was no explicit, well-considered request.

    It must be concluded that this criterion is not fulfilled in a considerable part of life-terminating actions.

Incompetent patients
Here it should be pointed out that, both by the medical profession and by the courts, life-terminating actions with incompetent patients have been accepted. The clearest examples are two severely handicapped and ill babies whose lives were intentionally ended. The attending physicians reported the case, were prosecuted, brought before a District Court (Alkmaar[6] and Groningen[7]), and later before a Court of Appeal (Amsterdam[8] and Leeuwarden,[9] respectively). We will briefly present these cases.[10]

    In the Alkmaar case the baby had spina bifida, hydrocephalus, a spinal cord lesion, and brain damage. The specialists decided not to operate on the spina bifida, because of the bad prognosis. The baby appeared to be suffering severe pain that was difficult to treat. The parents did not want the baby to suffer and asked for the termination of her life. Three days after the baby was born, she was killed by the attending gynecologist, P., after consultation with other specialists who had examined the baby. She died in the arms of her mother.

    The Groningen case concerned a newborn baby with trisomie-13, a syndrome that manifested itself in a number of disorders (deformities of skull, face, and hands; heart and kidney malfunction; brain damage). The baby was diagnosed to be non-viable; death was to be expected at most within a year, and probably within six months. After the situation had stabilized to a certain point, the baby was taken home and looked after by the parents, who had learned to supervise the tube-feeding. After a week, some tissue (meninges) came out through an opening in the skull. This area appeared to be very sensitive and the baby appeared to be in pain. The family physician, K., gave pain treatment, but that did not appear to be fully effective. After a number of days, with the explicit consent of the parents (though whether it was at their request remains unclear), and after consulting the pediatrician who had seen the baby before, the life of the baby was ended by lethal injection.

Both doctors appealed to the defense of necessity. All four courts accepted this and released the physicians without punishment.[11]

In the court decisions two lines of reasoning can be traced, not uncommon in discourses that defend euthanasia. First, in cases of short life expectancy, stopping or not starting treatment while accepting that the patient will die is virtually morally equated with intentionally killing the patient. A decision not to treat is regarded as an intention to shorten the life of the patient. The second main reasoning of the courts is that proportional pain and symptom treatment with has as a side-effect a shortening of life is morally equated with intentional killing by the administration of lethal substances. So, the moral significance of the principle of "double-effect" is rejected. We will come back to this reasoning below.

The 1995 survey reports that over 1000 newborns die in the Netherlands before their first birthday and estimates that the lives of about 15 are actively and intentionally terminated by doctors.

I conclude that under specific circumstances the intentional shortening of life of incompetent patients is supported by the KNMG (Royal Dutch Medical Association), is practiced to a certain extent, and is justified by the courts. Therefore, the practice of intentional shortening of life is not following the condition of a voluntary, well-considered, free request.

Unbearable suffering without prospect
According to the physicians this clearly is the most important reason for patients to ask for euthanasia. It is, however an "open" criterion, i.e., not clearly defined and in no way standardized so far. It is the doctor who decides whether the suffering has become sufficiently unacceptable or unbearable to give in to a request for euthanasia.[12] With about two-thirds of the 9700 requests for euthanasia the physicians did not comply, probably because in their opinion the situation had not yet become intolerable.

On the other hand, according to the attending physicians there were medical alternatives to alleviate the suffering in about 17% of the cases of euthanasia and assisted suicide, but the patients rejected them and the doctors complied with the request. This is at odds with the requirement that there must be no (reasonable) alternative to alleviate the suffering. Euthanasia should only be used as a last resort. This condition (an application of the subsidiary principle) was supported by the former Cabinet and by the KNMG (Sorgdrager and Borst-Eilers 1995; Kastelein 1995). However, the present Cabinet appears to have reversed this position. The bill that has just been accepted establishes that the refusal by the patient of available treatment alternatives does not render euthanasia unlawful.[13] Furthermore, the integral terminal palliative care is not yet sufficiently developed in the Netherlands. Only in recent years the hospice movement is really gaining importance.[14]

I conclude that this requirement (a) is interpreted differently by different physicians and (b) is not always fulfilled in the sense that euthanasia is only used as a last resort. With respect to the latter, two contradictory movements can be distinguished. On the one hand, the acceptance that this condition can also be fulfilled when a patient refuses a reasonable medical alternative to alleviate the suffering and, on the other hand, the increasing possibilities and availability of integral palliative care that could take away a request for euthanasia. If both movements will gain force, euthanasia will become less a last resort and increasingly more a choice for a certain kind of death. This will confront society even more urgently with the question of how to regulate this and avoid abuse, and will confront the medical profession with the question whether it want to be the executioner of such deaths.

Consultation
The guidelines for permissible euthanasia and assisted suicide require the doctor, before agreeing to either, to engage in a formal consultation (*consultatie*), and not merely an informal discussion (*overleg*), with a colleague.

In cases of euthanasia and assisted suicide 92% of doctors had, according to the survey (Van der Wal et al. 1996, Table 10.1), discussed the case with a colleague but in 13% of these cases the discussion did not amount to a formal consultation. Hence, consultation took place in 79% of cases. However, other findings of the survey indicate that consultation occurred in 99% of the reported cases but in only 18% of unreported cases (id., Table 10.2). Since almost 60% of all cases of euthanasia and PAS were not reported (see next section) it can be calculated that consultation occurred in only around half of all cases. The discrepancy between this 50% and 79% is not exactly clear, but seems to be related to a certain bias in responses to some questions (id., 113).

In 97% of the cases of life-termination without explicit request there was no formal consultation, though in 43% the case was discussed with a colleague. Even when consultation did take place, it was usually with a physician living locally and the most important reasons given for consulting such a physician were his views on life-ending decisions and his living nearby: expertise in palliative care was hardly mentioned.

I conclude that even this relatively easy condition did not reach a truly high level of compliance.

Reporting
In 1995 41% of cases of euthanasia and assisted suicide were reported to the local medical examiner, as required by the reporting procedure. While this is an improvement on the figure of 18% reported in 1990, it means that a clear majority of cases, almost 60%, still go unreported. Furthermore, the increase in reporting from 18% (1990) to 41% should not lead to optimism, since the number of euthanasia cases increased between 1990 and 1995 (900 cases) with almost the same number as the

reported cases (980 cases). Moreover, the survey confirms that the legal requirements are breached more frequently in unreported cases, in which there is less often
a written request by the patient; a written record by the doctor; or consultation by the
doctor (Van der Wal 1996, Table 11.6).

The most important reasons given by doctors for failing to report in 1995 were
(as in 1990), the wish to save oneself and/or the patient's relatives the inconvenience
of an investigation by the authorities, and to avoid the risk of prosecution (though, as
the consistently tiny number of prosecutions indicates, this risk is very small
indeed).

The purpose of the reporting procedure is to allow for scrutiny of the intentional
termination of life by doctors and to promote observance of the legal and professional requirements for euthanasia. *The fact that a clear majority of cases still goes
unreported confirms the failure of the procedure to fulfill its purpose and belies any
claim of effective regulation, scrutiny and control.*

DEVELOPMENTS AFTER 1996

A New Procedure

In 1998 the Parliament accepted a new regulation of euthanasia *reporting* that
became effective on November 1, 1998.[15,16] This did not imply a change of the Penal
Code prohibition of euthanasia, but a change of the procedure by which euthanasia
should be reported. According to that procedure, the report of every euthanasia case
as well as the filled out form of the medical examiner, should no longer be sent
directly to the public prosecutor, but should be sent to one of five regional
euthanasia review committees. This committee, consisting of a physician, a lawyer,
and an ethicist, should evaluate the case in the light of the courts' decisions on life-
terminating actions thus far. The committee's opinion on the case is sent to the
public prosecutor, together with the reports of the attending physician and of the
medical examiner. The prosecutor has the freedom and duty to form his own opinion
on the case, but the opinion of the committee will be of major importance in the
decision of the prosecutor to prosecute or not.

Since in this procedure the legal authorities were placed at larger distance from
the euthanizing physician, the government hoped that a higher percentage of cases
would be reported. However, the first year report of these review committees
indicates that so far this new procedure did not result in a substantial increase of the
number of reported cases.[17] Furthermore, this report indicates that in the reported
cases the information of the attending physician on the existence of alternatives and
on the quality of the consultation was not always sufficient. In some cases additional
information was requested from the physician. Ultimately, in all cases the euthanasia
or assisted suicide was approved by the corresponding committee.[18]

A second change in the reporting procedure is that euthanasia (at the request of
the patient) and life-termination without explicit request should be reported by
different procedures. In this way the government wants to stress that the two kinds

of acting should not be morally equated. This does not mean, however, that unrequested life-termination is ruled out altogether. Actions of this kind should be reported to a national committee, which will give its evaluation and then send the case to the public prosecutor who will then decide whether the case should be brought before court. Also in these cases the opinion of the national committee as well as the court decisions on such cases will significantly influence the decision of the prosecutor. But also in these cases there is no guarantee that they will be reported in the first place. So far this national committee has not yet been established.

A New Bill

In 1999 the Cabinet introduced a new Bill on euthanasia and assisted suicide. The Bill provides:[19]

1) Euthanasia must be performed in accordance with "careful medical practice." Requests must be voluntary, well-considered, persistent, and emanate from patients who are experiencing unbearable suffering without hope of improvement and the doctor and the patient must agree that euthanasia is the only reasonable option. At least one other independent physician must be consulted who must see the patient and give a written opinion on the case.

2) All cases must be reported to and evaluated by the established regional euthanasia review committees consisting of a lawyer, a physician and an ethicist or another professional who is used to deal with ethical issues. So this law gives a legal basis to the regional euthanasia review committees mentioned above.

3) Euthanasia and physician-assisted suicide will not be punishable if performed by a doctor who has complied with the requirements listed in (1) and who has reported the case to the local medical examiner.

4) The local medical examiner must send his or her report as well as the physician's report to the regional committee. At the same time the local medical examiner informs the public prosecutor about the case of euthanasia or assisted suicide. If on the basis of this report or for any other reason any serious infringement is suspected, the prosecutor will not give permission for burial or cremation until a further investigation has been conducted.

   The report to the regional review committee must demonstrate that all the requirements have been met. If in the committee's opinion the requirements were met, the case is done. Only if the committee judges that the requirements were not met, this opinion, together with the physician's report, is sent to the public prosecutor who will then decide on further prosecution.

In addition to these criteria, the law contains the following provisions concerning children and advance directives:

5) A physician may grant a euthanasia request by a child between 12 and 16 only with the parents' consent. Requests by children aged 16-17 do not require parental consent, though parents should be involved in the decision-making process.

      The provision that euthanasia requests of minors between 12 and 16 years in exceptional cases could be granted without the parents' consent was dropped by the Cabinet in response to critical questions by members of Parliament.

6) The proposal also establishes a legal basis for advance euthanasia declarations via a type of "living will" in which an incompetent patient would request euthanasia in the event he or she became mentally incompetent. Though such a statement does not imply that a physician has a duty to perform euthanasia at any moment, it provides the legal opening to intentionally end the life of an incompetent patient who had signed such a document.

This Bill was approved by the Second Chamber of the Dutch Parliament on November 28, 2000 and by the First Chamber on April 10, 2001.

Is this law likely to ensure effective control of voluntary euthanasia? The enactment of this law does not imply a major change in the requirements and circumstances under which euthanasia will be approved. But it can be interpreted as a further acceptance and institutionalization of euthanasia in society. The relaxation of the reporting procedure by the introduction of interdisciplinary committees as a buffer between the doctor and prosecutor, as well as the enactment of the law in itself, which would remove any lingering doubts about the legal permissibility of voluntary euthanasia, may assuage the fear of prosecution and encourage reporting. Yet it is doubtful whether this regulation would lead to a far higher reporting rate of euthanasia and assisted suicide. Also under the new legislation euthanasia is allowed only under certain conditions and circumstances. When the physician has not fulfilled the requirements, the case has to be investigated and possibly brought before court. Physicians show themselves quite reluctant to legal control.[20] There is no reason to expect that this would change. Hence, it is to be expected that precisely those cases in which the requirements have not been fulfilled, will still not be reported. Government and physicians have brought themselves into a prisoner's dilemma. When all the reported cases will be scrutinized critically, physicians will hesitate to report precisely those cases in which the requirements were not fully observed. To obtain a higher reporting rate legal scrutiny should be relaxed, but then one may doubt its value. Therefore, the fundamental problem that the legal authorities are not able to effectively control the life-terminating actions by physicians is not likely to be solved in the new legal situation. It might well increase the permissiveness towards euthanasia among physicians and society at large and further an increase in the number of cases.

Furthermore, the legal approach will be different. So far euthanasia is prohibited in the Penal Code and in individual cases the physician must be able to prove that he fulfilled the requirements in order to successfully appeal to the defense of necessity.

But under the proposed regulation of euthanasia in the Penal Code the public prosecutor must be able to prove that the physician has *not* fulfilled requirements in order to start prosecution. So the burden of proof is shifted from the physician to the prosecutor. This may be a reason for public prosecutors to forego prosecution in cases in which they foresee difficulty in proving noncompliance with the requirements. Probably effective control and prosecution of unacceptable euthanasia will become even more difficult.

ETHICAL CRITIQUE OF EUTHANASIA

Conceptual clarifications revisited

In the presentation of the three categories of actions that may have an impact on the moment of death of the patient, but should be distinguished from euthanasia and assisted suicide the concepts of proportional, or medically useful treatment, the principle of double-effect and the distinction between "bringing about death" and "letting die" have been used. These concepts constitute complicated and much debated issues involving philosophical problems with respect to causality and the relation between a natural course of events and technical interventions. It is not possible in this context to discuss these issues extensively. But they often play an important role in the debate as was apparent in the reasoning of the courts in the cases of the two babies whose life was ended intentionally. Therefore, I want to briefly present my view on these issues that, I think, is very much in line with traditional medical ethical reasoning and with the way many physicians experience their practice.

A central question is the relation between life and health and human (technical) intervention. It is important to recognize that human life and health are natural phenomena that have an existence and dynamic of their own. Health is a regulative idea, indicating a state of wholeness that cannot be fully objectified and can never be *produced* by medicine. Disease is to be understood as a disturbance of a state of (better) health. Medical interventions try to stop, to push back or take away pathological processes that constitute the disturbance in order to facilitate the natural processes of life to restore the ill person to a state of equilibrium that underlies health (Schipperges 1978, 487; Gadamer 1996, 31-44, 70-82, 103-16). Today's medicine, no doubt, has enormous possibilities to (temporarily) control physiological processes making it possible to maintain patients alive that would otherwise have died. Yet these interventions cannot produce health. Their possibilities are limited. They entail, on the contrary, always the risk and often the certainty that they will cause damage and more suffering to the patient. The principle of proportionality states that a medical intervention can be justified only if for the patient the benefits of that intervention outweigh its burdens and risks. It is important to note that this principle involves the assessment of the positive and negative effects of the intervention for the patient. It should not involve a value judgment of the life of the patient as a basis for a decision of the value of a treatment.

What do these observations mean? The observation that medicine cannot fully control life and health implies that the physician cannot be held responsible for whatever happens to the patient. The course of the disease and the dying process can be and ultimately always will lie beyond medical possibilities and often even sooner beyond ethically justifiable medical intervention. Therefore, a decision to let a patient die when no proportional treatment exists any longer cannot be considered in any morally meaningful way a decision to *bring about* death (Callahan 1992; 1989). The former decision implies the acceptance that the justifiable possibilities to maintain life are exhausted and that the death of the patient is unavoidable. The position that stopping disproportional life-prolonging treatment and letting a patient die is morally equal with actively bringing about death presupposes a degree of control over life that does not exist. From an ethical perspective such reasoning evaluates an action solely on the basis of its outcome, since those two kinds of action are only equal in the outcome: the death of the patient. But such a narrow consequentialist reasoning neglects the moral importance of the nature of the act itself and of the intention of the actor. It is possible that a physician experiences the condition and the care for a dying patient as a whole in which "natural processes" and his interventions can hardly be distinguished. This does not rule out the importance of maintaining those distinctions in order to prevent intentional killing from becoming a medical solution to difficult situations. For if this occurs, effective protection of patients' lives becomes impossible, as we saw above.

The principle of double-effect, which has a long history in western ethics, is a central ethical principle in medicine (Reich 1978). It is based upon the distinction between the intended and unintended foreseen results of an action. This distinction is common in medicine. The intended result of the administering of chemotherapy is the healing of the patient from cancer. A foreseen but unintended result can be the loss of hair. This is accepted because of the (expected) intended effect on the cancer. A condition for an acceptable treatment is that the intended effects outweigh the foreseen unintended effects.

In the case of pain and symptom treatment sometimes the shortening of life can be an unintended side effect. The principle of double-effect can be used to explain why this can be ethically justifiable.[21] The administering of palliative drugs is not bad in itself – otherwise it would not be acceptable in spite of possible good results. Second, the intention of the physician is (or should be) relief of pain; the (possible) shortening of life is an unintended side effect. This should be clear from the dosages that are administered: these are appropriate for the situation and not "overdoses." Third, the good effect (relieve of pain) is not produced by the side effect (shortening of life). Fourth, there is a good reason to allow the bad side effect: the suffering of a patient in the last phase of life (Keown 1999). In other words, the intended effects morally outweigh the side effects. The objection that it is impossible to distinguish between the intended and foreseen unintended effects of a physician is not convincing. Physicians constantly make this distinction in their treatment decisions and these decisions are justified with the use of this distinction. Denying the possibility to make this distinction would mean that physicians are intending the negative side effects of their treatments in the same way as the positive effects. In the care for terminal seriously suffering patients physicians may find it difficult to clarify their intentions. But this does not deny the importance of the distinction under discussion for the moral assessment of physician's actions. Furthermore, we have seen that Van der Wal and Van der Maas categorize the different kinds of "medical decisions near the end of life" partly on the basis of the intention of the physician to shorten life, thus underlining the importance of it in the moral evaluation of medical actions.

In the following I will present and criticize some of the main arguments in favor of euthanasia and present some additional arguments against euthanasia.

Arguments for Euthanasia Presented and Discussed [22]

Autonomy
A major argument of those who approve of euthanasia is the principle of autonomy or self-determination. Autonomy is seen as a fundamental anthropological characteristic and respect for autonomy as an ethical principle. In both meanings of the concept, a person is considered to be an individual agent who on the basis of rational deliberation is able to choose his values and norms as well as voluntarily behave in accordance with them. Because a human being is viewed as an autonomous being, each individual's autonomy should be respected. The important principle of informed consent is based upon the principle of autonomy (Beauchamp and Childress 1994, chap. 3).

These considerations on autonomy also apply to the euthanasia issue, because an important argument used in favor of euthanasia is the respect for the autonomy of the patient who asks for it. Increasingly, the right to self-determination is extended to the moment and the way one is going to die (Leenen 1989; Willems 1987). Death should not be awaited with fear and trembling while one's physical and mental condition is deteriorating, but death may be brought about willfully by a "free decision" of an autonomous person who has chosen to "step out of this life."[23] The relatively wide support for a proposal from an elderly, well-respected man of law illustrates the rise of this mentality in the Netherlands. According to this proposal, people from a certain age should be able to obtain a "euthanasia-pill" from their GP by request. This would allow the elderly to terminate their life at a self-chosen moment (Drion 1992).[24]

Autonomy revisited
A number of objections can be raised against the use of the autonomy concept as an argument for euthanasia:
a) The before-mentioned concept of autonomy forms part of a very individualistic view of mankind. Society is considered to be a collection

of individuals in which relationships are accidental and peripheral to human existence. What makes a human being really human is, in this view, his reasoning. This anthropology fails to recognize the importance of the spiritual, historical and social dimensions of mankind. It reasons in terms of rights and duties and not in terms of care and responsibility. This view implies an impoverishment of human life and relationships. It threatens to throw the individual back on himself and tends to further loneliness.

b) The concept of man as an autonomous individual, who freely and rationally chooses his norms and values and decides to act according to them, is unrealistic. And, particularly, so in health care. It is unlikely that a person who would fulfill the criteria for euthanasia would at the same time reach the high ideal of autonomy that would have to justify euthanasia. Illness, as soon as it gets serious, always implies a reduction of the freedom of reasoning and of choice because of the discomfort, pain and anguish the illness involves. A Dutch neurologist has argued that the very fact of being in a terminal state, as well as the medication that is often given in those situations, make a clear and normal functioning of the brain almost impossible (Schulte 1988; Reitman 1995). Furthermore, the patient is completely dependent upon others, who through their attitude, gestures, tone of voice, etc., can suggest the patient to ask for euthanasia. This is especially true for the physician and can even happen unconsciously if in the doctor's opinion euthanasia is considered to be a good "solution" in some cases. [25]

One need not deny the theoretical possibility of a truly voluntary request by a seriously ill patient to have his life ended, to realize that in practice it would be difficult to make sure that only such requests will be granted, once euthanasia is accepted and legalized. Recent research demonstrated that 58.8 % of patients with a persistent wish to be killed, had a depression, whereas of a comparable group of patients who did not have a depression 7.7 % had such a death wish. Such a depression is often not recognized while once diagnosed it demonstrates to be well treatable in many cases (Verhagen 2000).

The above is not to deny the importance of the agency and the wishes and needs of the patients and to respond to them. On the contrary, it is to argue that the principle of respect for autonomy is a very feeble basis for intentionally ending the life of a patient. It should be pointed out here that the interpretation of self-determination as a ground for physician-assisted death is not broadly supported internationally. [26]

c) The autonomy of the *patient* cannot logically be the main reason for accepting euthanasia or assisted suicide, simply because the *physician* is performing it. Euthanasia also requires the autonomous consent of the physician. When performing euthanasia the physician freely and intentionally kills the patient. But the free and intentional killing of another person, even with good intentions, cannot fail to influence the

attitude of the physician towards people, life and death in general, and
to very ill patients, competent and incompetent, in particular. Our acts
influence our attitude (e.g., think of the phenomenon of cognitive
resonance). This also applies to physicians. Performing and justifying
euthanasia will make them different people. This will affect medical
care and therefore society as a whole.

d) The whole system of review and control that is necessary to restrict
legalized euthanasia to the permitted cases contradicts the ideal of
euthanasia on the basis of individual autonomy (Cameron 1996).

Mercy

In modern medicine there is a pursuit to gain control over life and health. The
intention is to maintain life and health as much as possible (Jaspers 1989, 262;
Editorial 1970; Callahan 1990, chap. 2). This has led to formidable results in
medical possibilities to treat diseases and disorders. However, the enormous
technological facilities and the lure of technological control also make it possible to
prolong patients' lives until a situation is generated which turns out to be unbearable
for the patient (Cassell 1993). Terrible suffering and agony near the end of life
(which, of course is not only generated by medical technology) is experienced as
dehumanizing. This should not be tolerated; humans have a right to die with dignity,
it is stated. Ending the patient's life is then considered an act of *mercy* of the
physician and of *beneficence* towards the patient. (Note that a traditional term for
euthanasia is mercy-killing).

Mercy revisited

Many who accept euthanasia do so in the name of mercy or compassion. And mercy,
without doubt, will in general be a motive when euthanasia is performed. I find this
understandable and wherever any kind of hastening death would be allowed it would
be on the basis of compassion. I also realize that in the course of the ages in extreme
cases (some) physicians have consciously given high doses of drugs knowing that
death would thereby be hastened. Whether they had the clear intention to bring
about death is difficult to know and whether it was the right thing to do remains to
be seen. But apart from that I think that today's euthanasia practice in the
Netherlands clearly extends beyond such extreme cases, particularly in view of the
modern possibilities of palliative medicine.[27]

In addition to mercy euthanasia in my opinion is motivated by a pursuit for
control on the part of the patient and of the physician. This aspiration for control
pervades modern medicine. It primarily concerns control over the quality of life, but
under circumstances this requires control over death. For when the *quality-of-life*
cannot be maintained at the desired level, the *situation* can be brought back under
control by ending the patient's life.[28] Therefore, euthanasia can be seen as a
culmination and not so much as a correction of a medicine that is becoming more
and more an instrument for control by technical means. The acceptance of

euthanasia may well further such a medicine, because if the efforts to control the quality of life fail but leave the patient in a miserable condition, euthanasia still provides a way out. This is not to say that physicians consciously experience that aspiration for control. It has become part of the ethos of modern society and therefore also of medicine.

It may be granted that when the life of a patient is intentionally terminated, death itself is not the good sought for but the elimination of suffering. However, ending a patient's life to stop the suffering, implies the judgment that such a life is fully marked by the suffering and that the good of eliminating suffering justifies the elimination of the good of the person's life. Medically and ethically I find this unacceptable. Medicine should fight suffering because of the person and should not turn against the person because of his suffering. In practice this distinction may seem to become vague, e.g., in situations in which the suffering can only be made tolerable by such high doses of drugs that the patient's consciousness is pushed back and his life is (almost certainly) shortened. Yet it maintains its fundamental ethical importance because of the fundamental goods of the patient's life and of the physician's position in society (see further). The general acceptance of euthanasia may cause to atrophy our ability to deal with our human condition of vulnerability and openness to suffering in another way than by medical technology.

Some Additional Arguments Against Euthanasia

Palliative care
A request for euthanasia is frequently a sign of insufficient or inadequate care or attention. The patient feels uncomfortable, has pain, is afraid of what still may happen, and is worried about relatives or relationships. Such a request should be taken as an appeal to our responsibility to provide the care that is needed. This is precisely what hospice care intends to do. Good care and attention by physicians and nurses, attention by relatives, adequate pain treatment (nowadays almost always possible) in most cases by far make a request for euthanasia disappear (Zylicz 1998; Danhof 1999; Baar 1999; Twycross 1995). Exceptions to that rule may exist, but exceptions should neither determine morality nor legislation.

Life termination without request
Voluntary euthanasia will provoke "non voluntary euthanasia." From the data of Dutch euthanasia practice it can be concluded that, not so much the request, but rather the condition of the patient is the fundamental reason for physicians to terminate a patient's life. Yet, in cases where the condition of the patient gives reason to perform euthanasia on request, the patient will often no longer be able to make a free request (I forego the complicated issue of advanced directives). However, when the patient's condition functions as the fundamental ground for euthanasia, the fact that there is no request could then be considered insufficient reason *not* to perform euthanasia. In this way the ideological background of

voluntary euthanasia also provides a justification for life termination without request. Against this background it is not astonishing that courts have also accepted the latter under specific circumstances and that the Cabinet has announced in the Memorandum of explanation to the new law on euthanasia and assisted suicide another regulation for dealing with life termination of incompetent patients.[29]

Doctor-patient relationship
Several authors have expressed their concern that the acceptance of euthanasia will in the long run seriously change the doctor-patient relationship for the worse (Singer and Siegler 1990; Doyal 1993). "There is an inherent conflict of interest between the beneficent healing/comforting role in which life and wholeness is sought, and one in which death is caused intentionally" (Reichel and Dyck 1989, 1322). The AMA and other American organizations of health care professionals express this same concern:

> The ethical prohibition against physician-assisted suicide is a cornerstone of medical ethics. Its roots are as ancient as the Hippocratic Oath that a physician 'will neither give a deadly drug to anybody if asked for it, nor ... make a suggestion to this effect' and the merits of the ban have been debated repeatedly in this nation since the late nineteenth century. (....). Physician-assisted suicide remains 'fundamentally incompatible with the physician's role as healer, would be difficult or impossible to control, and would pose serious societal risks.'[30]

The acceptance of euthanasia as part of medical practice is likely to lead to a situation in which optimal care and protection of vulnerable patients is undermined. Fenigsen has given several examples of a new mentality that is becoming manifest in The Netherlands (giving up patients unnecessarily, or withholding medically indicated treatment since the patient is "just living by himself," or is already 70 years of age etc.,) (Fenigsen 1989; 1995). When euthanasia is a generally accepted option, then for terminal patients who get into difficulty it will be more and more an option *not* to ask for euthanasia.[31] When such patients do not feel really appreciated and respected they may start asking for euthanasia, maybe in the first place to get the care and attention they would appreciate. Such a society would become an inhabitable place for patients who need permanent care (Mayo and Gunderson 1993). We should avoid our society developing into that direction. I am not saying that acceptance and legalization of euthanasia would immediately produce such a society. In The Netherlands there are still many people and institutions committed to care for care-dependent people. Yet, during the months in which the new bill on euthanasia was debated it was noted that there are unacceptably long waiting lists for instance for psycho-geriatric patients.[32] And I notice that it is becoming more difficult to explain to the new generations of medical students why euthanasia is a moral problem at all.

CONCLUSION

Juridical Evaluation

The empirical evidence on euthanasia practice in the Netherlands, not least that of the two major surveys, is far from reassuring. Advocates of voluntary euthanasia have long claimed that tolerating it subject to "safeguards" would allow it to be "brought into the open" and effectively controlled. The data discussed above prove this to be only partly true. The reality is that an increasing number of cases of euthanasia and assisted suicide are indeed reported. But it is not certain that all the reports contain complete and correct information on the case in question. And, more seriously, most cases of intentional life terminating actions – at least until very recently a clear majority – have gone unreported and unchecked. In view of the undisputed fact that in these cases there has not been even the opportunity for official scrutiny, claims of effective regulation are unwarranted. The way in which euthanasia has been brought into the open and been dealt with by political and legal authorities and leaders of the medical and ethical elite has in my opinion led to quite a permissive attitude towards intentional life-terminating actions in a situation of seriously inadequate control.

In the light of the assertion made in the introduction that any form of legal regulation of euthanasia is only acceptable if it would guarantee that the state would in principle be able to assess each case, this development constitutes a strong argument against legalization of euthanasia.[33]

Ethical Evaluation

In addition to the before-mentioned juridical argument, I see several serious medical and social ethical arguments against the legalization of euthanasia. They mainly revolve around a concern for a health care and a society in which all human beings, independent of their capacities and condition are protected, respected, and cared for. Even in our age with a technologically advanced medicine, our human condition may confront us sometimes with appalling suffering for which there seems to be no alleviation. I realize very well that *only* saying that euthanasia is or should be forbidden, is unconvincing. Those situations present a real challenge to our societies and health care. Just a few remarks have been made on this issue, since this contribution is not the place to elaborate this topic. But I am convinced that so far the Dutch have not found the proper response to those challenges.

Henk Jochemsen, Lindeboom Chair of Medical Ethics, Free University of Amsterdam, and Director, Prof. dr. G.A. Lindeboom Institute, Ede, The Netherlands.

NOTES

[1] This requirement can be founded on the European Convention on Human Rights (1950), especially article 2, which requires the state to protect the life of every citizen by law. So even if the law would allow for the intentional termination of the life of a citizen it can do so only under the condition that the legal authorities have the possibility to control in each case whether the requirements that the law demands had been fulfilled. The Belgian jurist, Velaers (1996, 573), concludes his thorough study of this issue as follows:

> Only when it is demonstrated that medical deontology and the control system of the judiciary and of the disciplinary court can guarantee satisfactorily that euthanasia will be kept within acceptable limits, the lawgiver can take its responsibility. However, if it is not possible to provide that guarantee, in our view the lawgiver is not allowed to create any space, even not under very restrictive conditions.

[2] Standpunt van het Kabinet inzake medische beslissingen rond het levenseinde [Position of the Cabinet with respect to "Euthanasia and other medical decisions concerning the end of life"] (Den Haag: Ministerie van Justitie, 8 Nov 1991).

[3] The new legal regulation (bill no. 22 572) amends the law on the Disposal of the Dead (article 10), which provides the legal basis for a form used by physicians to certify natural death. The law as amended provides that both the form to certify natural death and the form to notify all cases of euthanasia, assisted suicide and a life-terminating action without request will be set out in an enactment ['Algemene Maatregel van Bestuur', AMvB]. This AMvB has been published in the Staatsblad [official publication of the government] of 28 Dec 1993, as: *Besluit van 17 Dec 1993, Stb 688*. This law became formally effective on June 1, 1994.

[4] For summaries of the research in English see: Van der Maas et al. (1996) and Van der Wal (1996).

[5] The report does not give exact data at this point, but from the data that are given it can be estimated that in at least a few hundreds of those cases there was no explicit request (Van der Wal et al. 1996, 77-8, 80, 92-3).

[6] Vonnis Arrondissementsrechtbank te Alkmaar d.d. 26 april 1995 in the case against P. *Tijdschrift voor Gezondheidsrecht* No.5 (1995) p.292-301.

[7] Vonnis Rechtbank te Groningen d.d. 13 november 1995 in the case against K. *Pro Vita Humana* 3, no.1 (1996) p.29-32.

[8] Arrest Gerechtshof te Amsterdam d.d. 7 november 1995 in the case against P. *Pro Vita Humana* 3, no.1 (1996) p.25-28.

[9] Arrest Gerechtshof te Leeuwarden, d.d. 4 april 1996 in the case against K. *Tijdschrift voor Gezondheidsrecht* 5 (1996) p.284-91.

[10] For a more extensive discussion of these cases, see Jochemsen (1998a).

[11] The Leeuwarden court formulated the following conditions that at least must be fulfilled by the physician to be able to successfully appeal to the defense of necessity (see *supra* note 9, section 12.10):
- no doubt must exist about diagnosis and prognosis,
- the doctor must consult colleagues,
- death must be brought about in a careful and correct way,
- the doctor must report the case,
- the parents must consent to the killing of the baby

[12] In a recent publication, the secretary of the KNMG medical ethical committee, argued that the physician must indeed make a judgment on the suffering of the patient and conclude that this is unbearable before he can perform euthanasia; the moral justification of euthanasia is not just the request of the patient but as much his suffering (Dillmann 1998).

[13] *Wet Toetsing levensbeëindiging op verzoek en hulp bij zelfdoding [Law on Assessment of Life Termination on Request and Assisted Suicide]*. Tweede Kamer 1998-1999, 26691, nr. 3, 10.

[14] I speak of hospice *movement* since it is rather a kind of approach to terminal care than recognizable institutions or the availability of certain medical "know how" as such. I do not deny that palliative care and palliative medicine has been practiced in the Netherlands for many years. However, the application of the most modern pain and symptom control, in the context of integral palliative care according to the hospice philosophy, is not widespread and still depends very much on local initiatives. An inventory in 1997 demonstrated that there were 35 initiatives in palliative care, six of which are (mostly small) hospice institutions and the others concern intramural palliative care within other health care institutions like nursing homes. In addition to these 35 there were about 120 initiatives of activities that show some overlap with hospice care, like the functioning of a special pain team or specific caring facilities for terminal patients (Francke et al. 1997). More recently the government and the medical profession are making considerable efforts to improve the situation (See: http://www.palliatief.nl/_e/n0/fr.htm, last accessed Aug 12, 2001).

[15] These changes are announced and described in: Kabinetsstandpunt naar aanleiding van de evaluatie van de meldingsprocedure euthanasie. *Brief van Minister van Justitie en van de Minister van Volksgezondheid, Welzijn en Sport aan de Tweede Kamer, d.d. 21 januari 1997 (kenmerk 603400/97/6).* [Position of Cabinet with respect to evaluation of reporting procedure of euthanasia. Letter of Ministers of Justice and of Health, Welfare and Sports, d.d. January 21, 1997].

[16] *Staatscourant 1998,* 101 and 103.

[17] Regionale toetsingscommissies euthanasie. *Jaarverslag 1999.* [Regional euthanasia review committees. *Annual report 1999*]. Den Haag 1999, p.8ff. This report indicates that, in 1999, 2216 cases of euthanasia or assisted suicide were reported, whereas in 1998 2241 cases were reported. (Jaarverslag Openbaar Ministerie 1998 inzake euthanasie en hulp bij zelfdoding. *Pro Vita Humana* 7, no 1 (2000), p.34). In April 2001 the Department of Justice announced that in 2000 the number of reported cases decreased to 2123 cases.

[18] See *Annual report 1999* of Regional euthanasia review committees, *supra* note 17 at 8.

[19] See *Wet Toetsing levensbeëindiging op verzoek en hulp bij zelfdoding [Law on Assessment of Life Termination on Request and Assisted Suicide]*, *supra* note 13. See also Jochemsen (2001).

[20] See *supra* note 17 and accompanying text.

[21] The American Medical Association in this context explicitly refers to this principle as well as to the distinction between withdrawing or withholding a treatment and assisted suicide, in its statement presented at the Hearing before the Subcommittee on Health and Environment of the Committee on Commerce of the House of Representatives of the USA, d.d. March 6 1997 (Serial No. 105-7), p. 55.

[22] For another discussion of arguments pro and contra euthanasia and assisted suicide see Kimsma and van Leeuwen (1998) and Jochemsen (1998b).

[23] Interesting in this context is that according to the 1995 Survey, 52% of the attending physicians, 83% of the medical examiners and 73% of the public prosecutors (fully) agree with the statement that "Everyone has the right to self-determination about one's own life and death" (Van der Wal et al. 1996, 174).

[24] In an interview a few days after the approval of the euthanasia bill in the First Chamber, the minister of health. Mrs. E. Borst voiced the opinion that it would be good to start the debate on the provision of such a "suicide pill." (See *NRC-Handelsblad* April 14, 2001, p. 35.) Later, she declared she does not intend to come with a proposal in this respect.

[25] Cf. the opinion of a number of American medical and nurses' organizations: "...experience to date provides little basis for confidence that health care professionals can reliably determine whether patients have provided truly informed consent for assisted suicide. Frank, sensitive, and extended conversations between physicians and patients are presumptively antecedents to such a determination" (p.15). "The well-established phenomena of transference and countertransference further complicate the problem of relying upon physicians and nurses to identify voluntary requests" (p.15,16). See: Brief of the American Medical Association, the American Nurses Association, and the American Psychiatric Association, et al. as Amici Curiae in support of Petitioners. *In the Supreme Court of the United States, October term 1996;*

State of Washington, Christine O. Gregoire (petitioners) versus H. Glucksberg, A. Halperin, T.A. Preston, P. Shalit (respondents). Doc. No.96-110, d.d. 12 november 1996. See also: Muskin (1998).

[26] Cf. The Appleton International Conference (1992), and also Recommendation 1418 on "The protection of the human rights and dignity of the terminally ill and the dying," adopted by the Assembly of the Council of Europe on 25 June 1999. This recommendation states: "recognizing that a terminally ill or dying person's wish to die cannot of itself constitute a legal justification to carry out actions intentionally to bring about death" (2000 *Pro Vita Humana* 7(1): 34).

[27] See e.g. the antropological study of Pool (1996).

[28] "Health care professionals also experience great frustration at not being able to offer the patient a cure. For some, the ability to offer the patient the 'treatment' of assisted suicide may provide a sense of 'mastery' over the disease and the accompanying feelings of helplessness." Brief of the AMA, *supra* note25, at 16.

[29] See *Wet Toetsing levensbeëindiging op verzoek en hulp bij zelfdoding [Law on Assessment of Life Termination on Request and Assisted Suicide], supra* note 13, at 4.

[30] Brief of the AMA, *supra* note25, at 5.

[31] Cf. "Were physician-assisted suicide to become a legitimate medical option, then a decision not to select that option would make many patients feel responsible for their own suffering and for the burden they impose on others." Brief of the AMA, *supra* note25, at 17.

[32] See e.g. Task force aanpak wachtlijsten. Analyse landerlijke inverntarisatie wachtlijstgegevens. Verpleging en verzorging. [Analysis national inventory waiting list data. Nursing and caring homes]. Den Haag: Ministerie van VWS [Health department] 2000; Idem for: Geestelijke gezondheidszorg [mental health]. Den Haag: Ministerie van VWS, 2001

[33] Cf. the conclusion of the British Select Committee of the House of Lords (1994, 48) on the question whether euthanasia should be legalized: "Ultimately, however, we do not believe that these arguments are sufficient reason to weaken society's prohibition of intentional killing. That prohibition is the cornerstone of law and of social relationships. (...) We acknowledge that there are individual cases in which euthanasia may be seen by some to be appropriate. But individual cases cannot reasonably establish the foundation of a policy which would have such serious and widespread repercussions." This report received remarkably little attention in the Netherlands.

REFERENCES

Baar, F.P.M. 1999. Palliatieve zorg voor terminale patienten in verpleeghuizen [palliative care for terminal patients in nursing homes]. *Pro Vita Humana* 6(6): 169-75.

Beauchamp, T.L., and J.F. Childress. 1994. *Principles of biomedical ethics,* 4th ed. Oxford: Oxford University Press.

Callahan, D. 1989. Can we return death to disease? *Hasting Center Report* 19(1): 4-6.

———. 1990. *What kind of life? The limits of medical progress.* New York: Simon & Schuster.

———. 1992. When self-determination runs amok. *Hastings Center Report* 22(2): 52-5.

Cameron, N.M. de S. 1996. Autonomy and the "right to die." In *Dignity and dying: A Christian appraisal,* eds. J.F. Kilner, A.B. Miller, and E.D. Pellegrino, 23-33. Grand Rapids (MI): Eerdmans.

Cassell, E.J. 1993. The sorcerer's broom. *Hastings Center Report* 23(6): 32-9.

Danhof, E. 1999. Palliatieve zorg is gebaat bij een bundeling van krachten tot een loket. [Palliative care would benefit from concentrating efforts in one desk]. *Pallium* 1(5): 12-14.

Dillmann RJM. 1998. Euthanasie: de morele legitimatie van de arts [Euthanasia: the moral legitimation of the physician]. In *Levensbeëindigend handelen door een arts: tussen norm en praktijk [Life terminating acting by a physician: between norm and practice],* eds. J. Legemaate, and R.J.M Dillmann, 11-25. Houten: Bohn Stafleu Van Loghum.

Doyal, L. 1993. Making sense of debates about euthanasia. [Editorial comment] *Current medical literature. Anaesthesiology* 7(2): 35-9.

Drion, H. 1992. Het zelfgewilde einde van oude mensen [The selfchosen end of elderly people]. Amsterdam: Balans.

Editorial. 1970. A new ethic for medicine and society *California Medicine* 113(3): 67-8.

Fenigsen, R. 1989. A case against Dutch euthanasia. *Hastings Center report* 19(1): S22-S30

———. 1995. Physician-assisted death: Impact on long-term care. *Issues in law & medicine* 11(3): 283-97.

Francke, A.L., A. Persoon, D. Temmink, and A. Kerkstra. 1997. *Palliatieve zorg in Nederland*. Utrecht: Nivel.

Gadamer, H.G. 1996. *The enigma of health- the art of healing in a scientific age*. Stanford: Stanford University Press

Jaspers, K. 1989. The physician in the technological age. *Theoretical Medicine* 10(2): 251-67. 262.

Jochemsen, H. 1994. Euthanasia in Holland: an ethical critique of the new law. *Journal of Medical Ethics* 20 (1994), 212-7.

———. 1998a. Dutch court decisions on non voluntary euthanasia critically reviewed. *Issues in Law & Medicine* 13(4): 447-458.

———. 1998b. The range of objections to euthanasia, In *Asking to die. Inside the Dutch debate about euthanasia*, eds. D.C. Thomasma, T. Kimbrough-Kushner, G.K. Kimsma, and C. Ciesielski-Carlucci, 227-40. Dordrecht: Kluwer Academic Publishers.

———. 2001. Update: The legalization of euthanasia in the Netherlands. *Ethics & Medicine* 17(1): 7-12.

Jochemsen, H., and J. Keown. 1999. Voluntary euthanasia under control? Further empirical evidence from the Netherlands. *Journal of Medical Ethics* 25: 16-21.

Kastelein, W.R. 1995. *Standpunt hoofdbestuur KNMG inzake euthanasie*, [Position General Committee KNMG on Euthanasia]. Utrecht, August.

Keown, J. 1995. Euthanasia in the Netherlands: Sliding down the slippery slope? In *Euthanasia examined*, ed. J. Keown, 261-96. Cambridge: Cambridge University Press.

———. 1999. Double effect and palliative care: A legal and ethical outline. *Ethics & Medicine* 15(2): 53-4.

Kimsma, G.K., and E. Van Leeuwen. 1998. Euthanasia and assisted suicide in the Netherlands and the USA: Comparing practices, justifications and key concept in bioethics and laws. In *Asking to die. Inside the Dutch debate about euthanasia*, eds. D.C. Thomasma, T. Kimbrough-Kushner, G.K. Kimsma, and C. Ciesielski-Carlucci, 35-71. Dordrecht: Kluwer Academic Publishers.

Leenen, H.J.J. 1989. Dying with dignity: Developments in the field of euthanasia in the Netherlands. *Medicine and Law* 8: 517-526.

Mayo, D.J., and M. Gunderson. 1993. Physician assisted death and hard choices. *Journal of Medicine and Philosophy* 18: 329-41.

Muskin, P.R. 1998. The request to die. Role for a psychodynamic perspective on physician-assisted suicide. *Journal of the American Medical Association* 279(4): 323-8.

Pijnenborg, L. 1995. *End-of-life decisions in Dutch medical practice*. PhD. Diss. Erasumus University. Rotterdam.

Pool, R. 1996. *Vragen om te sterven. Euthanasie in een Nederlands ziekenhuis. [Asking to die. Euthanasia in a Dutch hospital]*. Rotterdam: WYT Uitgeefgroep.

Reich, W., ed. 1978. Acting and refraining. In *Encyclopedia of Bioethics*, 33-5. New York: Free Press.

Reichel, W., and A.J. Dyck. 1989. Euthanasia: A contemporary moral quandary. *The Lancet* ii: 1321-3.

Reitman, J.S. 1995. The debate on assisted suicide- redefining morally appropriate care for people with intractable suffering. *Issues in law & medicine* 11(3): 317-8.

Schipperges, H. 1978. Motivation und Legitimation des Arztliche Handelns. In *Krankheit, Heilkunst, Heilung*, eds. H. Schipperges, E. Seidler, and P.U. Unschuld, 447-489. München: Alber.

Schulte, B.P.M. 1988. Over euthanasia en in het bijzonder over de relatie tussen hersenfunctie en vrijwilligheid bij euthanasie [On euthanasia and in particular on the relation between brain functioning and voluntariness of euthanasia]. *Vita Humana* XV(1): 9-12.

Select Committee of the House of Lords (UK). 1994. *Report of the Select Committee on Medical Ethics*. House of Lords, Session 1993-94. London: HSMO.

Singer, P.A., and M. Siegler. 1990. Euthanasia – a critique. *New England Journal of Medicine* 32(26): 1881-3.

Sorgdrager, W., and E. Borst-Eilers. 1995. Euthanasie. De stand van zaken. *Medisch Contact* 50(12): 381-4.

The Appleton International Conference: developing guidelines for decisions to forgo life-prolonging medical treatment. 1992. *Journal of Medical Ethics* 18(suppl.): 3-5.

Twycross, R. 1995. Where there is hope, there is life: a view from the hospice. In: *Euthanasia examined. Ethical, clinical and legal perspectives*, ed. J. Keown, 141-68. Cambridge: Cambridge University Press.

Van der Maas, P.J., et al. 1991. *Medische beslissingen rond het levenseinde*. Den Haag: SDU Uitgeverij. [Published in translation as *Euthanasia and Other Medical Decisions Concerning the End of Life*. Amsterdam: Elsevier 1992.]

Van der Maas, P.J., G. Van der Wal, I. Haverkate, C.L.M. de Graaff, J.G.C. Kester, B.D. Onwuteaka-Philipsen, A. van der Heide, J.M. Bosma, and D.L. Willems, Dick L. 1996. Euthanasia, physician-assisted suicide, and other medical practices involving the end of life in the Netherlands 1990-1995. *New England Journal of Medicine* 335: 1699-1705.

Van der Wal, G., and P.J. Van der Maas. 1996. *Euthanasie en andere medische beslissingen rond het levenseinde. De praktijk en de meldingsprocedure*. [Euthanasia and other medical decisions concerning the end of life. Practice and reporting procedure.] Den Haag: SDU uitgevers.

Van der Wal, G., P.J. Van der Maas, J.M. Bosma, B.D. Onwuteaka-Philipsen, D.L. Willems, I. Haverkate, and P.J. Kostense. 1996. Evaluation of the notification procedure for physician-assisted death in the Netherlands. *New England Journal of Medicine* 335: 1706-11.

Velaers, J. 1996. Het leven, de dood en de grondrechten – juridische beschouwing over zelfdoding en euthanasie [Life, death and fundamental rights – juridical discours on suicide and euthanasia. In *Over zichzelf beschikken? Juridische en ethische bijdragen over het leven, het lichaam en de dood [Self determination? Juridical and ethical essays on life, the body and death]*, 469-574. Reeks: Het recht in de samenleving, van het Centrum Grondslagen van het Recht, Universiteit Antwerpen. Antwerpen: Maklu uitgevers.

Verhagen, S. 2000. Behandeling depressiviteit: altijd zinvol, vaak effectief [treatment of depression: always useful, often effective]. *Pallium* 2(2): 13-7.

Willems, J.H.P.H. 1987. Euthanasie en noodtoestand. *Nederlands Juristenblad* 22: 694-8.

Zylicz Z. 1998. Palliative care: Dutch hospice and euthanasia. In *Asking to die. Inside the Dutch debate about euthanasia*, eds. D.C. Thomasma, T. Kimbrough-Kushner, G.K. Kimsma, and C. Ciesielski-Carlucci, 187-203. Dordrecht: Kluwer Academic Publishers.

CHAPTER SIX

GEORGE P. SMITH, II

MANAGING DEATH

End-of-Life Charades and Decisions

This chapter examines the decisions and charades, or symbolic actions and pretensions, that the medical profession follows in death management – primarily to avoid legal consequences for negligence, malpractice, or homicide, but very often also because there are inappropriate guidelines for suitable treatment of terminal illnesses. Through use of resuscitative (e.g., full code) or non-resuscitative (e.g., Do Not Resuscitate) codes, orders are given by physicians regarding the courses of medical action to be taken when a patient suffers a cardiopulmonary arrest. Sadly, all too often full resuscitation is undertaken in situations where it is clearly contraindicated.

One way to resolve the confusion and complexity in death management is to achieve a wider acceptance of the doctrine of medical futility and thereby allow health care providers to agree on uniform responses to end-of-life conditions often involving intractable pain as, for once treatment is deemed futile, a physician is freed from the moral, medical and legal duty to provide treatment. Coupled with this acceptance is a need to educate the public to the medico-legal value of permitting terminal sedation to be validated as reasonable palliative care when medical conditions are determined to be futile.

Earlier this year, *The Archives of Internal Medicine* published a first of a kind study comparing the attitudes of randomly selected physicians in the United States and The Netherlands regarding requests by terminally ill patients for assistance in hastening death (Williams et al. 2000; Cavanaugh 1998). The impact of this study on care at the end of life is yet to be determined. What is interesting for purposes of this essay is the commonality of compassion, sound medical judgment, and good sense that appears to exist globally in the medical profession on this complex and contentious issue.

That said, however, current clinical realities oftentimes are unable to immediately influence historically driven practices. Later, I shall turn to those practices which are themselves often nothing more than confused euphemisms or

D.N. Weisstub, D.C. Thomasma, S. Gauthier & G.F. Tomossy (eds.), Aging: Decisions at the End of Life,
91-106.
© *2001 Kluwer Academic Publishers. Printed in the Netherlands.*

"substitutions of an agreeable or inoffensive word or expression for one that is harsh, indelicate, or otherwise unpleasant taboo."[1]

CLINICAL RESPONSES TO DEATH

The *Archives of Internal Medicine* study was conducted during 1995-96 and drew from the responses of 67 physicians from The Netherlands and 152 from the state of Oregon. Physicians were presented with four clinical vignettes regarding hypothetical patients who were dying. The first vignette, for example, described a patient with widespread cancer that had invaded his bones, causing excruciating pain which was uncontrolled by morphine. The physicians were then asked whether it would be appropriate, even if premature death would likely occur, to increase the morphine. In the other vignettes, the doctors in the survey were asked about terminally ill patients who, neither depressed nor in pain, requested life-ending injections because they were debilitated, felt that life was without purpose, or feared becoming a social-economic-emotional burden on their families. American and Dutch doctors were in general agreement regarding the issue of increased levels of morphine, almost always favoring it for patients grouped within the first vignette, but less often for those outside this clinical profile.

On the issue of physician assisted suicide (PAS), both groups of doctors tended to agree on its acceptability where a physician helps or assists a terminally ill patient obtain life-ending drugs. In cases when patients are suffering from excruciating pain, a majority of the respondents favored PAS – but fewer favored it in cases outside this specific vignette (Williams et al. 2000).

Interestingly, the Dutch physicians were much more likely to support acts of euthanasia where they participate in the death of their patients by the administration of drugs. The survey found a resistance among the Oregon physicians to render final life-ending action – yet, found a willingness to assist their patients who wish to end their life.

The Dutch physicians in the survey were found to be more willing to participate in acts of PAS for those patients finding life meaningless than for those concerned about becoming a burden. There were no such differences among the Oregon doctors. Indeed, the researchers found that, among American physicians, the fear of being a burden was an acceptable justification for assistance with dying more often than for their counterparts in The Netherlands (Williams et al. 2000).[2]

The Art Of Coding: Order-Code Or No Code

There are but two basic responses to individuals in cardiopulmonary arrest: order-code or no code (Saunders and Valente 1986, 62). Thus, to code a patient, in essence, means to commence cardiopulmonary resuscitation (CPR). A no code – most commonly, DNR – means no aggressive assistance will be given to a patient in medical distress (Saunders and Valente 1986; Hashimoto 1983). Many consider a CPR order to be a "bad prognostic sign" because, put simply, few code survivors leave the hospital (Saunders and Valente 1986, 63). Indeed, an in-hospital survival rate of fifty percent is considered quite impressive (id.).[3] Even for no code patients, frequent re-assessments, perhaps every seventy-two hours, should be made in an effort to re-evaluate a physician's orders and give supportive or palliative care to the distressed patients (Saunders and Valente, 1986, 64; Bernabei et al. 1996; Smith 1985a). Communication of this no code should be given, as well, to all members of the patient's health management team (Saunders and Valente 1986, 64).

CPR was developed originally to preserve life, restore health, relieve suffering, and limit disability of persons who unexpectedly went into cardiac arrest. It is a desperate invasive procedure that was not intended to delay the impending death of patients who are suffering from terminal illnesses. Despite this, health-care institutions have classified CPR as an "emergency" procedure for which patients' consent is presumed – absent a pre-directive to the contrary (Boozang 1993, 24).[4] This classification has led to the expanded use of CPR, well beyond the select group of patients for which it was intended, and has thus become a pervasive, indiscriminate, and often contraindicated use by health-care workers (Annas 1995). In fact, some states have enacted legislation codifying the expectation of resuscitation in an attempt to ensure the administration of CPR to hospital inpatients (Boozang 1993, 24; Sabatino 1999).[5]

A further cause of the contraindicated use of CPR has been the rise in patient autonomy and informed consent. This rise precipitated the shift in the decision-making power regarding the withdrawal or refusal of advanced life-saving medical technologies. Traditionally, the patient relied upon the professional judgment of physicians. Today, ideally, the medical model is patient-driven and, thus, when patient wishes are communicated or discerned, should be directive (Pellegrino and Thomasma 1988; Katz 1984; Symposium 1998; Orentlicher 1998). As a result of these changes, physicians and other health-care workers often administer CPR in situations in which they feel its use is contraindicated. Their decisions to use or withhold CPR are influenced by the fear of litigation, inappropriate or unclear guidelines, or the misguided directions of the patient or his family. These influences cause, predictably, many physicians' objective judgments to become clouded by factors other than the patient's best interests (Smith 1995).

While it is always the preferred course – and the one advocated by the American Medical Association (1991) – to test the standard of futility according to individual patient values and goals, practice has shown physicians simply do not discuss routinely CPR with their adult patients who are admitted for medical and surgical care (Lo 1991). Indeed, there is a fuller discussion of DNR with AIDS and cancer patients than with patients without diseases with poor prognoses (e.g., coronary artery disease, cirrhosis) (ibid.). Similarly, in order to avoid ambiguities in actual DNR orders, procedure specific orders should be given – with full documentation of the rationale for the order being set out (Mittelberger et al. 1993).

FRAMING THE DNR DEBATE

In the absence of a written Do Not Resuscitate Order, it is standard hospital practice to require resuscitation for every patient suffering a cardiopulmonary arrest and, in essence, force them to die in a code. For the house staff physicians and interns in most hospitals, a DNR order in a patient's medical record lingering near death typically limits the type of medical assistance provided to the patient when death itself approaches. The absence of a DNR order obligates the residents to approach any and all complications aggressively – treating the patient as though he were acutely ill and not dying. The practical result of a situation of this nature is that the interns may very well expect to invest considerable time – in the middle of the night – "working up" either a very sick patient or coding one who is dying. The consequence of this is that the interns are diverted from their other hospital tasks and patient supervisions. Not only are such actions a waste of time and energy, they are not cost effective (Muller 1992, 894-5).

When there are no demonstrable benefits to a medical intervention which maintains an expiring patient other than the act of survival itself, the best interests of the patient may not be served by a resuscitation (id., 894). Accordingly, in those cases where a physician – more likely a resident, but occasionally an attending – is convinced of the futility or potential harm of further treatment, an intermediate code may be negotiated. This action may well have the effect of over-riding the wishes of the patient, family or even private or attending physicians but allows the resident physician a convoluted way to avoid hospital policies. Stated otherwise, such a course of action, "allows them a means of restricting their therapeutic activity when they confront the possibility of having to provide treatment they not only thought was futile but could also inflict significant harm on the patient" (ibid.). The intermediate or limited code has the ultimate effect, then, of providing a means by which resident hospital physicians guard themselves not only against in-house disciplinary action and legal liability but control as well the extent to which they are forced to pursue futile drains of their time and hospital resources (id., 895). It is an artful euphemism, to be sure, and a charade at best.

While it is understandable that hospitals have guidelines for physician behavior in the use of resuscitation, these codes of operating procedures fail to make allowance for patient variation and contemporary clinical practices (id., 896). Designed in response to unnecessary acts of resuscitation, DNR orders have given rise to yet another moral dilemma: the negotiation of slow codes. Stated otherwise, the slow or intermediate codes – while being a cultural response to unworkable circumstances arising in clinical practice – bring with their implementation and use a troubling ethical dilemma. They bypass the "very intent of the resuscitation guidelines: to honor the principle of patient (or surrogate) autonomy and to prevent physicians from making unilateral decisions about resuscitation by requiring a joint decision-making process" (ibid.).

Hakim et al. (1996) found that, in deciding whether to issue a DNR order, physicians unduly relied upon a patient's age and short-term prognosis – giving those two factors a weight that goes beyond their actual ability to predict life expectancy *and* quality of life. Their study of 6,802 seriously ill patients with an average life expectancy of six months (with illnesses such as coma, heart failure, and cancer) yielded some interesting findings. First, compared with other medical specialists, surgeons waited nearly twice as long to write a DNR for a very sick patient while intensive care specialists and lung specialists were most likely to issue a DNR and cardiologists the least likely. As to age as a factor in the issuance of DNR orders, it was found that only 22% of the patients under age 55 had DNR orders, compared with 56% of those over 85. DNR orders were written most rapidly for patients older than seventy-five.

All too often, ambiguities are found in DNR orders. A rather simple effort to issue a procedure specific DNR order and include the rationale for the order has been found to be an effective way to lessen both confusion and eliminate ambiguity (Mittelberger et al. 1993).

UTILIZING CODES: CODIFYING EUPHEMISMS

When there are no orders written that specify what resuscitative measures should be taken with particular patients, hospital policies may well dictate, as observed, that a full code should be called or, in other words, resuscitation be initiated (Quigley 1988; Ross and Pugh 1988). Yet, circumstances may arise in which it is just as appropriate, instead of calling a code, to initiate minimal resuscitative measures which do not rise to the level of being a full blown code (Quigley 1988) and might be termed a "short code." This type of code is sometimes referred to as a *show code* and allows the health care personnel to initiate resuscitation and then proceed to stop their actions either after a few efforts or a predetermined period of time (ibid.). This code is largely regarded as a symbolic gesture designed to reassure or placate the family of a patient, or the health care personnel themselves, that "everything was done" (Muller 1992, 890, 896).

The *show, soft, slow, partial, limited* or *light blue* codes are all termed intermediate codes. Each designation conveys pertinent information concerning not only the type but the extent of response to be followed in the event of a patient suffering cardiopulmonary arrest. Thus, a partial, limited, or soft code is taken commonly to set forth those circumstances where either drugs might be administered without chest compressions or where resuscitation is initiated but drugs or intubation would be withheld (id., 890).

Very often the uses of intermediate codes arise from informal arrangements negotiated verbally, often at night, between residents and nurses who reach an agreement – before a patient goes into cardiac arrest – regarding the courses of action or inaction to be taken. Attending physicians are sometimes consulted – with the ultimate decision regarding the use of a limited code being negotiated between the attending physician and the house staff. Although not ordinarily formalized in writing, occasionally the stated medical reasons for selecting one resuscitative technique over another are in fact written in the patient's record. Clearly, then,

intermediate codes are, in reality, little more than clinical deviations from
established hospital protocols and regulations. As such, they are neither easily
observed nor are they occurrences which are publicly acknowledged (ibid.).[6]

Types of Partial Codes

Medical realities and patient desires shape the parameters for issuance of partial
codes. It is within this spectrum that marked confusion and disagreement occur
primarily because of a failure by hospital personnel to differentiate partial codes
according to their intents and purposes. Thus, it is essential to first seek to
distinguish between partial codes for patients who are monitored and those who are
not (Ross and Pugh 1988). Then, distinctions between patients partially coded with
and those without specific medical conditions (e.g., chronic obstructive pulmonary
disease) viewed likely to result, in the event of full CPR, with ventilator dependence
are drawn (id., 6).[7]

Partial codes may be distinguished, additionally, by the particular intent of the
orders. The question then, becomes: are the code orders being motivated by patient
autonomy or by a health care provider's standard of beneficence. While the
autonomy, the patient's right to direct that course of medical treatment consistent
with his life values is acknowledged, not every patient (or family) preference can or
should be followed. Under the principle of beneficence, a health care provider must
ascertain whether his actions contribute to a patient's well-being. Thus, a conflict
often arises when these two foundation principles come into play (Ross and Pugh
1988, 4).

There are – essentially – four situations where partial codes may be written. The
first arises when, for example, a partial code (e.g., "basic CPR only") is ordered as a
result of a family refusal to consent to a DNR order. Most often, the physician is
responding in such a case to what is believed to be an irrational family response in
the first instance. Such an order may also be given in writing when discussion of
approaching death with either the patient or the family is unfeasible yet the attending
physician wishes to convey an obviously deceptive impression "that all that can be
done is being done." Because of external problems in such a situation, a preferred
DNR cannot be given. Yet, because the physician has determined the best medical
interests of the patient will be served by no resuscitative effort, a partial code may be
ordered. Although such a use is motivated properly by beneficence, issuing a partial
code under these circumstances is considered – under most circumstances – to be
unethical (id., 7).

A second type of partial code is seen in cases where patients are on monitors.
Here, for example, a "chemical code only" (or the use of cardiac drugs) is ordered
most commonly when either a patient or his family has requested a DNR order or
has consented to its issuance. Because the physician and, even possibly, the patient
and/or the family are of the opinion that either the patient's actual condition or
present quality of life are stabilized to such a degree that desirable efforts should be
pursued to prevent an actual arrest, at the same time there is an understanding not to

reverse it if it were in fact to occur. Thus, in a case of this nature, preventing an arrest is sought by simply treating pre-arrest symptomatologies (ibid.).

When the patient or family expresses an unambiguous and reasoned decision not to allow ventilator use to prolong life, a third situation arises where a partial code such as – "do not intubate" – may be written. Typically, this situation presents itself when the physician reveals to all concerned parties that, because of the patient's condition (e.g., chronic obstructive pulmonary disease), once respirator care is initiated there will be little chance of weaning the patient from the respirator. Although the physician places reliance upon an accurate prognosis, the patient or family makes the actual decision for the code (ibid.).

The fourth and most difficult of all partial codes for health care providers is seen in cases where an order such as, "do not intubate," is written at the request of a competent patient who, for personal reasons, does not wish specific parts of the code performed. Normally, the basis of these reasons is to be found in fears that the patient will become either a vegetable or an untubated appendage to a machine. Simply because a competent patient makes a request of this nature, does not mean it must be respected; for, it does not comport with a sound medical judgment by a health care provider, it will not be executed (id., 7). Beneficence will usually trump autonomy.

The inherent problem with partial codes is that patients and their families simply fail to realize that there are not discrete elements in resuscitation plans. Thus, from a medical perspective, a patient should submit to full resuscitation or no resuscitation (id., 8). Medical realities dictate ultimately that euphemistic requests for less than effective full treatment options will not always be followed.

When patient requests are made for a full resuscitative effort, "except intubation and ventilation," they are made normally without a full understanding that to undertake such an effort, intubation for maximal oxygenation will be required normally and ventilator assistance will be necessary to stabilize the patient. If it is determined subsequently that ventilator support is required for the longer term, the patient may thereupon request the withdrawal of support (id., 7).

Similarly, when patients view CPR decisions they may request, for example, chest massage and mask ventilation and nothing more. Such a decision fails to take into consideration the fact that CPR is a synergistic process – with each component building upon the effectiveness of other components. Thus, the first stage of CPR typically includes chest massage and artificial respiration – followed, if necessary, by defibrillation and, finally, with cardiac drugs. While appearances might suggest the process is indeed severable, if the patient is in a monitored unit, cardiac drugs might well be used first, with defibrillation beginning before chest massage. And, for unmonitored patients, while resuscitation begins with chest massage and artificial respiration, it is undertaken because there is no other equipment readily available. Resuscitative efforts which use equipment are recognized as more effective than either chest massage and artificial respiration (id., 8).

FUTILITY AS AN OPERABLE STANDARD

Legislative definitions may be proffered for what is a terminal medical condition and include incurable and irreversible conditions "within reasonable medical judgment" which will either cause death "within a reasonable period of time" or merely extend the dying process.[8] Depending of course upon individual patient profiles and disease etiologies, medical judgment will vary regarding when specified conditions are terminal.

One approach to resolving this quandary is to be found in wider acceptance of the doctrine of medical futility. By utilizing one of five operative standards under this doctrine, a physician could conclude a patient's condition is indeed terminal and proceed to search a wide range of palliation options – with terminal sedation being given central consideration (Smith 1998).[9] Accordingly, in cases where a cure is physiologically impossible, continued treatment is non beneficial, a desired or positive benefit is unlikely to be achieved, a particular treatment option – although regarded as plausible – has yet to be validated or, a determination is made that a course of treatment is either quantatively or qualitatively futile, a physician is freed ethically from pursuing further medical treatment (Smith 1995). Bolstered by wide professional approbation of this doctrine, then, physicians – exercising their best medical judgment – would be allowed to withhold CPR from patients in futile conditions, without actual consent (Tomlinson and Brody 1990; Blackhall 1987; Council on Ethical and Judicial Affairs 1991). This humane action would, of course, only be undertaken when it would be in the patient's best medical interests not to have a hopeless, non-qualitative existence continued.

The American Heart Association's Guidelines for Cardiopulmonary Resuscitation and Emergency Cardiac Care allow resuscitation to be discontinued in pre-hospital (or field) settings when, after an adequate trial of advanced cardiac life support, the patient remains non resuscitable (Delbridge et al. 1996). Because of familial non-acceptance of field termination of unsuccessful out-of-hospital cardiac arrest, however, emergency medical services continue the medical charade of transporting pulseless patients to hospital emergency departments, knowing fully such actions must be considered futile (ibid.). As a consequence of the significant neurological damage associated with resuscitation by aggressive advanced life support, many questions remain as to initial decisions to initiate the procedure in the first instance. A high patient price, with significant family distress, is quite often the end result of such efforts (Callahan and Madsen 1996).

The probability that resuscitative efforts will provide more than a marginal benefit appears to be the emerging, and most acceptable, basis for determining whether to initiate aggressive treatment. Under this standard, cardiopulmonary patients who are near death and unlikely to survive after CPR is administered are clearly an identifiable group who would benefit marginally by such resuscitative efforts (Murphy and Finucane 1993). In essence, then, under this standard, what is seen is a clear example of the codification of a futile treatment and a validation of the medical principle of *triage* (Smith 1985b). Thus, for example, CPR has been labeled a futile act for patients with metastatic cancer – this simply for the reason that survival after CPR is reported to be zero (Lo 1991; Schneiderman and Jecker 1993).

A Moral Calculation

Today, futility may well be seen as less an objective medical judgment, made only by physicians, than a template for decision-making for the attending physician, his patient or health care surrogate. As such, patient goals, values, and beliefs should be evaluated in an effort to determine whether the total good of a patient is achieved by recognizing a withhold/withdraw decision (Pellegrino 1999, 4-5).

Pellegrino suggests an individualized or situational use of futility as a prudential guide in end-of-life cases (id.). Accordingly, for him a balance is sought, not with mathematical certainty, but rather by moral calculation based upon a clinical assessment of three criteria: effectiveness, benefit, and burden. This balance ideally "gives a weight to each of these three dimensions in relationship to the other and, ultimately, to the patient's good" (id., 6).

Under the Pellegrino approach, subjective and objective components are integrated with expertise and the authority proper to the major participants in end-of-life care: physicians, nurses, patient, or surrogate. Thus, a declaration of futility cannot be a unilateral decision. Rather, it requires a joint determination and agreement as to what is total patient good (id., 7).

The "good" of the patient is, indeed, complex. As Pellegrino and Thomasma (1988) suggest, it includes essentially four components arranged hierarchically. The medical good is seen as the lowest goal and considers both the psycho-social and the physical functioning of the patient as a human organism. At the next level, the patient's individual assessment (or that of his surrogate decision-maker) of his condition is taken – his goals, preferences, and wishes regarding the continuation of his life. The third level uses the natural law as a reference point and evaluates "the good of the patient as a human person" by assessing "what is proper to the life of humans as humans" (id., 8). Finally, at the fourth or highest level, the spiritual good (of these end-of-life decisions) is to be considered. Here, the point of reference is Scripture, tradition, and church teaching – all of which are indefinable by either patient or physician. This is, unquestionably, the most difficult, ignored or entirely negated element in the assessment. Both Pellegrino and Thomasma acknowledge this and express concern over its lack of real-life pertinence because, they agree, every patient, physician, or surrogate will surely have either some faith commitment or faith rejection.

This insightful and humane template for decision-making redefines futility, then, "as a prudential guide to moral assessment of the good of the patient and to the moral permissiveness of withholding or withdrawing particular treatments in seriously ill or dying patients" (Pellegrino and Thomasma 1988). It surely adds, at a minimum, a new set of values to consider in the ongoing debate of this issue.

Simplicity in the Final Decisions

The implicit, but rarely articulated, question in resuscitative decisions is whether and when to end the life struggle against death (Nolan 1987, 11). Coded resuscitative efforts are, as has been seen, nothing but a subterfuge which all too often rob the patient at the end of life of his dignity and autonomy yet protect his family members and the assisting medical team from experiencing fully their own failures in preventing the patient's death (id., 11). Unable to undertake a surgical procedure with a statistical measure of success being assured, a coded patient becomes, in reality, little more than a metaphoric euphemism for a systemic failure of not only the patient, himself, but of the medical establishment as well (ibid.; Smith 1997).

Only by developing a common understanding of and shared vocabulary for the dying process (Nolan 1987), thereby promoting a free line of communication among all affected parties (Teno et al. 1997), can inhumane resuscitative efforts be stopped and codes eliminated altogether. In the final analysis, to understand a phenomenon, one must understand the language and its symbols (Nolan 1987, 11). Heavy emotional connotations flow from such terms as persistent vegetative state, CPR, and do-not-resuscitate. Because of the varied connotations – and, indeed, ambiguities arising from the use of these words, transmitting the precise information needed for a patient or his family to understand the consequences flowing from their use is fraught with misunderstanding (Peppin 1995, 23).

If a comprehensive acceptance and use of the doctrine of medical futility could be achieved as a pivotal measure to define when cardiopulmonary resuscitation should be withheld (Miles 1995), other aids and constructs could be developed for guiding humane decision-making in end-of-life cases. Thus, for example, greater utilization of and reliance upon hospital ethics committees could be promoted when confusion or disagreement arises among patients, their families, and attending physicians (Smith 1990). Fuller discussion of a patient's code status before an episode of distress arises (Paris et al. 1993, 1695), together with unambiguous documentation (Miles 1995, 896) in the patient's medical record of those medical reasons for selecting the particular techniques for resuscitation (or non-resuscitation, as the case may be), would also go far to dispel the need for slow or intermediate codes (Tomlinson and Brody 1988).

PAIN MANAGEMENT AND PALLIATIVE CARE

One report has suggested that more than 50% of patients with terminal cancer have physical suffering during the last days of their life controlled, as such, only by sedation (Fainsinger et al. 1991). Another report shows that 40% of all dying patients in the United States die in pain (Lynn et al. 1997). Recently, the Institute of Medicine found that anywhere from 40% to 80% of patients with terminal illness report that their treatment for pain is inadequate and prolongs the very agony of death (Field and Cassell 1997).

The emancipation principle of palliative care states clearly that no scientific or chemical efforts should be spared to enable dying persons to escape from pain that "shrivels their consciousness" and prevents them from maintaining dignity in their final days (Roy 1990). Indeed, the goal of continually adjusted care demands that those who are hopelessly ill be given whatever medication is needed to control pain (Wanzer et al. 1989). While there can be little dispute about the validity of this principle, there is a widening gap between what *can* be done and what in fact *is* done to implement the emancipation principle (Mount 1990). Stated otherwise, although pain can be managed, the central problem remains how to deal with situations where pain management is merely palliative and the disease symptomatology giving rise to the pain itself continues a malignant progression toward death – resulting in an exceedingly low or even nonexistent quality of life for the patient and unendurable and refractory episodes of dyspnea, delirium, myoclonus, vomiting, and intractable pain (ibid.; Zucker and Zucker 1997).

An Alternative or a Continuum of Care?

While it has been argued that palliative care is the principle alternative to euthanasia, others contend palliation and euthanasia are but a continuum of medical treatment (Ogden 1994). Indeed, some physicians maintain that providing final assistance for the hopelessly ill upon request is a professional responsibility and sound medical practice as such (Wanzer et al. 1989, 848). Still others suggest that in specific contexts, terminal sedation "is covert physician assisted suicide or euthanasia" (Rousseau 1996).

The very integrity of acceptance and use of sedating pharmacotherapy is tied inextricably to two principles: informed consent and double effect. Before sedation is prescribed and initiated for control of refractory symptoms or those that include a terminal disease with impending death, all other types of palliative treatment should be exhausted. Additionally, there should be mutual agreement by the patient and his family of the need for terminal sedation and a full knowledge of the double and ultimate effect of the actions together with the execution of a valid do not resuscitate order (Orentlicher 1997).

Others have suggested that intravenous barbiturate administration is a preferable alternative to extended use of narcotics, which always runs a risk of severe toxicity (e.g., depression, constipation, nausea, dysphoria, and drug tolerance). By use of a single hypnotic agent, somnolence or pharmacologic hypnosis may achieve the same sedating effect, thereby dissociating patient consciousness from refractory symptomatologies (Stiefel and Morant 1991). Benzodiazepines may be used as a second drug to alleviate noxious side effects or used simply as singular agents in these circumstances (Rousseau 1996).

More and more, as palliative care management develops a national if not, indeed, an international praxis, it can be hoped that terminal sedation will in time be understood as but a continuum of proper treatment. Efforts must be undertaken to assure that terminal sedation does not fall into a quagmire of taxonomical confusion.

If viewed as an action that validates personal autonomy or self-determination, this type of palliative care will no longer be seen incorrectly as either euthanasia or physician-assisted suicide. Rather, with this reclassification or clarification in terminology will come an understanding of a medically proper way to assure a modicum of dignity at death.

A NEW MEDICO-LEGAL RIGHT

Although a unanimous United States Supreme Court held in June, 1997, that there was no federal or fundamental right to commit suicide or, thus, to have assistance in effecting it,[10] two concurring opinions would appear to legally validate the medical right to terminal sedation as an efficacious form of palliative treatment for intractable pain. Indeed, Justice Sandra Day O'Connor opined that those individuals suffering from a terminal illness accompanied with great pain may presently, in the states of New York and Washington, obtain whatever level of medication determined professionally by a physician will "alleviate that suffering, even to the point of causing unconsciousness and hastening death."[11]

Justice John Paul Stevens went further in his concurring analysis to probe the full extent of one's liberty interest in defining a personal concept of existence. He concluded that since palliative care cannot alleviate every degree of pain and suffering for all patients, there may be situations in which a competent person could make an informed judgment or "a rational choice for assisted suicide."[12] Legal approbation was given specifically to the unequivocal AMA endorsement of the policy allowing pain-killing medication for terminally ill patients – up to and including use of terminal sedation – even if death is hastened.[13]

For Justice O'Connor, the central issue in this area of consideration is finding a definition of terminal illness and, similarly, safeguarding the voluntariness of patient decisions that have the effect of hastening death.[14] Although sound, reasonable judgment will always be the touchstone for a determination of terminality. With each disease etiology shaping this decision, recognition of the validity of the tests for determining medical futility would go far in resolving this conundrum. With this would come, it is to be hoped, an additional realization that wider use of terminal sedation as sound palliative care would provide a more acceptable patient choice for self-determination than recourse to both the idea and practice of assisted suicide or of euthanasia (Stolberg 1997). There is both a moral imperative and a political mandate for national health policies to provide more humane end-of-life care for the dying (Sulmasy and Lynn 1997). To those disposed to tendentiousness, the suggestion of a taxonomical change of assisted suicide terminology in order to recognize the right of competent, terminally ill individuals to exercise autonomy or self-determination through use of terminal sedation in palliative management would be viewed as but a shallow ruse (Smith 1989). The process of public education needed to effect a significant change here is admittedly complex. Indeed, society may not be equipped to grasp the full consequences of such an educative dialogue on this topic. It therefore remains the primary responsibility of the medical profession – supported by law – to provide the leadership needed to rethink the standards of humane care for treatment at the end-of-life (Quill 1991).[15]

By accepting and applying standards of medical futility to come to grips with a more uniform approach to and understanding of terminal illness, a ready willingness in turn will be seen to accept the use of terminal sedation as a part of palliative treatment and, thus, part of a more comprehensive right to die with dignity and without intractable pain and suffering (Symposium 1996). Indeed, as has been argued, this form of care can be viewed truly as part of a continuum of healthcare to which every individual should be entitled (Ogden 1994).

With the recent effort undertaken by Justice O'Connor and Justice Stevens in the U.S. Supreme Court,[16] the nation is now being led in a re-evaluation of the ethical validity, medical propriety, and legal correctness of terminal sedation as a normative standard of humane conduct and palliative treatment at the end of life.

George P. Smith, II, Professor of Law, The Catholic University of America, U.S.A.

NOTES

1   *Webster's New International Dictionary*, 3d ed., s.v. "euphemism."

2   See also Ganzini et al. (1998) and Sullivan, Hedberg, and Fleming (2000), concluding that physicians are proceeding carefully under the Oregon Act and further documenting the fact that the poor, uneducated, mentally ill, or socially isolated are not seeking disproportionately (or getting) lethal prescriptions of drugs under this law.

3   The Council on Ethical and Judicial Affairs of the American Medical Association (1991) has reported that in approximately one third of some two million patient deaths occurring in hospitals in the United States each year, CPR is attempted in approximately one-third of this population. Of those receiving CPR, one-third survive and one-third of these individuals survive, in turn, until discharge from the hospital. Ultimately, the success or failure of CPR resuscitation depends upon the nature and severity of a patient's major illness before arresting.

4   New York has codified a presumption in favor of resuscitation: "(E)very person admitted to a hospital shall be presumed to consent to the administration of cardiopulmonary resuscitation . . . unless there is consent to the issuance of an order not to resuscitate as provided in this article" (N.Y. PUB. HEALTH LAW § 2962). In addition, this 1988 statute establishes the lawfulness of DNR orders if issued in compliance with the statute itself (Golden 1988).

5   See, for example, N.Y. PUB. HEALTH LAW § 2962 discussion at note 4.

6   Some physicians have even ordered slow codes when they are in disagreement with a patient's request for resuscitation (Peppin 1995, 20). See also Lo (1991, 1875), who noted physicians' use of limited or partial DNR orders with patient or surrogate decision-maker consent.

7   The President's Commission for the Study of Ethical Problems in Medicine and Biomedical and Behavioral Research (1983) concluded that "partial codes" and "slow codes" may be used interchangeably. Yet, others consider partial codes as either a sub-category or a totally different form of "slow codes" (Ross and Pugh 1988, 4).

8   See REV. CODE WASH. ANN. § 70.122.020(9) (West 1996-97).

9   Other treatments might include antibiotics, transfusions, and intensive care (Lo 1991; Meisel 1999). Interestingly, the 106th Congress considered the Pain Relief Promotion Act of 1999 (H.R. 2260, S. 1272) in an attempt to amend the Controlled Substances Act and thereby promote pain management and palliative care without permitting assisted suicide and euthanasia (Orentlicher and Caplan 2000).

[10] *Washington v. Glucksberg*, 521 U.S. 702 (1997); *Vacco v. Quill*, 521 U.S. 793 (1997).

[11] *Id. Washington v. Glucksberg*, 521 U.S. at 737.

[12] *Id. Washington v. Glucksberg*, 521 U.S at 748.

[13.] *Id. Washington v. Glucksberg*, 521 U.S at 751.

[14] *Id. Washington v. Glucksberg*, 521 U.S at 738.

[15] Humane treatment for those individuals with end-stage dementia should be hospice or palliative care and – under proper conditions – I argue, terminal sedation (Morrison and Siu 2000).

[16] See *supra* notes 10, 11. This final section of the paper derives in part from Smith (1998).

REFERENCES

Annas, G. 1995. New York's do-not-resuscitate law: Bad law, bad medicine, and bad ethics. In *Legislating medical ethics: A study of the New York do-not-resuscitate law*, eds. Baker, R., and M.A. Strosberg, 141-155. Boston: Kluwer Academic Press.

Bernabei, R., G. Gambassi, K. Lapane, F. Landi, C. Gatsonis, R. Dunlop, L. Lipsitz, K. Steel, and V. Mor. 1998. Management of pain in elderly patients with cancer. *JAMA* 279(23): 1877-82.

Blackhall, L.J. 1987. Must we always use CPR? *New England Journal of Medicine* 317(20): 1281-5.

Boozang, K.M. 1993. Death wish: Resuscitating self-determination for the critically ill. *Arizona Law Review* 35: 23-85.

Callahan, M., and C.D. Madsen. 1996. Relationship of timeliness of paramedic advanced life support interventions to outcome in out-of-hospital cardiac arrest treated by first responders with defibrillators. *Annals of Emergency Medicine*. 27(5):638-48.

Cavanaugh, T. 1998. Currently accepted practices that are known to lead to death and PAS: Is there an ethically relevant difference? *Cambridge Quarterly on Healthcare Ethics* 7: 375-81.

Council on Ethical and Judicial Affairs, American Medical Association. 1991. Guidelines for the appropriate use of do-not-resuscitate orders. *JAMA* 265(14): 1868-71.

Delbridge, T.R., D.E. Fosnocht, H.G. Garrison, and T.E. Auble. 1996. Field termination of unsuccessful out-of-hospital cardiac arrest resuscitation: acceptance by family members. *Annals of Emergency Medicine* 27(5): 649-54.

Fainsinger. R., M.J. Miller, E. Bruera, J. Hanson, and T. Maceachern. 1991. Symptom control during the last week of life on a palliative care unit. *Journal of Palliative Care* 7(1):5-11.

Field, M.J., and C.K. Cassell, eds. 1997. *Approaching death: Improving care at the end of life*. Washington, D.C.: National Academy Press.

Ganzini, L., H.D. Nelson, T.A. Schmidt, D.F. Kraemer, M.A. Delorit and M.A. Lee. 2000. Physicians' experience with the Oregon Death with Dignity Act. *New England Journal of Medicine* 342: 557-63.

Golden, S.M. 1988. Do not resuscitate orders: A matter of life and death in New York. *Journal of Contemporary Health Law and Policy* 4: 449-467.

Hakim, R.B., J.M. Teno, F.E. Harrell, W.A. Knaus, N. Wenger, R.S. Phillips, P. Layde, R. Califf, A.F. Connors, and J. Lynn. 1996. Factors associated with do-no-resuscitate orders: Patients' preferences, prognoses and physicians' judgments. *Annals of Internal Medicine* 125: 284-93.

Hashimoto, D.M. 1983. A structural analysis of the physician-patient relationship in no-code decisionmaking. *Yale Law Journal* 93: 362-83.

Katz, J. 1984. *The silent world of doctor and patient*. New York: Free Press.

Lo, B. 1991.Unanswered questions about DNR orders [Editorial]. *JAMA* 265: 1874-5..

Lynn, J., J.M. Teno, R.S. Phillips, A.W. Wu, N. Desbiens, J. Harrold, M.T. Claessens, N. Wenger, B. Kreling, and A.F. Connors Jr. 1997. Perceptions by family members of the dying experience of older and seriously ill patients. SUPPORT Investigators. Study to Understand Prognoses and Preferences for Outcomes and Risks of Treatments. *Annals of Internal Medicine* 126(2):97-106.

Meisel, A. 1999. Pharmacists, physician-assisted suicide, and pain control. *Journal of Health Care Law and Policy*. 2: 211-242.

Miles, S. 1995. Futility and medical professionalism. *Seton Hall Law Review* 25(3): 873-82.

Mittelberger, J.A., B. Lo, D. Martin, and R.F. Uhlmann. 1993. Impact of a procedure-specific do not resuscitate order form on documentation of do not resuscitate orders. *Archives of Internal Medicine* 153: 228-232.

Morrison, R.S., and A.L. Siu. 2000. Survival in end-stage dementia following acute illness. *JAMA* 284(1): 47-52.

Mount, B. 1990. A final crescendo of pain? [Editorial] *Journal of Palliative Care* 6(3): 5-6.

Muller, J. H. 1992. Shades of blue: The negotiation of limited codes by medical residents. *Social Science and Medicine* 34: 885-898.

Murphy, D., and T.E. Finucane. 1993. New do-not-resuscitate policies: A first step in cost control. *Archives of Internal Medicine* 153(14): 1641-8.

Nolan, K. 1987. In death's shadow: The meanings of withholding resuscitation. *Hastings Center Report* 17(5): 9-14.

Ogden, R. 1994. Palliative care and euthanasia: A continuum of care? *Journal of Palliative Care* 10(2): 82-5.

Orentlicher, D. 1997. The Supreme Court and physician-assisted suicide: Rejecting assisted suicide but embracing euthanasia. *New England Journal of Medicine* 337(17): 1236-9.

———. 1998. The alleged distinction between euthanasia and the withdrawal of life-sustaining treatment: Conceptually incoherent and impossible to maintain. *Illinois Law Review*, no. 3: 837-859.

Orentlicher, D., and A. Caplan. 2000. The Pain Relief Promotion Act of 1999: A serious threat to palliative care. *JAMA* 283: 255-8.

Paris, B.E., V.G. Carrion, J.S. Meditch Jr., C.F. Capello, and M.N. Mulvihill. 1993. Roadblocks to do-not-resuscitate orders. A study in policy implementation. *Archives of Internal Medicine.* 153(14):1689-95.

Pellegrino, E.D. 1999. Decisions at the end of life: The use and abuses of the concept of futility. Paper presented to The Pontifica Academia Per La Vita in Rome, Italy, February 26, 1999.

Pellegrino, E.D., and D.C. Thomasma. 1988. *For the patient's good: The restoration of beneficence in health care.* New York: Oxford University Press.

Peppin, J.F. 1995. Physician neutrality and patient autonomy in advanced directive decisions. *Issues in Law and Medicine* 11: 13-27.

President's Commission for the Study of Ethical Problems in Medicine and Biomedical and Behavioral Research. 1983. *Deciding to forego life-sustaining treatment: A report on the ethical, medical, and legal issues in treatment decisions.* Washington, D.C.: The Commission.

Quigley, F. 1988. Legalities of the no code/slow code. *Pennsylvania Nurse*, 15 Oct.

Quill, T.E. 1991. Death and dignity. A case of individualized decision making. *New England Journal of Medicine* 324(10):691-4.

Ross, J.W., and D. Pugh. 1988. Limited cardiopulmonary resuscitation: The ethics of partial codes. *Quality Review Bulletin* 14(1): 4-8.

Rousseau, P. 1996. Terminal sedation in the care of dying patients. *Archives of Internal Medicine* 156(16): 1785-6.

Roy, D. 1990. Need they sleep before they die? [Editorial] *Journal of Palliative Care* 6: 3-4.

Sabatino, C.P. 1999. Survey of state EMS-DNR laws and protocols. *Journal of Law, Medicine and Ethics* 27(4): 297-315.

Saunders, J.M., and S.M. Valente. 1986. Code/No code? The question that won't go away. *Nursing* 16: 60-64..

Schneiderman, L.J., and N. Jecker. 1993. Futility in practice. *Archives of Internal Medicine* 153(4): 437-41.

Smith, II, G.P. 1985a. Death be not proud: Medical, ethical and legal dilemmas in resource allocation. *Journal of Contemporary Health Law and Policy* 3: 47-63.

———. 1985b. Triage: Endgame realities. *Contemporary Journal of Health Law and Policy* 1(1): 143-151.

———. 1989. All's well that ends well: Toward a policy of assisted rational suicide of merely enlightened self-determination. *University of California-Davis Law Review* 22(2): 275-419.

———. 1995. Utility and the principle of medical futility: Safeguarding autonomy and the prohibition against cruel and unusual treatment, *Journal of Contemporary Health Law and Policy* 12: 1-39.

————. 1990. The ethics of ethics committees. *Journal of Contemporary Health Law and Policy* 6: 157-170.

————. 1997. *Final exits: Safeguarding self-determination and the right to be free from cruel and unusual punishment.* Chicago: Northwestern University Medical School.

————. 1998. Terminal sedation as palliative care: Revalidating a right to a good death. *Cambridge Quarterly on Healthcare Ethics* 7: 382-387.

Stiefel, F., and R. Morant. 1991. Morphine intoxication during acute reversible renal insufficiency. *Journal of Palliative Care.* 7(4): 45-7.

Stolberg, S. 1997. Cries of the dying awaken doctors to a new approach. *New York Times*, June 30, sec. A1.

Sullivan, A.D., K. Hedberg, and Fleming DW. 2000. Legalized physician-assisted suicide in Oregon – The second year. *New England Journal of Medicine* 342(8): 598-604.

Sulmasy, D., and J. Lynn. 1997. End of life care. *JAMA* 277: 1854-5.

Symposium. 1996. Appropriate management of pain: Addressing the clinical, legal, and regulatory barriers. *Journal of Law, Medicine and Ethics* 24(4): 288-368.

————. 1998. Physician-assisted suicide: Facing death after *Glucksberg* and *Quill*. *Minnesota Law Review* 82: 885-1101.

Teno, J., J. Lynn, N. Wenger, R.S. Phillips, D.P. Murphy, A.F. Connors, N. Desbiens, W. Fulkerson, P. Bellamy, and W.A. Knaus, William A. 1997. Advance directives for seriously ill hospitalized patients: effectiveness with the patient self-determination act and the SUPPORT intervention. *Journal of the American Geriatrics Society* 45(4): 500-7.

Tomlinson, T., and H. Brody. 1988. Ethics and Communication in Do-Not-Resuscitate Orders. *New England Journal of Medicine* 318(1): 43-6.

————. 1990. Futility and The Ethics of Resuscitation. *JAMA* 264: 1276-1280.

Wanzer, S.H., D.D. Federman, S.J. Adelstein, C.K. Cassel, E.H. Cassem, R.E. Cranford, E.W. Hook, B. Lo, C.G. Moertel, and P. Safar. 1989. The physician's responsibility toward hopelessly ill patients. A second look. *New England Journal of Medicine* 320(13): 844-9.

Williams, D.L., E.R. Daniels, G. van der Wal, P.J. van der Maas & E.J. Emmanuel. 2000. Attitudes and practices concerning the end of life: A comparison between physicians from the United States and from The Netherlands. *Archives of Internal Medicine* 160: 63-8.

Zucker, M.B., and H.D. Zucker. 1997. *Medical futility and the evaluation of life-sustaining interventions.* New York: Cambridge University Press.

CHAPTER SEVEN

ROGER S. MAGNUSSON

CHALLENGES AND DILEMMAS IN THE "AGING AND EUTHANASIA" POLICY COCKTAIL

> I had a demented aunt; she was demented for seven or eight years, in a nursing home, virtually catatonic. I had not had much of a relationship [with] her but I went out there with my partner to visit her. She's sitting in a chair with a table up front of her and every so often she would fall forward and hit the bridge of her nose on the edge of the table, and they had taped a piece of foam rubber on there [to protect her]. My partner looked at me and he said "you know, we wouldn't treat urchin cats and dogs like this, we would put them out of their misery; the whole place just warehoused with people like this..."
> — Mark, clinical psychologist

INTRODUCTION AND METHODOLOGY

There can be few more volatile policy debates than that of aging and euthanasia. If AIDS has come to be seen as the disease that "most justifies" the right to die (Magnusson 2002, 49), then old age must surely be one of the most difficult contexts in which to consider a policy of assisted death. Apart from the usual arguments about personal choice, pain and suffering, and compassion, versus arguments about palliative care, the role of medicine and the "slippery slope" (Miller 1996), looms the troubling issue of health care rationing (Callahan 1996). As the costs of chronic and end-of-life care continue to rise, some fear that even a well-intentioned policy permitting voluntary assisted death would become corrupted (Santamaria 1996). Depression and dementia, where evident in later life, further complicate the policy equation. Tobin predicts that doctors would become emboldened "to act on ascribed 'autonomous wishes', that is to say, on 'what the patient would have wanted if only he [or she] were able to tell us'" (Tobin 1995). Together, these factors generate a bleak (albeit imagined) scenario: one where death in old age becomes a subtly "persuaded act," a choice disproportionately made by the vulnerable and self-effacing, those wishing to conserve the family silver and to spare loved ones the agony of waiting.

Difficult or not, western societies will increasingly be forced to engage with the aging and euthanasia debate. The aging population, the rise in chronic diseases, the

D.N. Weisstub, D.C. Thomasma, S. Gauthier & G.F. Tomossy (eds.), Aging: Decisions at the End of Life, 107-137.
© *2001 Kluwer Academic Publishers. Printed in the Netherlands.*

growing costs of medical technology, and the drawn-out nature of the dying process
are just some of the factors motivating discussion (Magnusson 1997, 6-8). The
literature exploring the euthanasia policy debate is vast and this chapter does not aim
to provide a comprehensive review. Rather, it aims to use interview data, in
interaction with the literature, to illustrate some of the complexities, dilemmas, and
ironies that muddy the arguments on both sides and make this such a difficult
debate. In a number of places, I have drawn upon the work of Daniel Callahan, the
former director of the Hastings Center in New York, both to sharpen the issues and
to act as a "sparring partner" in the discussion.

   Recent research into illicit assisted death has tended to focus upon quantifying
levels of involvement. National surveys suggest that a significant minority of
medical practitioners have assisted their terminally ill patients to die (Table 1 at
p.136). Several studies have investigated why patients seek medically assisted death
(Emanuel, Fairclough and Emanuel 2000; Chochinov et al. 1999; Seale and
Addington-Hall 1994), as well as the decision-making processes of informal caregivers that result in illicit assistance (Cooke et al. 1998). However, there has been
very little interview-based research into the processes that operate at the bedside,
and the "culture" within which assisted death occurs. Interview research is an
effective way of exploring the social circumstances that lead to assisted death,
illustrating the tensions and dilemmas of end-of-life care, and sharpening debate
around issues that are all too frequently presented in abstract terms (Magnusson
2002).

   This chapter draws upon detailed, semi-structured interviews conducted with 49
doctors, nurses, and therapists, 37 of whom have reported participating in specific
episodes of assisted suicide and euthanasia (Table 2, at p.137). The interviews were
conducted primarily in Sydney, Melbourne, and San Francisco. Recruitment for the
study took a variety of forms (Table 3, at p.137). Twenty interviewees volunteered
to be interviewed after hearing about the study during the course of seminars and
conference papers that I presented. Nineteen interviews came about following
referrals from other interviewees or contacts. Certain interviewees played a pivotal
role here, referring me to "key players" within informal euthanasia networks, and
assuring potential contacts that the study was "legitimate" and that I was
"trustworthy." A "snowball" sampling technique is well suited to investigating
populations that are scattered, hidden, or lack formal organization (Burgess 1982,
77). In contrast to most health care work, assisted death is practiced surreptitiously.
Health care workers who participate are concerned about anonymity and keen to
avoid exposure. The interviews in this study were conducted on a pseudonymous
basis, under a research protocol monitored by the Human Research Ethics
Committee of The University of Melbourne, Australia. The methodology of the
study is described in more detail elsewhere (Magnusson 2002, 282-4).

   Although the interviews focused primarily upon the experiences and
involvements of health carers with HIV/AIDS-related assisted death, the experiences
and professional careers of respondents extended more widely to include cancer and
geriatric care. This chapter draws upon those broader experiences in an attempt to

explore some of the variables in what I have called the "aging and euthanasia" policy cocktail. As suggested by the structure of this chapter, while the variables in the euthanasia policy debate are complex and inter-related, it is helpful to distinguish between "bedside," and "systemic" issues. "Bedside" issues relate to the patient's subjective experience of illness and his or her reasons for wishing to die. "Systemic" issues, on the other hand, relate to the nature of the health care system within which suicide talk takes place, the broader cultural and legal environment, and the possible flow-on effects of a right to die.[1]

THE EUTHANASIA POLICY COCKTAIL: SOME "BEDSIDE" CHALLENGES

Advocates of a moral and legal right to assisted death typically base their claim upon the right to personal autonomy and self-determination, and secondly, upon the values of compassion and mercy in the face of unbearable suffering. Opponents of the right to die frequently reply that these reasons provide no logical stopping-point: "if either [respect for personal] autonomy or the merciful termination of an unendurable existence is the basis for this right, why limit it to the terminally ill?" (Kamisar 1995, 234; Callahan 1992, 54). It is true that defining limits to a lawful right to die could be a difficult line-drawing exercise. However, the tendency to respond to "bedside" issues with "systemic" arguments (the "slippery slope" argument, for example), is unsatisfactory. The factors generating demand for euthanasia must be confronted on their own terms. Both prohibition and legalization involve social costs: wise policy, therefore, involves balancing both burdens and benefits at both bedside and systemic levels.

An honest policy debate, therefore, must begin at the beginning. It must recognize that the elderly have a voice, and that some are very passionate about their "right" to an assisted death. It is not only bodies that are ravaged by disease, but personalities and personal narratives. As Justice Brennan of the U.S. Supreme Court has pointed out: "Dying is personal. And it is profound. For many, the thought of an ignoble end, steeped in decay, is abhorrent. A quiet, proud death, bodily integrity intact, is a matter of extreme consequence."[2] In Australia, before the Northern Territory's *Rights of the Terminally Ill Act* came into effect,[3] newspapers reported that two desperately ill people traveled to Darwin, searching in vain for a doctor who would assist them. Max Bell, a 66-year-old retired taxi driver with stomach cancer, told his doctor, Philip Nitschke, "If my dog was lying in a pool of his own shit, they'd put him down – they're treating me worse than an animal" (Alcorn 1996, A4). Seventy-year-old Marta Alfonso-Bowes, a retired poetry lecturer from Monash University, Melbourne, also arrived in the Territory. Suffering severe pain from tumors in her colon and intestines, she later suicided "alone, like an old dog," from an overdose of tablets in a Darwin hotel room (Kennedy 1995, 1, 4).

An honest policy debate must also be sensitive to the changing context of bedside care. Whereas once patients would die at home, "surrounded by praying family and friends and provided with the consolations of religion," now they die "in sterile hospitals, narcotized and frequently alone in the middle of the night with the television set or Muzac blaring" (Coomaraswamy 1996, 145). It has been estimated that, in the United States, more than 80% of people die in hospitals, hospices, and convalescent institutions, in contrast to less than 50%, fifty years ago (McCue 1995, 1041).[4]

"Opiophobia" and Clinical Disrespect

Recalling his nursing days, John, a funeral director and former nurse, tells of an "image that sticks in my mind." John was the charge nurse on night duty, caring for a man in his late 70s or early 80s who had cancer and diabetes. "Because [of] his diabetes," says John, "he had gangrene. We went to do a dressing on his toe. We'd given him either pethidine or morphine beforehand so he [could] stand the pain of the dressing being done." As John unwrapped the dressing, the man's "right great toe quite literally fell off into the dressing." The man was in great pain, so John telephoned the registrar, who was asleep (but on call), to prescribe further painkillers. When the registrar refused, "I told him he was a cruel mongrel," John recalls, and "that he really should get over to the ward to check the patient out [and that] perhaps he'd change his mind." "I ended up getting the nursing supervisor to get him out of [his] quarters to check the patient out, and he still refused to [prescribe anything]." "His excuse was, 'I'm a Christian, if I give this man any more drugs we will be killing him *and you don't understand this, nurse'*."

"I just thought then that it was one of the cruelest things I'd ever seen happen," says John, "from a profession supposedly there to help people, but they've gotten their help mixed up with the prolonging of life." "That's when I first formed my opinion of things," he says.

A number of other interviewees gave similar accounts of pain, medical indifference, and clinical disrespect. Amanda, a Melbourne nurse, tells of trying to persuade nurses at a Catholic hospital to give her grandmother adequate pain relief. "She was in great pain," says Amanda, "[but] the nurses refused to give her more morphine, they said that she just had some three hours ago and it wasn't time for her next dose." Struck by the "cruelty and insensitivity" of the hospital staff, Amanda replied "Oh, do you think she'll get addicted; I mean, she's [only] got 12 hours to live!" It is important to note that, out of experiences like these, attitudes are cemented and euthanasia careers are born. Despite advances in pain management, unrelieved pain, and "opiophobia" remain widespread (Billings 2000, 555-6).

The Psychosocial Dimensions of Pain

Inadequate pain relief is only one of many challenges in elder care. Pain itself is a complex and ambiguous notion, merging with anxiety, fear, fatigue, loneliness, confusion, frustration, guilt, depression, grief at loss of independence, companionship, and lifestyle, and many other emotions (Magnusson 2002: 68-99). Nurse interviewees, in particular, were alive to the emotional components of pain. Robert, a physician, admits that "the use of morphine as an emotional palliative is often more important than as a pain palliative," while acknowledging that "the ambiguity of the indication" means that some nurses can be reluctant to administer the charted dose.

The psychosocial dimensions of pain force health carers to confront their assumptions about the kinds of suffering that justify pharmacological intervention, and those that do not. However, if the distinction between physical pain, and psychosocial suffering has a disruptive impact upon pain relief regimes, it promises to be even more problematic within the context of euthanasia. Stanley, a therapist, and seasoned administrator of assisted death, comments that:

> I certainly know a number of people who are chronically depressed and their pain, their anguish – even though it's mental pain and anguish – is as pronounced and as articulate as any other people's pain...I think mental anguish is as legitimate a reason [for euthanasia] as physical anguish. I think it's a bit disingenuous for us to say "that doesn't count."

The logical consequences of this view are illustrated in the *Chabot* case in the Netherlands. Bereaved by the death of her husband and two sons, Mrs Chabot refused psychiatric help and wished for death. A Dutch court held that she fulfilled the criteria of unbearable suffering, with no prospect of improvement, and that despite the absence of somatic disease, it was lawful for her physician to assist her death (Griffiths 1995).[5]

To summarize, therefore, inadequate pain relief is a significant challenge in end-of-life care, and witnessing inadequate care can act as a potent incubator for pro-euthanasia attitudes. At the same time, the "holistic" nature of pain and suffering raises the broader question of what forms of suffering medicine should recognize as "justifying" relief. These value judgments would be implicit in any statutory right-to-die which required "unbearable suffering" as a pre-condition to assistance (in contrast, for example, to a law merely requiring a terminal diagnosis).[6]

Too Much Medicine

If lack of medical care is one risk for elders, "too much medicine" is another. Erin, a nurse, recalls his experiences working in an intensive care unit. "We kept them alive, that respirator was blowing air into them, sixteen breaths a minute, but the alarm was going off every three minutes and they were uncomfortable and being suctioned and tormented. I was *doing harm*, you know. And for the first time I took those principles back out and really looked at them, as a care giver...but also trying to understand what the patient was going through."

Later, a "revolving door pulmonary patient" asked Erin for assistance to die. Erin recalls:

> [The patient said] "Do you know what it's like to be on a respirator?" (He'd been intubated a number of times). He said "Do you know the terror I have when I try to go to sleep at night knowing that I have a DNR [a Do Not Resuscitate Order] on my record, [hoping that] they are not going to resuscitate me, [but] knowing somebody is going to slip in there and say "Oh well, he didn't really mean that, we're going to...go

ahead and put him back on the machine." He knew, and I knew, that that had happened.
He had TB, so everybody was wearing masks and he said "Is this quality of life?"

The patient went on to talk of doctors cracking jokes at the bedside, unable to see his
face through the mask, and of the man's disappointment that he could no longer see
his grandchildren. That night, Erin placed the cardiac monitor on standby and dialled
down the oxygen. The next morning, he fended off an intern intent on resuscitation,
despite the DNR and the fact that the man was gray and had been dead for hours.

In several cases, the mindless pursuit of "vitalism" reached epic proportions.
Mark, a clinical psychologist, describes being with a patient whose partner had died
at home. After unsuccessfully trying to report the death to the police, and to the local
hospital, Mark's patient was advised to call "911." The patient did so, explaining
that this was not an emergency and that his partner had died in his sleep four hours
previously. "Ten minutes later," Mark recalls, "a fire truck, a police car, a paramedic
unit all come screeching out in front of the house." There were "four or five firemen
all dressed up in thick rubber boots and yellow rubber coats and fire hats all carrying
boxes of medical equipment." They came "bursting into the house." The dead man
had previously signed a "Do Not Resuscitate" order. Despite this, the medics were
"absolutely convinced that they were under a legal requirement to attempt to
resuscitate him, they attached all kinds of medical equipment to him, gave him
oxygen, tried heart paddles, the whole business. *The corpse was cold.*"

The ideology of the palliative care movement provides a welcome corrective to
both the excesses of technological medicine, and the lack of adequate care. Palliative
care is often contrasted with curative medicine, and lauded as an important counter-
balance to the medicalization and bureaucratization of the dying process. It aims to
emphasize quality of life despite the inevitable trajectory of disease, by "seeking to
meet the differing physical, emotional and social needs of patients on a holistic [and
individual] basis" (Hart, Sanisbury and Short 1998, 69). Despite the fact that
euthanasia (a "gentle or harmonious death") is "central to the proper function of the
hospice" (Gillett 1994, 263), the dominant philosophy – if not ideology – of the
palliative care movement is that euthanasia is neither morally nor legally acceptable
(Farsides 1998, 150; cf Hunt 1994). However, the extent to which palliative care
provides an "answer" to pro-euthanasia claims based upon personal autonomy, and
upon compassion for suffering, is bitterly contested.

Palliative Care and Personal Autonomy

> I believe in self-determination, some people are going to mess up, some people are
> going to make mistakes, some people are going to abuse that, but to my mind self-
> determination – I'd rather err on that side. That's just my feeling on the matter.
>   – Stanley, therapist.

Interview research highlights the fragile environment within which notions of
"personal autonomy" and "self-determination" operate. The meanings of "suicide
talk" are complex (Magnusson 2002, 68-99). In a recent study, Chochinov and
colleagues tracked the "will to live" of elderly, terminally ill cancer patients. They

found that will to live is highly unstable, and appeared to be influenced by anxiety, depression, shortness of breath, and sense of well-being, with each of these variables assuming greater prominence at different time intervals leading up to death (Chochinov et al. 1999; Emanuel, Fairclough, and Emanuel 2000).

The respondents in the present study acknowledged that requests for assistance to die may be a metaphor for current anxieties, future fears, pain, grief, loss of control, and many other unresolved needs. Requests are seldom taken at face value, but usually precipitate a process of exploring the unmet needs and motivations that lie behind the request. Patients may be "testing you out, wanting to know to what extent I can trust you," or seeking reassurance about the future. Kerry, a consultant nurse, points out that patients will "commonly" talk about suicide "in the context of their fears about what the future will bring." According to Amanda, a community nurse:

> So many people will ask you about euthanasia or assisted suicide…and often you find that there are other reasons – like people say they're afraid of the pain, they're not afraid of dying but they're afraid of the pain. And if you give them some assurances that you [will] at least attempt to make it as comfortable as possible, then…you can often talk people through it.

Like several other nurse interviewees, however, she added, "some you can't."

As they anticipate the progression of their disease, patients will not uncommonly say that, when they reach a certain point of illness, or disability, that they do not wish to go on living. Fitzgerald argues that implicit in such statements is the cultural stereotype of "the life not worth living," which can powerfully constrain the judgments and perceptions of ill and disabled people (Fitzgerald 1999, 271-2, 281). Callahan (1992, 53) is even more forthright, asserting that:

> The degree and intensity to which people suffer from their diseases and their dying, and whether they find life more of a burden than a benefit, has very little directly to do with the nature or extent of their actual physical condition…suffering is as much a function of the values of individuals as it is of the physical causes of that suffering. Inevitably in that circumstance, the doctor will in effect be treating the patient's values.

The fact that perceptions may change and that, over time, patients may come to bear the unbearable was an important theme in interviews. Ruth, a community nurse, referred to this phenomenon as the "shifting goalposts." The goalposts keep getting shifted because when the patient reaches the point where they had previously decided life was intolerable, it has been a gradual process, "it's sort of snuck up on them, and it's not as bad as they thought it was going to be." "People…draw lines in the sand," says Stanley. They say "'if I can't walk then I'm going to kill myself.' [But] when they get to the point where they can't walk, they find another reason for drawing another breath, and then they move on."

In a related theme, interviewees felt that the process of "lingering on," and living through the dying process was itself a rewarding experience for some patients, even those who had sought assistance to die in vociferous terms. Emma, a community nurse in San Francisco, provides one example:

The patient that really influenced me the most was an elderly man with cancer. [He] was an angry, depressed man who had refused all treatment, and in the most direct way every day that I saw him would say, "Why can't you just kill me, I want to die." I kept saying, you know, "as a nurse I can't do that," and I really had mixed feelings about it. I was his [only] contact with the health care system, me, and the hospice social worker that used to come to his house. He [also] had these poor home health aids, he screamed at them, he abused them, he called them racist names, I felt really bad for these women that were locked up with this guy for 12 hours at a time. This was a guy who used to introduce himself – he'd say, "Hello, my name is Rotten Richard," and that was how he saw himself.

[To] make a long story short – it was a very interesting process, which I don't think happens very often – but in the last two months of his life, he underwent this incredible emotional and spiritual transformation. It was just an incredible thing to watch this unfold, and by the time he died, he was telling everybody how much he loved them, and how grateful he was, and by living [those] last two months in a lot of physical pain and discomfort he really had an opportunity to experience himself as a loving and a lovable and considerate person, in a way that I don't think he had in 70 years. Seeing that really made me much more cautious about this issue, because I really felt that if I had been able to assist this guy legally I would have done it, it seemed very clear cut; you know, all the criteria that people talk about: repeated requests, rational person, no hope for a cure, palliative care.... clearly he was more at peace with himself when he died than he would have been if he had [had] euthanasia. Despite the fact that his physical existence was very unpleasant during that time, he was able to experience something that was very positive for him.

The fluid nature of "the life not worth living" is an issue that must be presented with care, so as not to patronize suffering patients, minimize their suffering, or neglect it altogether. For some interviewees, the value of "lingering on" lay not in the possibility of a sudden change of heart about the meaning and value of life, but in the opportunities provided by the dying process for family and loved ones to enter into an intense, closing relationship with the dying person. Robert, a physician, regretted that "a lot of people, in my experience, play charades," pretending that the patient isn't really dying, because they are afraid of dying themselves, and afraid of confronting the death of a loved one. "I think that euthanasia often is the short cut, it's the quick way out [of] a difficult situation [that] people don't want to confront." He observes how families that have used this time as a "closing experience" in their relationship with the patient, suffer less after the death.[7] Robert's experience is echoed in the literature. Byock, for example, claims that "with the skillful application of psychosocial and spiritual support, dying often becomes an extremely important, valuable and, not rarely, a privileged time in life" (Byock 1994, 6; Koch 1996, 58).

Not everyone, however, feels comfortable discounting patient autonomy or asserting the benefits of "lingering on." In a statement encapsulating North American individualism, Kyle, a physician and Vietnam War veteran, comments: "I've always believed quality of life is an issue that only the individual can really measure. I cannot measure pain and I cannot measure loss of vanity and I cannot measure suffering and I cannot measure depression. I think people always should be able to make their own decisions and I think it makes a stronger society when people have to assume that responsibility."

Autonomy, Depression and Dementia

Depression and dementia are complicating factors for those who would ground the right to die in personal autonomy. It has been estimated that up to 45% of terminally ill cancer patients consider suicide, and that up to 60% of those who express a sustained desire for euthanasia have a clinically diagnosable depression (Block 2000, 209-18). Despite this, interviewees were divided over whether the risks of depression and dementia warranted psychiatric assessment. A number expressed confidence in their own assessment skills, disliked third party interference, or felt that psychiatric assessment interfered with patient autonomy. Joseph, a hospital physician, said "I don't send my patients for psychiatric assessment when they ask me for this because I think *it's the patient's...right.*" Even so, several respondents who were psychiatrists, psychologists or therapists reported receiving "referrals" from trusted doctor colleagues who sought a "second opinion" on whether it was appropriate to proceed toward euthanasia.

Those involved in end-of-life nursing were, in general, more sensitive to the life-crippling effects of depression. Kerry, a consultant nurse, reflected that "I often hear nurses saying things like: 'I don't know why they're hanging around, I don't know why they don't just give up.' The danger is in thinking: 'Yes it *is* terrible, I've seen a hundred people die that way; *yes – die!',* instead of listening to what the person is saying which may be about 'I'm frightened of something, I'm scared of suffering'." An important theme in the literature (as in the interviews) is that health care workers can become so "dazzled" by the tragedy of their patients' lives that they fail to distinguish clinical depression within life events that otherwise make sadness appear "rational" (Billings 2000, 556; Mullen 1995, 126). Worse still, they may agree with the patient's sense of futility in a way that legitimates and encourages suicidal ideation.

Nevertheless, weaving its way through many of the responses was the assumption that a certain degree of depression is appropriate: "it's unusual for people *not* to have a sense of sadness, a sense of loss or regret; I mean, *depressing things are happening.*" "You or I would probably respond in the same way if we were in their position," said Damian, a psychiatrist, "so I can't say it's abnormal or pathological – their desire to kill themselves...has to be seen as rational." The clinical imperative for Damian, therefore, in assessing the "suitability" of a patient for euthanasia, was not to exclude depression, per se, but to satisfy himself that the patient was making the decision in a "clear state of mind" and was not suffering from a "treatable depression." This involved excluding dementia "at a level that would seriously impair" judgment, and ensuring that the patient was receiving adequate pain relief. "It would be something the patient would have to request [and] the patient would have to be in a state where I thought [the request] was fairly reasonable," said Damian.

Interestingly, Damian's approach does not involve denying that patients who wish to die may be depressed; rather, it distinguishes between "appropriate and rational" depression, and "abnormal or pathological" depression. After attempting to control for the factors that might distort reasoning, Damian believes that euthanasia can be an appropriate response to "rational" depression because "it is important that people have that right." Nevertheless, what combination of circumstances qualifies as a rational reason for seeking euthanasia "seems destined to remain...entirely subjective" (Ryan 1995, 583).

A more conventional approach to psychiatry would be less willing to enter into this kind of distinction. For Block, the relevant distinction is between "normal" and "inevitable" grief and adjustment reactions on the one hand, and clinical depression on the other (Block 2000). While there is no "bright line" separating the two, depression is diagnosed according to the range and intensity of symptoms. Once diagnosed, the assumption is that major depression can and should be treated with supportive therapy and/or psychopharmacology, even in the terminally ill. It is doubtful, however, whether this approach should be considered to be any more "value neutral" than Damian's. Just as the values of the health carer will influence whether opioid pain relief is seen to be an appropriate treatment for emotional suffering, so they will influence perceptions of whether suicide can ever be rational, or whether treatment with antidepressants is the only humane response to a patient who is terminally ill, suffering, and seeking assistance to die.

If depression raises doubts about the "autonomy" of voluntary euthanasia, then the rising prevalence of dementia raises squarely the issue of non-voluntary euthanasia.[8] Interviewees expressed differing views about whether it is appropriate for health carers to euthanize demented patients. One doctor felt that it was reasonable "to inform the patient and let them take a decision for themselves before the dementia is [so] advanced that they're not capable of making a rational decision." Another doctor thought, "I suppose it depends on the degree of dementia, but if we're talking [about] a matter of days or weeks, [it] hardly seems to matter."

Unlike voluntary euthanasia, the doctor who hastens the death of a patient who is burdened by delirium or dementia can claim no mandate in the patient's personal autonomy. One person interviewed reported receiving urgent requests from a patient who was becoming progressively more demented who begged her, "help save me from playing with my shit." Beyond a certain point, however, as one hospital nurse pointed out, patients "don't even talk about wanting to suicide...[they] become a lot more oblivious to what's going on around [them]."

If voluntary euthanasia were to be legalized, the issue of "advance euthanatic directives" ("AEDs") would soon follow. After all, the right to *withdraw* life-sustaining treatment already encompasses advance directives and surrogate decision-making in many countries. Like advance psychiatric directives, the moral authority of AEDs depends, at least in part, upon theories of "personhood" (Hughes 2001).[9] Less abstractly, an AED would purport to be an expression of the competent patient's deeply-held values, previously explored with the treating doctor within the context of the patient's progressing illness and anticipated mental incompetence. Absent the crucial safeguard of personal advocacy by the patient concerned, however, AEDs are potentially "instruments of power and control" in the hands of physicians (Widdershoven and Berghmans 2001, 95). The risks are greater still when there is no advance directive, and where the doctor is acting solely on their own perceptions of the patient's suffering, or on the patient's wishes as mediated by family and friends. Interview research confirms that this form of euthanasia does occur, and with few real safeguards (Magnusson 2002, 231-8).

Palliative Care, Suffering and Compassion

> She woke up in pain, she asked for something and I said "what do you want? I gave you an injection an hour ago, we really can't give you anything more", and she said "I want IT done, I do not want to see morning." She asked for a cup of tea, so I made her a cup of tea and we settled and we discussed it. I said "you realize that if I do this you will not see morning," and that [was] what she wanted. She'd been in so much pain it was cruel to watch her go on. She had cancer of the cervix and it had disseminated through the spine and gone to the brain, so she was in a pretty bad way.
>
> When she'd had her cup of tea she asked for the injection. She had no bloody skin, she was just skin and bone, she had no muscle left by that time. I found a reasonable spot on her upper thigh and injected her, and then sat and talked. I said "this wasn't in the contract when I met you – that I'd be doing this; this wasn't part of the deal when I met you." Her final words were "well next time you'll read the small print you little bastard." She was always one for a smart line, and they were her last words. She just lapsed into unconsciousness, and [I] cradled her.   – John, nurse.

The euthanasia debate has placed palliative care under the spotlight. While some argue that the hospice movement is moving towards a position of "caution," rather than "outright rejection" of voluntary euthanasia (Hunt 1994, 128-32), assisted death nevertheless represents an assault upon its "core business" of helping patients to live through their dying process. The challenges to hospice come as much from those in the profession who are ignorant of symptom control techniques, as from euthanasia activists. The former cause patients to suffer unnecessarily, thereby legitimizing the claims of euthanasia advocates. The latter disregard the "value inherent in the dying experience for the patient and the family" (Byock 1994, 2), and support a policy that implicitly devalues the contribution of palliative care as a speciality. The perceived "legitimacy" of the anti-euthanasia view, therefore, is linked to the capacity of palliative care to relieve suffering and to provide a good death.

While qualitative research cannot resolve broad empirical questions about the quality of palliative care services within a society or country, it can illustrate the diversity of opinion among health care workers themselves. Responding to the suggestion that palliative care is capable of resolving the issues that motivate requests for euthanasia, Amanda, a community nurse, burst out:

> What a load of crap! I think that it's a gross generalization to put on everybody [who] is dying...that they *must have* [palliative] care. I think that carers [try] to make *themselves* feel better [but] there's no recognition that...[some patients] just cannot tolerate their life the way it is. I think that there is room for good palliation, don't get me wrong...but there's certainly a group of people [for whom] good palliation is never enough.

Similarly, Ruth, a nurse, reflected that "Some people just get to a point when they say, 'I just don't have any more energy left to fight this disease, I've been doing it now for 12 years, and I've run out, and all I want to do is die.'"

The measure of success of palliative care is partly a function of what can reasonably be expected of it. Woodruff asserts that pain in terminal cancer can be "well controlled in 95% of patients" (1995: 157),[10] although the appropriateness of using morphine to deeply sedate distressed patients whose symptoms cannot otherwise be controlled, remains a contested aspect of this debate (Billings and Block 1996, 27; Syme 1999a; Syme 1999b; Hunt 1994, 119-23; O'Connor, Kissane, and Spruyt 1999). The issue was nicely put by Robert, a physician, who observed, "If palliation includes rendering the patient unconscious, *it always works.*"

Outside of pain, palliative care experts tend to be more circumspect. Pollard (1994, 40, 41) states:

> Even good palliative care can be less effective in relieving emotional suffering or anguish because mental suffering has many possible causes arising from factors apart from the patient's medical state…Many of the causes of misery in the dying, especially the elderly, are due to factors in their social environment, their personal relationships or their psychology…It would be both unreasonable and unfair to expect palliative care to be able to control factors beyond its ambit.

Immobility, weakness, fatigue, dependency, and incontinence are common problems in elder care. In so far as palliative care, as a speciality, attracts criticism, it may be precisely because of its ambitious, "holistic" focus upon the "multiple psychosocial aspects of suffering" (Byock 1994, 2). Ironically, as Harper points out, the quality of nursing and medical care can itself prolong the dying process: "some residents become stabilized at a level of bodily animation at which they can, and often do, remain for years – often senile, doubly incontinent, immobile and totally dependent. In the end, it could be said that they "cease to exist," rather than die…" (Harper 1995, 102).

It remains to be seen whether the limitations of palliative care in achieving a good death will lead to a gradual shift in emphasis in the philosophy of hospice care. Brody, for example, argues that assisted death has a role to play as a compassionate response to "the failure of medical interventions to arrange a good death" (1992, 1386). If this is perceived as confronting, it may have as much to do with the way in which a lethal infusion implicitly questions the aspirations and expertise of palliative care as a speciality, as with the moral issue of killing.

Quite apart from tensions in the guiding philosophy of palliative care, interview research suggests that keeping assisted death "on the table" as an option of last resort provides an important security blanket for some patients. The following comments, from a hospital physician and a family physician (both with impressive euthanasia credentials) illustrate the point:

> I think it tends to make people *less depressed* to know that they have access to [assistance] if that's what they want…[it] lifts some sort of a weight from their shoulders and they can go ahead and live their life without fearing that they're gonna have a miserable death that they have no control over.   – Joseph

> Very few have used it…very few have felt obligated, but they've been at peace, knowing that it was an option, it was a treatment plan or modality that was out there.
>   – Kyle

Heilig and Jamison (1996, 114) go even further, arguing that the availability of euthanasia, if needed, acts as a potent suicide *prevention* strategy: "where patients feel "empowered" to end their own lives with or without assistance, they often do not do so in the end, and in fact enjoy a longer life than if they felt compelled to take action out of fear of becoming unable to do so at a later date."[11] On this view, therefore, a legal right to die has the potential to enhance, rather than to degrade, the journey that hospice seeks to make with the terminally ill. In a similar vein, Dworkin and colleagues argue that the legalization of assisted death would assist in *expanding* the delivery of hospice care services to the terminally ill, since assessment by a palliative care team would be a condition precedent to receiving lawful assistance to die (Dworkin et al. 1997, 42).

THE EUTHANASIA POLICY COCKTAIL: SOME "SYSTEMIC" CHALLENGES

In this part of the chapter I want to turn from the "bedside" context to some of the broader, systemic challenges that must be confronted in the euthanasia policy debate. If debate at the bedside revolves around the limits of personal autonomy, suffering and compassion, at the health care system level it tends to focus upon the proper role of medicine and the fear of a "slippery slope." There is also the phenomenon of "underground" assisted death, a frequently forgotten variable in the euthanasia debate.

Healing, Palliating, and Killing: Perspectives on the Doctor's Role

> I'm not here to cure everybody...Death is part of the natural cycle of life, and as a physician I have an opportunity to grant people peace and comfort in their death, and it's something that I'm not unwilling or afraid to do.   – Joseph, HIV/AIDS Specialist

> You don't [give] euthanasia instead of offering palliative care, you offer euthanasia as an option, as part of that whole package.   – Chris, hospital nurse

Involvement in assisted death challenges participants to justify the role of killing within the caring paradigm.[12] Interviewees attempted to do this with greater or lesser degrees of finesse. Zane, a palliative care physician, said:

> I believe that there comes a time when you acknowledge all your attempts of treatment have failed and once the patient says they want to go and they want to go quickly and you are sure that they have fulfilled all their family wishes and have had reconciliation, seen their relatives for the last time, then I feel it is the physician's role to hasten a person's death.

For Kyle, a family physician, the process was even more straightforward. "When I have to make a decision like that I place myself in the position [of the patient]. What would I want done to me? How would I want to be treated?" "I'm fairly much at comfort with myself and my conscience," he said. "It's not a big issue that causes a lot of internal debate."

For others, however, ambiguity and mixed feelings were a strong theme. Amanda, a community nurse, admitted that "my personal beliefs [around] euthanasia conflict a great deal with my professional identity."[13] Richard, one of the few "opponents" of euthanasia interviewed in this study, worried about the non-accountability of doctors who participate in illicit assisted death. "What would one trust a doctor to do unsupervised?" he asked. "Should killing someone require less supervision than taking their tonsils out?" "If you're going to medicalize [assisted death] and [give] doctors all this power, then it needs to be...subject to scrutiny."

In an interesting parallel to the religious origins of the hospice movement, a number of interviewees who had participated in assisted death referred to their involvement as a "ministry." Merril, a physician, commented: "I feel strongly that my commitment of service to the patient represents a...ministry. Though many people...may grow stronger through suffering in pain, there is a point at which it becomes intolerable. I don't see [assisted death] as a sin. I don't see it as damnation of myself, or of the patient who decides to do this. It sort of sits on fragile ground for me." Erin, a former Catholic brother, felt that he had "never been more priestly" than when he assisted patients to die. Stanley, a therapist and former priest, had even carved out a role as a creator and celebrant of ritual at pre-death "wakes."[14]

In bright contrast to the perspectives above, Callahan argues that "Medicine should try to relieve human suffering, but only that suffering which is brought on by illness and dying as biological phenomena, not that suffering which comes from anguish or despair at the human condition." He adds: "The great temptation of modern medicine, not always resisted, is to move beyond the promotion and preservation of health into the boundless realm of general human happiness and well-being" (1992, 55).

Callahan's point is that doctors who assist suicide are inevitably joining with the patient in declaring that the patient's life is no longer worth living, not merely making a judgment about the patient's treatment or health. Medicine is not a slave to patients' suicidal urges, and so a doctor who assists is not merely honoring the patient's right of self-determination, but participating in the judgment that the patient is better off dead (1992, 52-3). However, such a judgment is not a clinical one, argues Callahan, and doctors lack the moral and technical competence to make it (1992, 53, 54-5).

There is much in the philosophy of palliative care that invites, perhaps even requires, carers to explore existential issues which are, strictly speaking, beyond medicine's domain.[15] Maddocks points out that palliative care "attends to the physical, emotional and spiritual needs of the patient and family. It addresses fear and anger as well as pain, recognizing the emotional associations of physical distress" (1996, 62). Gillett suggests that palliative care "demands more of us as persons than as technicians" (1994, 267), while Barbato invites doctors to embrace a model of care that permits participation in the patient's "inner dialogues," a care that embraces "the social, psychological and spiritual dimensions of our patients' lives" (1998, 297). Inevitably, the physician who forges an intimate connection with the dying patient will become a participant in a dialogue about meaning:[16] the invitation is there, it seems, for palliative medicine to encompass a "ministerial" role that moves beyond the physical to the existential.

To suggest that medicine should restrict itself to "biological" challenges, therefore, appears incompatible with the aspirations of palliative care. The aims of palliative care may also provide a response to the argument that medicine lacks suitable criteria upon which to judge when it is appropriate to assist a patient to die. As a clinical speciality, palliative care cannot be judged according to the aims of curative medicine, but only according to its own distinct aims. The palliative care movement encourages patient participation in decision-making, nurtures a "holistic" relationship with the patient, is sensitive to the complex causes of suffering, aims to relieve pain and distress during the final stages of life, and to facilitate a gentle death. If these aims – relief of suffering and a gentle death – cannot reasonably be achieved except by hastening death, in circumstances where this is the patient's wish, then the provision of assistance would not seem to be out of step with the aims of palliative care (Brody 1992). If killing in these circumstances is nevertheless wrong, then it must follow that the aims of palliative care are themselves misguided or need qualification.[17]

This still leaves Callahan's objection that since "unbearable suffering" is a subjective notion heavily influenced by the patient's values, doctors will inevitably be abandoning themselves to the values of the particular patient whom they choose to assist. There are several responses to this. First, just because the patient's values influence the patient's experience of suffering and, just because that suffering cannot be weighed on any "objective" scale, does not lessen the importance of the aims of palliative care. The priority for the palliative care physician is to reduce suffering and to fashion a good death for each individual, not to dictate the patient's values or to discount their experience. Secondly, it is the aims of palliative care – to relieve suffering during the course of terminal decline and to produce a good and meaningful death – which frame the doctor's role, not abstract judgments about when death is preferable to life (a matter about which reasonable minds may differ).[18] These aims keep the focus where it should be: providing comfort and preserving meaning through a good death (Little 1999). Thirdly, it is inevitable that conflict may arise between the doctor and the patient over whether the patient's desire is "reasonable." Doctors, after all, have their own personal values, and perceptions of the relative burden of the patient's suffering. According to Warren, medicine may need to resolve this conflict by bowing to the autonomy of the patient:

> The critical thing is what the client's deeply thought out belief system [is]. If the client lives in a world where one of the worst things is pain, and he doesn't see any reason for enduring it, then pain is the critical thing. If, on the other hand, he believes in the afterlife...then the role of pain is completely different, and therefore, the role of the professionals surrounding him is completely different...What we have to be doing here is respecting the ethical integrity of the client...Beyond a certain point, what we have to do as professionals is to get down on our knees and bow to the autonomy of the client.

Similarly, Damian admits, "what I might consider a reasonable quality of life may not be [for] one particular patient...One has to respect that some people just have higher standards or higher expectations of what life should provide."

Not every doctor would feel comfortable with such a ringing endorsement of patient autonomy. Different doctors may have differing "thresholds" in terms of the evidence of suffering they would need to see – in spite of palliative care – before they felt comfortable assisting a patient's death. Some would not feel comfortable under any circumstances. Doctors would need to resolve any disparities between their own consciences and their commitment to providing a good and meaningful death (much as they do now). Ultimately, however, if relief from suffering is a worthy aim, then it must revolve around the subjective experiences of patients themselves.

These factors provide one answer to Callahan's criticism that any framework for "impartial oversight" of assisted death would be ideologically biased, since it would presumably exclude doctors who opposed euthanasia, or who "believe that pain and suffering can be relieved in all but the rarest cases" (Callahan 1994, 1656). The bias argument cuts both ways. For example, medicine has traditionally regarded patients who wish to end their lives as depressed: "If this is true, then no patient who expresses such thoughts could meet standards for decision-making capacity" (Tulsky, Ciampa and Rosen 2000, 495). The "no euthanasia" view is not value free: it embodies the value judgment that no amount of suffering can ever justify an assisted death.

A final argument, based upon the lack of "objective" criteria to determine when assisted death is appropriate, points to the "creeping influence" of the doctor's own values, which may potentially mislead or even pressure the dying patient. Richard, a psychiatrist, fears that a physician who has "strong beliefs about the right of individuals to self-determination may form a different judgment regarding euthanasia than the therapist who typically has a more parental or paternalistic style. Such factors clearly influence the therapeutic process and would be particularly critical in the area of judgments about euthanasia." Similarly, Emma fears that health professionals, "stressed out by the amount of suffering they have observed," could choose to assist "out of their own need to not have [sic] to witness the patient suffering, rather than the patient's need not to suffer." These risks should not be dismissed out of hand. They point to the risks of legalizing assisted suicide in the absence of a statutory framework requiring independent assessment and multi-disciplinary review. While opinions and values may vary, independent review provides a process for questioning whether all that might be done for the patient has been done, and for ensuring that misconduct does not shelter within the sanctity of the doctor/patient relationship.

Euthanasia and Money

The economics of euthanasia is a familiar theme in the policy debate. Periodically, one reads discussion of euthanasia as an explicit strategy to restrain rising health care costs (Callahan 1996) or, more vaguely, of the apprehension that "decisions concerning euthanasia...may be influenced by resource constraints" (Little 1998, 56).[19] Occasionally there is also mention of remuneration for those who assist death (Magnusson 2002, 238-9; Ogden 2001, 396; Little 1998, 56). Echoing familiar themes in the literature, Terri, a palliative care nurse, feared that if euthanasia were legalized, elders might have their wishes abused, especially by family members keen to receive an inheritance. Perhaps the most prominent theme, however, is that lack of money to pay for expensive care and treatment may motivate the elderly and chronically ill themselves to seek euthanasia. The broader, cost-constrained, health care environment is seen as a major factor "coercing" the inadequately insured to make this choice.[20]

The risks are particularly great in countries such as the United States which lack a universal health care system (Quill 1998, 552-3). Emma, a San Francisco community nurse, pointed out that in the case of AIDS, health care services were mobilized for the benefit of AIDS patients as a result of "empowerment and activism." Emma fears that the opposite may also occur: that those who are most vulnerable, who lack connections to activist communities, those with the least money and education, could be "subtly or not [so] subtly encouraged to take advantage of assisted suicide if it were legal because nobody wants to take care of them." Another interviewee, a hospice physician, noted that he would ask patients who raised the topic of euthanasia, "Are you worried that you're going to run out of funds? Are you worried that your estate is going to dwindle and you're not going to have anything to leave to your heirs?"

Into this already complicated picture comes the specter of age-based health care rationing. Although an opponent of assisted death, Callahan has nevertheless advocated health care rationing for the elderly as a solution to the American Medicare crisis. He argues:

> The key issue for the future is whether this society will continue its present course of using more marginally beneficial technologies to improve health care for the elderly – refusing, in effect, to accept the historical decline in health associated with aging – or whether society will take a population-based approach by putting more resources into improving health in younger years and thus increasing the likelihood of better health in later years...Only the latter course seems based in common sense.
>
> (Callahan 1999a, 2077)

While Callahan does not advocate denying elders "good long-term and home care, and basic primary medical care" (Callahan 2000, 10), he does advocate limiting access to "high technology, even life-extending technology" which could yield only marginal health benefits in the elderly, given their shorter life expectancy.

If age-based health care rationing were ever introduced, it is hard to believe that palliative care would be immune, especially the expensive, multi-disciplinary, round-the-clock kind of care that palliative care physicians argue is most effective in managing end-of-life suffering.[21] If ever anything were calculated to make elders feel that they were no longer wanted, and that they owed society a duty to die quickly and quietly, then surely it is age-based rationing and its likely corollary: inadequate funding for quality palliative care.

We live in societies where there is no shortage of examples of the elderly and sick dying in misery and distress. It has been estimated that in the United States less than 15% of patients receive hospice care (Cheyfitz 1999, 14). Ganzini and colleagues have shown that, in Oregon, palliative interventions can significantly alter the patient's desire for a lethal prescription, in circumstances where this would be legal (Ganzini, Nelson, Schmidt 2000, 560). It would be a tragedy if society's lack of financial commitment to palliative care, and to a decent standard of health care for elders, were to act as an incentive for euthanasia. That would be a tragedy: a tragedy of health care funding policy. The concern of opponents of euthanasia, however, is that legalizing assisted death would make killing themselves an easier option for those denied adequate care. But is this an appropriate objection? If society will not adequately fund palliative care, why should dying, elderly patients be denied an escape from inevitable suffering and misery? "Systemic" concerns are an important factor in the euthanasia policy equation. But they are hardly a trump card against those advocates of assisted death who base their arguments on "bedside" suffering. In a cost-constrained health care environment, the challenges of the "bedside" and the interests of the patient need to be carefully distinguished from the "funding crisis", and other "systemic" concerns.

Nasty Visions: Euthanasia and the "Slippery Slope"

The "flow-on effects" of legalizing euthanasia are a constant source of concern for opponents of assisted death. Some, like Ria, fear that "the cost of [a] lethal injection" may be reduced funding for palliative care and research. Pollard and Winton warn that "Killing the failures of medical or social care would be negative, in that it would not contribute to finding solutions to their problems" (Pollard and Winton 1991, 428). Others fear that euthanasia may become the easy way out for doctors who are not professionally or emotionally equipped to meet the needs of their patients (Miles 1994). Others, still, are alarmed at the potential for an "easy marriage between respect for autonomy" and "the need for greater cost-containment" (Reichel and Dyck 1989, 1322). For Callahan, the greatest danger of legalizing assisted death is "the social legitimation of suicide as a way of dealing with the suffering and sorrows of life" (1999b, 7). "Make euthanasia legal," writes Glover "and it becomes something we're all forced to consider" (Glover 1999, S2).[22] Finally, one cannot but mention the concern that legalizing euthanasia will lead to non-voluntary and involuntary euthanasia, as some believe it has in the Netherlands (Keown 1995).

Those who point to the slippery slope are convinced that bad things will happen if euthanasia is legalized. One limitation of their accounts, however, is that they tend to point to the bottom of the slope, rather than providing a coherent account of how we would inevitably end up there. As Frey argues, "Merely to fear the failure of safeguards is not itself to show the failure of any particular one" (Frey 1998, 47;

Burgess 1993). Interestingly, early data from Oregon suggest that the legalization of physician-assisted suicide under the *Death With Dignity Act* has stimulated physicians to improve their ability to care for the terminally ill, and has marginally increased the number of referrals to hospice (Ganzini et al. 2001).[23] On the other hand, research by Sullivan and colleagues in Oregon during 2000 led them to believe there was a significant increase in patients whose suicides were motivated by being a burden to family, friends and caregivers (Sullivan, Hedberg and Hopkins 2001). While the likelihood of choosing physician-assisted suicide appears to rise with level of education, it does not necessarily follow that any risks of abuse would be quarantined to the educated classes (cf Callahan 1999b).[24] Overall, however, the number of patients who have (legally) sought assistance to die in Oregon and the Netherlands remains small. Ironically, this has prompted some to argue that, if assisted death was intended as a response to terminal suffering, it has ended up being an "emotionally charged irrelevance" (Emanuel 2001, 1377).

Nevertheless, slippery slope theorists have played a valuable role in focusing debate upon the efficacy of safeguards, the value of guidelines, and the legal shape of the right to die (Quill, Cassell and Meier 1992; Baron et al. 1996; Amarasekara 1997). They also serve as a useful reminder of the cost of making a mistake (Caplan, Snyder and Faber-Langendoen 2000, 479). As Robert points out, "Once it's the law, people will then start to die, and you can't suddenly after six months say, 'well, we made a mistake.'"

Perhaps the greatest limitation of the slippery slope literature is that few opponents of euthanasia have faced up to the policy challenges of current, "underground" euthanasia practices.[25] Illicit euthanasia is a forgotten variable in the euthanasia policy debate. As Dworkin, Nagel, Nozick and colleagues pointed out in their "Philosopher's Brief" to the United States Supreme Court, "The current two-tier system – a chosen death and an end of pain outside the law for those with connections and stony refusals for most other people – is one of the greatest scandals of contemporary medical practice" (1997, 41).[26] Increasingly, the nature of "underground euthanasia" is itself changing. "Non-physician assisted suicide" is alive and well in North America and has the potential to cut medical practitioners out of the loop entirely (Ogden 2001a; Ogden 2001b).

Illicit Assisted Death: A Forgotten Variable in the Euthanasia Policy Debate

Survey studies suggest that a substantial proportion of those who die do so with intentional assistance. In the Netherlands, where doctors can avoid criminal liability for assisted death if certain procedures are followed, Van der Maas and colleagues estimated that 2.3% of all deaths in 1995 were the result of euthanasia, and 0.4% were attributable to physician-assisted suicide. In a further 0.7% of cases, life was ended without the patient's explicit request (Van der Maas et al. 1996). In Australia, where euthanasia is unlawful, Kuhse et al. (1997) estimated that in 1996, 1.8% of all deaths were the result of voluntary euthanasia or physician-assisted suicide, while an additional 3.5% of deaths involved the termination of the patient's life without an explicit request.

For many who work in end-of-life care, this will come as no surprise. Hospital-based interviewees variously reported playing "fast and loose" with the morphine during the final stages of life (a "nice soft fuzzy euthanasia death under another name," according to one), and choosing death-hastening strategies that were not instantaneous but mimicked, if possible, a natural decline.[27] The processes involved are both complex and covert (Magnusson 2002, 191-7). In addition to the stresses of involvement, participants complained that the prohibition on euthanasia stifled discussion. "It's very secret," said Kerry, a consultant nurse, "it's sort of whispered conversation." This, in turn, makes "any reflection or examination of processes exceedingly difficult."

A feature of "underground" euthanasia is that attempts are frequently "botched." Jamison notes that the "lack of models, experience, and training makes this an act that must be reinvented constantly. Every experience is new, fraught with its own fears, hesitancy, and ignorance, and nearly every one who participates is an actor with an unrehearsed script" (Jamison 1996, 241). Failed episodes of assisted death frequently end with the panicking health care worker injecting every available drug into the patient, followed by choking or suffocation. Other features of underground euthanasia include the erratic and idiosyncratic criteria according to which health carers "assess" their patients, hasty and rash participation in assisted death, the all-permeating culture of deception, non-consensual euthanasia, lack of professional distance, cavalier speech and actions, and burn-out (Magnusson 2002, 200-247). These features illustrate the very opposite of those attributes which characterise "medical professionalism."[28]

Elsewhere, Ballis and I have argued that these features of illicit euthanasia provide a *prima facie* argument in favor of legalization. From a harm minimization perspective, legislative guidelines provide a protective framework for end-of-life decision-making, and embody a series of norms designed to influence conduct. In contrast to the "anarchy of denial and secrecy" (Helig and Jamison 1996, 115), legalization is arguably a "conservative" response to illicit assisted death: it recognizes the failure of regulation, calls health care workers to account for their ad hoc involvements, and seeks to channel their actions within a more structured and humane framework (Magnusson and Ballis 1999, 328; Magnusson 2002, 268-71). While nothing is certain, a statutory euthanasia regime may reduce the incidence of illicit assisted death, and possibly, "pre-emptive" suicide.

If the aim is to minimize harm and suffering, while respecting the dignity and values of the dying, then the practice of illicit euthanasia must be factored into the euthanasia policy equation. That equation involves balancing the potential harm that could be caused by the devaluing of life, and by the potential abuse of elderly, vulnerable patients in an environment where voluntary euthanasia could legally be practiced in accordance with defined procedures, against the current abuses and indignities resulting from unrelieved suffering (the absence of euthanasia), "amateur" suicide, and illicit, "backyard" euthanasia.

CONCLUSION

In this chapter I have presented interview data in interaction with the literature in order to illustrate some challenges and dilemmas in the "aging and euthanasia" policy debate. The aging of the post-war, "baby boomer" generation, the extended deteriorative decline that characterizes modern dying (Battin 1995, 201, 225), and the failure of medicine to provide the kind of death that the dying wish for, are key factors that will stimulate this debate for many years to come.

We should, perhaps, surrender the idea that we can achieve a perfect policy. Legalization and prohibition both carry risks of harm. In weighing that harm, we need to identify and give full weight to the many variables in the policy equation, recognizing that in some areas we can only speculate. As interview research illustrates, medicine both under-treats and over-treats at the end of life. Both problems generate pro-euthanasia advocacy that is only partially offset by advances in palliative care.

It is sometimes assumed that while curative and palliative care are delivered in a "professional" capacity, according to the established norms of the profession, euthanasia involves moral judgments about the value of life that can only be made individually, on "subjective" grounds. Palliative care, however, is little different. Evaluating the degree of suffering (and the kind of suffering) that "justifies" opioid pain relief, the withdrawal of life prolonging measures, or even terminal sedation, involves moral judgments about the value of life and the "point" of continuing to live.

I have argued that within the philosophy of palliative care lies the seeds of a rival theory of terminal care that would permit euthanasia as an option of last resort in order to reduce unbearable suffering and to achieve a good death for each individual patient. A good death is a worthy aim for all patients, not only the 90-95% whose pain can currently be controlled. Provided that physicians remain fixed upon the aims of palliative care, assisting patients to die only when conventional interventions fail, then it may be possible to avoid the darker reaches of the inevitably subjective environment within which both patients and physicians operate. Ironically, the availability of euthanasia as an option may also alleviate anxiety and enhance the success of "traditional" forms of care.

At the same time, we must recognize that personal autonomy operates within a fragile environment in the dying. Autonomy may be challenged by depression, delirium, and dementia. Requests for assistance to die encode complex and ambiguous messages. In reality, the goalposts change, patients may be grateful for lingering on and for the opportunities for tenderness and intimacy for which illness provides a context. As health care funding comes under increasing pressure, there is the possibility that lack of adequate care may become a subtle incentive for the self-effacing and elderly to accelerate their death (Glover 1999). To oppose assisted death on the basis of these "systemic" constraints, however, may perpetuate, rather than reduce, the suffering experience of the dying. Finally, counterbalanced against these and other "nasty future visions" is the disturbing reality of current illicit euthanasia.

The "aging and euthanasia" policy cocktail is complex, and there is little agreement about how different variables should be weighted. My own qualified support for a legalized, but heavily regulated regime draws criticism from libertarians who point out that, while it is commonly assumed that legalizing physician-assisted suicide will enhance patients' autonomy, it will actually increase the power of doctors (Salem 1999).[29] It also draws criticism from physicians themselves, who doubt whether law is capable of responding sensitively to the "nuances" that inform clinical decision-making (Ellard 1998, 57).[30]

If euthanasia were to be legalized within a statutory framework, the monitoring of clinical discretion through law would effect a shift in the locus of control from the private sphere to the public realm, from health professionals to the state (Lewins 1998, 132). Herein lies a final irony in the euthanasia policy debate: the medical profession has an interest in "defending its turf," keeping end-of-life decisions firmly within the private realm and away from State control. This may explain why the significant proportion of those in the medical profession who support, or have indeed participated in, assisted death (Table 1, at p.136), remain a weaker force for change than might be expected.

Roger S. Magnusson, Senior Lecturer, Faculty of Law, University of Sydney, Australia.

NOTES

I am grateful for comments on an earlier draft of this chapter provided by my father, Associate Professor Eric Magnusson, Australian Defence Force Academy, University of New South Wales; and Professor George P. Smith II, Catholic University of America, Washington, D.C. Any errors are mine.

[1] In what follows I will also be assuming that, if assisted death were to be legalized, it must logically encompass both physician-assisted suicide (PAS) and active voluntary euthanasia (AVE). Some American commentators favor the former but not the latter on the basis that only PAS can assure the voluntariness of the act (Quill, Cassel and Meier 1992, 1381; Baron et al. 1996, 10). "Absent doctors" are a feature of many assisted deaths, prescribing the medication, or setting up an IV-line, before fleeing the scene for personal or legal reasons, often with shameful results (Magnusson 2002, 209-210, 224-6). As Brody recognizes, "letting the patient do the dirty work [is] an abrogation of responsibility rather than an exercise in professional integrity" (Brody 1992, 1386; Pellegrino 1995, 1675). Similarly, it makes little sense to argue that patients already have the right to die, as a function of their right to refuse life-preserving treatment. As Harper points out, nursing home residents are frequently on long-term, low maintenance therapy such as diuretics or cardiac medication. "This therapy cannot be discontinued without the sometimes severely distressing symptoms and complications recurring. While these complications might eventually be lethal, people do not like to drown in their own bodily fluids, or to experience cardiac distress" (Harper 1994, 102).

[2] *Cruzan v Director, Missouri Department of Health* 497 US 261 (1990), 310-1 (Justice Brennan, dissenting).

[3] The *Rights of the Terminally Ill Act* 1995 (NT) took effect on 1 July 1996. On 25 March 1997, however, the Commonwealth Parliament of Australia enacted the *Euthanasia Laws Act* 1997 (Cth). This federal statute withdrew the legislative powers of the Northern Territory to make laws with respect to euthanasia, drawing upon the constitutional powers of the Commonwealth Parliament to make laws with respect to Australia's Territories. During the intervening time, four patients died in accordance with the protocol outlined in the Act, using a computer-controlled injection device monitored by a general practitioner, Philip Nitschke (Kissane, Street and Nitschke 1998).

[4] Indeed, as Hoefler argues, the institutionalization of death has contributed to a culture in which death is decreasingly a part of common experience. In a culture characterized by individualism, faith in technology, increased longevity, and the decline of the multi-generational household, death may be part of public discourse, but it has little personal meaning. (1994, 10-42).

[5] One reason, perhaps, for restricting assisted death to situations where the patient has a terminal illness and is expected to die within a defined timeframe, is to ensure that euthanasia remains a way of easing an otherwise traumatic or lingering *death*, as distinct from escaping a *life* that would not otherwise be ending. The requirement of a terminal illness would exclude people with chronic illnesses and disabilities from obtaining assistance to die. However, this restriction limits the collateral damage euthanasia does to our intuitions about the value of life. At the same time, a law restricting assisted death in this way might be less effective in eliminating illegal practices than a law encompassing chronic illness.

[6] Both terminal illness and unbearable pain and suffering were requirements under the now-defunct Northern Territory legislation: *Rights of the Terminally Ill Act* 1995 (NT) s. 4. The Dutch model, by contrast, only requires unbearable pain or suffering, although treatment possibilities for relieving that suffering must have been exhausted or rejected by the patient. This ensures that there is no prospect for improvement (Griffiths, Bood and Weyers 1998, 101-2). This position has not changed as a result of the November 2000 amendments to the Dutch Penal Code, and the "due care" requirements under the *Termination of Life on Request and Assisted Suicide (Review Procedures) Act*. Oregon's "Measure 16" does not require "unbearable suffering," but does require an incurable and irreversible disease likely to cause death within 6 months: *Death with Dignity Act* (Oregon) ss. 1.01, 2.01.

[7] As Seale states: "In late modernity love is expressed through talk; if talk is absent there is nothing to reassure individuals that love is still there. Because unawareness involves silence it contains the threat of abandonment. Confessional talk provides more convincing proof of emotional accompaniment. With funeral rituals having lost much of their power, caring talk before death becomes, in the discourse of awareness, an act of anticipatory mourning that wards off the threat of abandonment and isolation, now feared far more than the older terrors of evil spirits" (Seale 1995, 611).

[8] It has been estimated that, as the percentage of Americans older than 75 years grows from 5.9% (in 2000) to an estimated 11.4% (in 2050), the prevalence of Alzheimer's disease will quadruple to afflict one in forty-five Americans (Kawas and Brookmeyer 2001, 1160).

[9] Hughes advocates what he terms a "situated-embodied-agent" ('SEA') view of the person, in which the person with dementia "is continuous with and connected to the person who signed the directive, by embodiment and by the situatedness that embodiment entails" (Hughes 2001, 89). He contrasts this with the "'Locke-Parfit view,'" in which personhood is signalled by "psychological continuity and connectedness." The SEA view suggests that advance directives retain moral authority whereas, with the LP view, a person with dementia may be regarded as "a different person from the one who wrote the directive, or not a person at all" (Hughes 2001, 89).

[10] In a ten-year prospective study of cancer pain relief in an anesthesiology-based pain service associated with a palliative care program, Zech and colleagues found that pain relief was good in 76% of patients, satisfactory in 12%, and inadequate in 12% (Zech et al. 1995).

[11] The same point has been made in the Dutch context: "In my professional work I often see the same response. When the decision for euthanasia is made, when patients know death is a possibility and in their control, then they are calmer and more reassured. As a consequence, the quality of their lives improves" (Anonymous 1996, 85).

[12] Elsewhere, Ballis and I have suggested a "typology" of attitudes and responses to assisted death (Magnusson and Ballis 1999; see also Magnusson 2002, 100-116).

[13] Having assisted the lethal overdosing of a friend in a *private* capacity, Amanda admitted that "the biggest problem I have now is that it sort of sets a bit of a precedent for me –it's like, I've done it once, how do you decide when you do or don't do it again?'

[14] Stanley expands: "Most of it's a coach[ing] kind of thing...it's helping people relax...creating...a peaceful, life-affirming environment, trying to keep the chaos that often surrounds death to a minimum. I've tried to help educate people who will survive and attend the dying person on the etiquette of being present at such a pivotal moment, *and there is etiquette*, it seems to me...Oftentimes it's helping the dying create a ritual to celebrate this transitional phase in their life...giving gifts and receiving them, giving blessings and receiving them, saying thank you, saying goodbye...all of those things, you know, [that] make for a peaceful, wise and good death."

[15] The same may be true of "curative" medicine, although to a lesser degree. In modern society, medicine has ousted the church as the "comforter at the deathbed;" "Family members who used to care for the dying now hand them over to the public hospital" (Walker, Littlewood and Pickering 1995, 581). Increasingly, it is the nurse who fulfills the confessional role and displays the virtues of "caring talk" and "emotional accompaniment" within the hospital (Seale 1995).

[16] Indeed, Little argues that euthanasia should not be based on compassion for the suffering of the patient, but on how assisted death might "augment or protect the meaning of the life to be ended" (Little 1999, 292).

[17] Exactly how and why opponents of euthanasia would choose to re-define the aims of palliative care may be revealing. As Tallis observes, "only those theologians who believe that there are more important things than human happiness feel that the dying should earn their death the hard way and go the whole distance along the tunnel of barbed wire" (1996, 3). He adds that one cannot feel safe at the hands of those for whom, after all, "excruciating pains have a deep and inalienable meaning as the 'kisses of Christ'."

[18] Whether a doctor who assists a patient to die should be regarded as making the implicit moral assertion that the patient's life is no longer worth living, depends very much upon how the doctor's action is framed. Arguably, a doctor whose sole concern is relieving the patient's suffering and ensuring a gentle death should no more be regarded as making judgments about the value of a patient's life than the doctor who, relying on the principle of double effect, intentionally administers death-hastening quantities of analgesia in order to relieve pain.

[19] In a recent study, Salmusy surveyed American physicians in six metropolitan areas and found a strong linear association between their willingness to participate in physician-assisted suicide (based upon responses to seven medical scenarios), and their choice of resource-conserving treatment options (Salmusy, Linas, Gold and Schulman 1998). However, there was no correlation between physicians' perceptions of the strength and direction of financial incentives in their own practices, and their willingness to prescribe lethal medication in the scenarios. The authors note that "This might suggest that physicians' attitudes regarding resource conservation and PAS are intrinsic characteristics of the physicians themselves and are not related to the financial environment of the practice." They conclude: "The association we report does not constitute proof that abuse of PAS will result from the legalization of this practice in a cost-constrained environment. Nevertheless, it suggests a sobering degree of caution in legalizing PAS in a medical care environment that is characterized by increasing pressure on physicians to control the cost of care."

[20] Little, for examples, notes that "Economic rationalism is a dominant philosophy still, and resource constraints have already entered medical and social thinking to a point where it is easily conceivable that the elderly and chronically ill may feel pressured to end their lives, rather than continue to burden families and the state" (Little 1999, 290).

[21] Byock, for instance, describes "authentic hospice care" in the following terms. "By authentic I mean care that is interdisciplinary and family-centered; care that manages the physical components of pain, and other distressing symptoms, that uses round the clock analgesia; and care that addresses the multiple psychosocial aspects of suffering and that provides the opportunity of looking beyond the psychosocial to

the spiritual" (Byock 1994, 2). In an environment of aged-based health care rationing, can it seriously be contended that this kind of care would remain available to the elderly dying?

[22] In a similar vein, Maddocks writes: "What I do not want...is to see a brisk new expectation arise in our society that a quick death is a noble one, that for the sake of the family, the health budget or the stress upon carers, it will be best to elect euthanasia. This is because I feel that some of the excellent sharing and exchanges of love that inspire and comfort the rest of us will be lost in accepting that fatal attraction, and our society will be the poorer for it" (Maddocks 1996, 70).

[23] Furthermore, reporting on the 27 patients who died in Oregon in 1999, during the second year of operation of the *Death With Dignity Act*, Sullivan and colleagues claim that "poverty, lack of education or health insurance, and poor care at the end of life were not important factors in patients' requests for assistance with suicide" (Sullivan, Hedberg and Fleming 2000, 602).

[24] Callahan argues that "it is members of the educated middle and upper classes who are most at risk. They are the ones who think control of one's body is the greatest value in life, and that dependence and bodily decay are a threat to their dignity" (Callahan 1999b: 7). For discussion of the demographic characteristics of patients whose deaths were assisted by Jack Kevorkian, see Roscoe et al. 2000.

[25] Instead, the common response has tended to be that if doctors already practice illicit euthanasia, then a statutory regime incorporating "safeguards" would be equally unlikely to influence doctors' practices. Callahan argues "If they now feel they can violate a long-standing moral prohibition in medicine to assist a suicide, why should we expect a sudden new respect for medical morality in the future?" (Callahan 1994, 1656). Similarly, Andrews states "If a person is willing to disregard a law which says lethal injections are never allowed, why would they be constrained by a law which says lethal injections are sometimes allowed? If some doctors are prepared to take the law into their own hands and reject the clear requirements for knowledge and consent, why would they change their attitude, especially when dealing with an incompetent patient?" (Andrews 1996).

[26] Similarly, defending criticisms of Dutch law, Griffiths, Bood and Weyers argue that "Those who invoke the hoary metaphor [of the slippery slope] to criticise Dutch legal developments rely on local taboos in their own countries *as if they described actual practice* and contrast such a mythical situation with the actual empirical data that exist for the Netherlands. Meanwhile, the Dutch are busy trying to take practical steps to bring a number of socially dangerous medical practices *that exist everywhere* under a regime of effective control" (1998: 303, emphasis supplied).

[27] Harvey notes that a similar process operates when life-support is withdrawn within intensive care units (Harvey 1997, 725-7).

[28] "In the euthanasia underground there is no specialised training. Participants remain ignorant about what is needed to achieve a gentle death. There is a proliferation of what participants refer to bluntly as 'botched attempts'. Accountability is also absent. The medical profession for the most part turns a blind eye to the practice of illicit euthanasia by its members. There are no norms or principles guiding involvement. Rather, participation is shrouded in secrecy and deception, triggered by highly idiosyncratic factors, with evidence of casual and precipitative involvements. In place of a tradition of disinterested service to patients, there is evidence of a complete lack of 'professional distance', sharp conflicts of interest, and examples of euthanasia without consent" (Magnusson 2002, 201).

[29] Salem argues that the legalization of assisted death corrodes personal autonomy because "it is precisely physicians who are in charge of freeing patients from medicine...What the patients' rights movement still struggles to recapture from medicine – control over the decision to die – is being returned to medicine through physician-assisted suicide. Eventually, people will have physician-assisted suicide not only because they want it, but because physicians agree they can have it" (1999, 35).

[30] As Robert, a physician, commented in interview, "it's a bit like trying to legislate for the treatment of heart attacks. At the end of the day I think we're looking at *clinical decisions*, and [that's why I am yet to be convinced that a law can] cover the range of clinical issues that arise out of that decision."

REFERENCES

Alcorn, G. 1996. Cancer kills man who sought NT euthanasia. *The Age* (Melbourne), 5 August, A4.

Amarasekara, K. 1997. Euthanasia and the quality of legislative safeguards. *Monash University Law Review* 23: 1-42.

Andrews, K. 1996. Second Reading Speech to the *Euthanasia Laws Bill* 1996, Australian House of Representatives, 28 October.

Anonymous. 1996. What is there to be frightened about? After all, it's not like I am going to the dentist! *Cambridge Quarterly of Healthcare Ethics* 5: 83-7.

Barbato, M. 1998. Death is a journey to be undertaken. *Medical Journal of Australia* 168: 296-7.

Baron, C., C. Bergstresser, D. Brock, G. Cole, N. Dorfman, J. Johnson, L. Schnipper, J. Vorenberg, and S. Wanzer. 1996. A model state act to authorize and regulate physician-assisted suicide. *Harvard Journal on Legislation* 33: 1-34.

Battin, M. 1995. *Ethical Issues in Suicide*. New Jersey: Prentice Hall.

Baume, P. and E O'Malley. 1994. Euthanasia: Attitudes and practices of medical practitioners. *Medical Journal of Australia* 161: 137-44.

Billings, J., and S. Block. 1996. Slow euthanasia. *Journal of Palliative Care* 12: 21-30.

Billings, J. 2000. Recent advances: Palliative care. *British Medical Journal* 321: 555-558.

Blaikie, A. 1999. Ageing: Old visions, new times?. *The Lancet* 354: siv3.

Block, S. (for the ACP-ASIM End of Life Care Consensus Panel). 2000. Assessing and managing depression in the terminally ill patient. *Annals of Internal Medicine* 132: 209-18.

Brody, H. 1992. Assisted death – A compassionate response to a medical failure. *The New England Journal of Medicine* 327: 1384-8.

Burgess, J. 1993. The great slippery-slope argument. *Journal of Medical Ethics* 19: 169-74.

Burgess, R. 1982. *Field research: A sourcebook and field manual*. London: George Allen & Unwin.

Byock, I. 1994. The hospice clinician's response to euthanasia/physician assisted suicide. *The Hospice Journal* 9(4): 1-8.

Callahan, D. 1992. When self-determination runs amok. *Hastings Center Report* 22(2): 52-5.

———. 1994. Regulating physician-assisted death (letter). *New England Journal of Medicine* 331: 1656.

———. 1996. Controlling the costs of health care for the elderly – Fair means a foul. *New England Journal of Medicine* 335: 744-6.

———. 1999a. Aging, death and population health. *Journal of the American Medical Association* 282: 2077.

———. 1999b. Good strategies and bad: Opposing physician-assisted suicide. *Commonweal* 126(21): 7.

———. 2000. On turning 70: Will I practice what I preach? *Commonweal* 127(15): 10.

Caplan, A., L. Snyder, and K. Faber-Langendoen. 2000. The role of guidelines in the practice of physician-assisted suicide. *Annals of Internal Medicine* 132: 476-81.

Cheyfitz, K. 1999. Who decides? The connecting thread of euthanasia, eugenics and doctor-assisted suicide. *Omega* 40(1): 5-16.

Chochinov, H., D. Tataryn, J. Clinch, and D. Dudgeon. 1999. Will to live in the terminally ill. *The Lancet* 354: 816-19.

Cooke, M., L. Gourlay, L. Collette, A. Boccellari, M. Chesney, and S. Folkman. 1998. Informal caregivers and the intention to hasten AIDS-related death. *Archives of Internal Medicine* 158: 69-75.

Coomaraswamy, R. 1996. Death, dying and assisted suicide. In *Suicide and depression in later life*, ed. G.J. Kennedy, 141-52. New York: John Wiley & Sons.

Dworkin, R., T. Nagel, R. Nozick, J. Rawls, T. Scanlon, and J. Thomson. 1997. Assisted Suicide: The Philosophers' Brief. *The New York Review of Books* 27 March 1997: 41-7.

Ellard, J. 1998. Euthanasia: Over the rubicon already? *Australian & New Zealand Journal of Medicine* 28: 57.

Emanuel, E. 2001. Euthanasia: Where the Netherlands leads will the world follow? *British Medical Journal* 322: 1376-7.

Emanuel, E., D. Fairclough, and L. Emanuel. 2000. Attitudes and desires related to euthanasia and physician-assisted suicide among terminally ill patients and their caregivers. *Journal of the American Medical Association* 284: 2460-8.

Farsides, B. 1998. Palliative care – A euthanasia-free zone? *Journal of Medical Ethics* 24: 149-50.

Fitzgerald, J. 1999. Bioethics, disability and death: Uncovering cultural bias in the euthanasia debate. In *Disability, divers-ability, and legal change,* eds. M. Jones and L.A. Basser Marks, 267-81. The Hague: Martinus Nijhoff.

Frey, R. 1998. The fear of a slippery slope. In *Euthanasia and physician-assisted suicide,* eds. G. Dworkin, R.G. Frey, and S. Bok, 43-63. Cambridge: Cambridge University Press.

Ganzini, L., H. Nelson, T. Schmidt, D. Kraemer, M. Delorit, and M. Lee. 2000. Physicians' experiences with the Oregon Death With Dignity Act. *New England Journal of Medicine* 342: 557-63.

Ganzini, L., H. Nelson, M. Lee, D. Kraemer, T. Schmidt, and M. Delorit. 2001. Oregon physicians' attitudes about and experiences with end-of-life care since passage of the Oregon Death With Dignity Act. *Journal of the American Medical Association* 285: 2363-9.

Giddershoven, G., and R. Berghmans. 2001. Advance directives in psychiatric care: A narrative approach. *Journal of Medical Ethics* 27: 92-7.

Gillett, G. 1994. Euthanasia from the perspective of hospice care. *Medical Law* 13: 263-8.

Glover, R. 1999. Why the vulnerable flock to dr death. *Sydney Morning Herald* 17 April: S2.

Griffiths, J. 1995. Assisted suicide in the Netherlands: the *Chabot* Case. *The Modern Law Review* 58: 232-48.

Griffiths, J., A.Bood, and H. Weyers. 1998. *Euthanasia & law in the Netherlands.* Amsterdam: Amsterdam University Press.

Harper, S. 1994. Terminal care in nursing homes. In *Willing to listen, wanting to die,* eds. H. Kuhse, 97-114. Ringwood, Victoria: Penguin Books Australia.

Hart, B.. P. Sainsbury, and S. Short. 1998. Whose dying? A sociological critique of the "good death." *Mortality* 3: 65-77.

Harvey, J. 1997. The technological regulation of death: With Reference to the Technological Regulation of Birth. *Sociology* 31: 719-35.

Heilig, S., and S. Jamison. 1996. Physician aid-in-dying: Towards a "harm reduction" approach. *Cambridge Quarterly of Healthcare Ethics* 5: 113-120.

Hoefler, J. 1994. *Deathright: Culture, medicine, politics, and the right to die.* Boulder, Colorado: Westview Press.

Hughes, J. 2001. Views of the Person with Dementia. *Journal of Medical Ethics* 27: 86-91.

Hunt, R. 1994. Palliative care – The rhetoric/reality gap. In *Willing to listen, wanting to die,* eds. H. Kuhse, 115-37. Ringwood, Victoria: Penguin Books Australia.

Jamison, S. 1996. When drugs fail: Assisted deaths and not-so-lethal drugs. In *Drug use in assisted suicide and euthanasia,* eds. M.P. Battin, and A.G. Lipman, 223-43. New York: Pharmaceutical Products Press.

Kamisar, Y. 1995. Physician-assisted suicide: The last bridge to active voluntary euthanasia. In *Euthanasia examined: Ethical, clinical and legal perspectives,* ed. J. Keown, 225-60. Cambridge: Cambridge University Press.

Kawas, C., and R. Brookmeyer. 2001. Aging and the public health effects of dementia. *New England Journal of Medicine* 344: 1160-1.

Kennedy, H. 1995. Plea from the grave. *Sunday Herald-Sun* (Melbourne), 8 October: 1, 4.

Keown, J. 1995. Euthanasia in the Netherlands: Sliding down the slippery slope? In *Euthanasia examined: Ethical, clinical and legal perspectives,* ed. J. Keown, 269-96. Cambridge: Cambridge University Press.

Kissane, D., A. Street, and P. Nitschke. 1998. Seven deaths in Darwin: Case studies under the rights of the Terminally Ill Act, Northern Territory, Australia. *The Lancet* 352: 1097-1102.

Kuhse, H., P. Singer, P. Baume, M. Clark, and M. Rickard. 1997. End-of-life decisions in Australian medical practice. *Medical Journal of Australia* 166: 191-6.

Lewins, F. 1998. The Development of bioethics and the issue of euthanasia: Regulating, de-regulating or re-regulating?. *Journal of Sociology* 34: 123-34.

Little, M. 1998. Euthanasia – Crossing the Rubicon or at the cross roads? *Australia & New Zealand Journal of Medicine* 28: 55-6.

————. 1999. Assisted suicide, suffering and the meaning of a life. *Theoretical Medicine and Bioethics* 20: 287-98.

McCue, J. 1995. The naturalness of dying. *Journal of the American Medical Association.* 273: 1039-43.

Maddocks, I. 1996. Hope in dying: Palliative care and the good death. In *An easeful death? Perspectives on death, dying and euthanasia,* ed. J. Morgan, 57-70. Sydney: Federation Press.

Magnusson, R. 1997. The sanctity of life and the right to die: Social and jurisprudential aspects of the euthanasia debate in Australia and the United States. *Pacific Rim Law & Policy Journal* 6: 1-83.

————. (forthcoming, February 2002). *Angels of death: Exploring the euthanasia underground,* Melbourne: Melbourne University Press.

Magnusson, R., and P. Ballis. 1999. The responses of health care workers to AIDS patients' requests for euthanasia. *Journal of Sociology* 35: 312-330.

Meier, D., C. Emmons, S. Wallenstein, T. Quill, R.S. Morrison, and C.K. Cassel. 1998. A national survey of physician-assisted suicide and euthanasia in the United States. *The New England Journal of Medicine* 338: 1193-1201.

Miles, S. 1994. Physicians and Their Patients' suicides. *Journal of the American Medical Association* 271: 1786-8.

Miller, R. 1996. Assisted Suicide and Euthanasia: Arguments For and Against Practice, Legalization and Participation. In *Drug use in assisted suicide and euthanasia,* eds. M.P. Battin, and A.G. Lipman, 11-41. New York: Pharmaceutical Products Press.

Mullen, P. 1995. Euthanasia: An impoverished construction of life and death. *Journal of Law and Medicine* 3: 121-8.

O'Connor, M., D. Kissane, and O. Spruyt. 1999. Sedation in the terminally ill – A clinical perspective. *Monash Bioethics Review* 18(3): 17-27

Ogden, R. 2001a. Non-Physician Assisted Suicide: The Technological Imperative of the Deathing Counterculture. *Death Studies* 25: 387-401.

————. 2001b. Nutech and non-physician assisted suicide: A reply to Werth. *Death Studies* 25: 413-8.

Pellegrino, E. 1995. Ethics. *Journal of the American Medical Association* 273: 1674-6.

Pollard, B. 1994. *The challenge of euthanasia.* Crows Nest, New South Wales: Little Hills Press.

Pollard, B., and R. Winton. 1993. Why doctors and nurses must not kill patients. *Medical Journal of Australia* 158: 426-9.

Quill, T., C. Cassel, and D. Meier. 1992. Care of the hopelessly ill: Proposed clinical criteria for physician-assisted suicide. *New England Journal of Medicine* 327: 1380-4.

Quill, T., D Meier, S Block, and J. Billings. 1998. The debate over physician-assisted suicide: Empirical data and convergent views. *Annals of Internal Medicine* 128: 552-8.

Russon, L., and D. Alison. 1998. Palliative care does not mean giving up. *British Medical Journal* 317: 196-7.

Reichel, W., and A. Dyck. 1989. Euthanasia: A contemporary moral quandary. *The Lancet* 2: 1321-3.

Roscoe, L., J. Malphurs, L. Dragovic, and D. Cohen. 2000. Dr. Jack Kevorkian and cases of euthanasia in Oakland County, Michigan, 1990-1998. *New England Journal of Medicine* 343: 1735-6.

Ryan, C. 1995. Velcro on the slippery slope: The role of psychiatry in active voluntary euthanasia. *Australian and New Zealand Journal of Psychiatry* 29: 580-5.

Salem, T. 1999. Physician-assisted suicide: Promoting autonomy – or medicalizing suicide? *Hastings Center Report,* 29(3): 30-6.

Salmasy, D., B. Linas, K. Gold, and K. Schulman. 1998. Physician resource use and willingness to participate in assisted suicide. *Archives of Internal Medicine* 158: 974-978.

Santamaria, B. 1996. Euthanasia's bell tolls for thee. *The Weekend Australian,* 13-14 July: 22.

Seale, C. 1995. Heroic Death. *Sociology* 19: 597-613.

Seale, C., and J. Addington-Hall. 1994. Euthanasia: Why people want to die earlier. *Social Science & Medicine* 39: 647-54.

Sullivan, A., K. Hedberg, and D. Fleming. 2000. Legalized physician-assisted suicide in Oregon – The second year. *New England Journal of Medicine* 342: 598-604.

Sullivan, A., K. Hedberg, and D. Hopkins. 2001. Legalized physician-assisted suicide in Oregon, 1998-2000. *New England Journal of Medicine* 344: 605.

Syme, R. 1999a. Pharmacological oblivion contributes to and hastens patients' deaths. *Monash Bioethics Review* 18(2): 40-3.

————. 1999b. Reply from Rodney Syme. *Monash Bioethics Review* 18(3): 34-40.

Tallis, R. 1996. Is there a slippery slope? *Times Literary Supplement* 12 January: 3.

Tobin, B. 1995. Letter, *The Australian* 24 May: 10.

Tulsky, J., R. Ciampa, and E. Rosen. 2000. Responding to legal requests for physician-assisted suicide. *Annals of Internal Medicine* 132: 494-9.

Walker, T., J. Littlewood, and M. Pickering. 1995. Death in the news: The public invigilation of private emotion. *Sociology* 29: 579-96.

Ward, B., and P. Tate. 1994. Attitudes among NHS doctors to requests for euthanasia. *British Medical Journal* 308: 1332-4.

Whitneym, S., B. Brown, H. Brody, K. Alcser, J. Bachman, and H. Greely. 2001. Views of United States physicians and members of the American Medical Association House of Delegates on physician-assisted suicide. *Journal of General Internal Medicine* 16: 290-6.

Woodruff, R. 1995. In *The last right? Australians take sides on the right to die*, eds. S. Chapman, and S. Leeder, 155-8. Port Melbourne, Victoria: Reed Books Australia.

Van der Maas, P., G. van der Wal, I. Haverkate, C. de Graaf, J. Kester, B. Onwuteaka-Philipsen, A. van der Heide, J.M. Bosma, and L. Willems. 1996. Euthanasia, physician-assisted suicide, and other medical practices involving the end of life in the Netherlands, 1990-1995. *New England Journal of Medicine* 335: 1699-1705.

Zalcberg, J., and J. Buchanan. 1997. Clinical issues in euthanasia. *Medical Journal of Australia* 166: 150-2.

Zech, D., S. Grond, J. Lynch, D. Hertel, and K. Lehmann. 1995. Validation of World Health Organization guidelines for cancer pain relief: A 10-year prospective study. *Pain* 63: 65-76.

APPENDIX

Table 1. Physicians' attitudes to assisted suicide and active voluntary euthanasia: selected studies

Study	Willingness to assist	Compliance with patient requests	Attitudes towards law reform
USA:			
Meier, Emmons, Wallenstein et al. (1998) • (national survey: 1902 physicians from 10 specialties) • Sample: 3102 (61% response rate)	• 11% would be prepared to (illegally) prescribe medication to assist a suicide; 36% would do so if it were legal. • 7% would be prepared to (illegally) give a lethal injection; 24% would do so if it were legal.	• 18.3% (320) had received requests for medication to assist in suicide; 16% (of 320) had written a 'lethal prescription' (3.3% of sample). • 11.1% (196) had received a request for a lethal injection. 4.7% (59) had given at least 1 lethal injection.	
Whitney, Brown, Brody et al. (2001) • (national survey: 658 physicians & 315 AMA members; combined response rate 74%)			• 44.5% favoured the legalization of physician-assisted suicide; 33.9% were opposed and 22% were unsure.
UK:			
Ward & Tate (1994) • (national survey: 312 general practitioners and hospital consultants) • Sample: 424 (73.6% response rate)	• 46% (142 of 307 doctors responding to the question) would be prepared to assist a terminally ill patient to die if it was legal to do so.	• 60% (163 of 273 doctors responding to the question) had been asked to hasten a patient's death; 124 had been asked to hasten death with active euthanasia. • 32% (38 of 119 doctors responding to the question, or 12% of all respondents) had complied with such a request.	• 47% (146 of 309 doctors responding to the question) thought that the law on euthanasia in Britain should be similar to the Netherlands.
Australia:			
Baume & O'Malley (1994) • (1268 NSW & ACT doctors) • Sample: 1667 (76% response rate)	• 59% of respondents agreed that it is sometimes right for a doctor to take active steps to bring about a patient's death.	• 46.4% of respondents had been asked by a patient to hasten his or her death; of those asked, 28% had complied with the request (12.3% of all respondents returning the survey); 7% had provided the means for suicide.	• 58% of respondents believed the law should be changed to allow active voluntary euthanasia.

Table 2. Occupational categories of interviewees (n=49).

Doctor (includes psychiatrists)	19
Nurse	17
Therapist (Psychologist/Counsellor)	7
Community Worker	5
Other (Funeral Director)	1

Table 3. Summary of Recruitment Methods (n=49).

Recruitment Category	Total (n=49)
Volunteers	
▪ Volunteered following public presentations	20
▪ Volunteered in response to a flier	2
▪ Volunteered in response to letter to community group	1
Referrals	
▪ Referred/recommended by interviewees	19
▪ Referred by non-interviewees	6
Approached directly by letter	1

CHAPTER EIGHT

LORRAINE SHERR & FABRIZIO STARACE

END-OF-LIFE DECISIONS IN TERMINAL ILLNESS

A Psychiatric Perspective

EUTHANASIA, SUICIDAL BEHAVIORS, AND TERMINAL ILLNESS

Suicide is a problem of major concern. Both completed suicide and suicidal acts resulting in self-harm carry particular burdens for mental health provision. The UK health of the nation White Paper recommends a 15% reduction in suicide rates but, without a complex understanding of the mechanisms underlying suicidal acts, this may not be easy to achieve. Suicide needs to be discussed in the total context of end of life and the gray area between suicide and euthanasia must also be considered.

Literature abounds attempting to record predictors, risks, and associates of suicide. Comprehensive overviews (Hawton 1995) show a number of clear risks that can inform practice and prevention. Schmidtke et al. (1996) note that across Europe gender and age risks reflect males at greater risk of suicide than females and an increasing incidence in younger age groups. Their large 16 center study across Europe also confirmed the risks associated with repeated attempts – over 50% of attempters had made more than one previous attempt, often within dose time proximity. Twenty percent of subsequent attempts occurred within the year. Of note was the association between attempts and social problems, particularly poverty. Many other studies confirm the age and gender findings, as well as the risks associated with repeated attempts. Medical contacts prior to the attempt have also been monitored with varying findings.

Obviously, this provides a potential input channel, given that contacts could provide a moment for prevention and intervention if the contacts are sufficiently trained or perceptive to anticipate the psychological state of the individual (Hawton et al. 1996). The role of mental health factors is continuously reported. Either suicide acts are associated with a history of mental health diagnosis or repeated suicide acts are associated with an absence of psychiatric assessment. Both these findings have implications for care. Initially, they need to be interpreted with

D.N. Weisstub, D.C. Thomasma, S. Gauthier & G.F. Tomossy (eds.), Aging: Decisions at the End of Life,
139-154.
© *2001 Kluwer Academic Publishers. Printed in the Netherlands.*

caution. Given that many units practice a policy of automatic referral to mental health specialists of all suicide attempts, subsequent attempts may include a reporting artifact where mental health diagnosis actually simply reflects the previous attempt rather than additional or different mental health problems. Global strategies to reduce the incidence of suicide may not be sufficiently tailored to the individual needs (Lewis et al 1997). The interlinking between mental health and suicide is well explored (Hawton 1995; Hawton 1998), although the direct role of terminal illness, and/or the mediating role of mental health impact of terminal illnesses is less often studied or articulated.

Another avenue for study has been the trend for increased attempts among young men (Hawton et al. 1997). Yet it is surprising that much effort is expended in understanding this increase with little correlating effort to explain the downward trend in the elderly. Indeed, studies have even excluded the elderly (Haste et al. 1998).

There seems to be strong evidence that study conducted of suicide should neither be done in isolation of attempted suicide (Hawton et al. 1998) nor, on the other end of the spectrum, of end of life issues and euthanasia.

SUICIDAL RISKS AND TERMINAL ILLNESS

Studies have described links between suicidal risks and terminal diseases (Stenager et al. 1994; Horte et al. 1996; Starace and Sherr 1998; Sherr and Starace 1998). This literature fails to provide clarity over why some medical conditions show a link whereas others do not. These could be seen as patient factors, illness factors, and prognosis factors. Patient factors relate to the individual make up of the person with the terminal illness, coping and adjustment, and levels of support. Illness factors play a differential role, often not connected to the physical impact or severity of impact. For example, cognitive impact is seen as different from muscle impact. Some diseases have a social aura of stigma or reaction that differs from others. Thus, population reactions to cancer and AIDS may differ dramatically to cardiac disease and diabetes. Prognosis factors are also of interest, given that they may be open to interpretation, and the arrival of new interventions may dramatically effect the suicidal burden of a given condition. The efficacy of such interventions for any individual may also well ameliorate their personal adjustment and reaction.

The nature and mechanisms underlying the links between terminal illness and suicide are not clearly understood. Some findings point to factors that may be associated with risk. Timing appears to be relevant with proximity to diagnosis emerging as a factor in suicidal risk. Gender, age of the patient, and deterioration of the condition are also associated factors which increase suicidal risk in terminal illness. Co-existence of mental health problems or previous suicide attempts are also seen as risks, with additional burdens.

Harris and Barraclough (1994) note that the association between suicide and medical disorders receives less attention than psychiatric disorder links. A comprehensive search revealed 235 reports of mortality studies where medical disorders were examined with a minimum of 24 months follow up. Utilizing statistical methodologies to observe ratios between observed and expected suicide rates in mortality, they found an increased risk for a number of conditions including HIV/AIDS, malignant neoplasms, head and neck cancers, Huntington disease, multiple sclerosis, peptic ulcer, renal disease, spinal cord injury and systemic lupus erythematosus. They noted insufficient evidence for conditions including amputation, heart valve replacement and surgery, intestinal diseases, hormone replacement therapy, alcoholic liver diseases, neurofibromatosis, Parkinson disease, and systemic sclerosis.

They also noted the protective factor of pregnancy and found no associations with any other conditions.

Table 1 provides a detailed summary of findings relating to an array of terminal illnesses where suicide and end of life have been explored. A number of diseases – most notably cancer and AIDS – have been studied in depth, while others have had less robust investigations. Although many studies do find an association, it is important to note that some do not. The clear message from this table is the need for care provision, particularly regarding end of life trauma for those diagnosed with any terminal illness.

Table 1. Studies of suicide and euthanasia associated with terminal illness

Condition	Studies	Findings
AIDS	Sherr (1995)	Suicidal burden greatest at time of diagnosis and at end-stage illness
	Starace and Sherr (1998)	Increased rate of suicide associated with HIV diagnosis.
	Onwuteaka et al. (1996)	7% of all euthanasia cases (84-93 – Netherlands) had AIDS
	Cooke et al. (1997)	12.1% of 140 people with AIDS received drug increase prior to death to hasten death
	Van der Wal et al. (1996)	9% of all euthanasia cases studied had AIDS. Euthanasia accounts for 13.4% of all AIDS deaths
	Laane (1995)	21 EU and AS deaths in Amsterdam (n=204)
	Back (1996)	AIDS Patients request PAS/EU (n=207): 13% AIDS
	Thompson et al. (1998)	The major causes of death were suicide 35%, AIDS 21%, drug toxicity 16%, natural causes 12%, and injury 7%.
Cancer	Allebeck & Bolund (1991)	59,845 cancer deaths, 24 suicides, 33 attempted suicide
	Grabbe (1997)	Risk associated with cancer and elderly suicides
	Van der Wal et al (1996)	78% of Euthanasia in Holland (84-93) had Cancer. Euthanasia accounts for 2.26% of all cancer deaths
	Grabbe et al. (1986)	Suicide decedents significantly more likely to have cancer history
	Verhoef et al. (1997)	85% Euthanasia and assisted suicide to patients with malignant neoplasms
	Suarex-Almazor et al. (1997)	People with terminal cancer showed an endorsement for legalization of euthanasia
	Sullivan (1997)	48 patients with metastatic cancer, 80% supported AS

Table 1 (cont'd). Studies of suicide and euthanasia associated with terminal illness

Condition	Studies	Findings
Cancer	Laane (1995)	69% of EU/AS in Amsterdam survey had cancer
	Back et al. (1996)	Patients requesting EU/PAS – 41% Cancer
	Haste et al. (1998)	No association with malignancy for either men or women
	Stenager et al. (1991)	Literature review showing moderately increased risk, greatest in first year after diagnosis.
	Suris et al. (1996)	N=162 (control n=865) adolescents. Cancer, asthma, diabetes, and seizures – no differences between conditions. Substantive suicidal ideation compared to controls, Females > Males
	Ginsburg et al. (1995)	N=52 newly diagnosed with lung cancer. 13% considered suicide
Stroke Cardio-vascular Disease	Stenager (1998)	Showed a strong link between stroke and suicide
	Laane (1995)	1% of Amsterdam EU/PAS had cardiovascular disease
	Haste et al. (1998)	No association for CHD or cerebrovascular disease
Epilepsy	Barraclough (1994)	Increased suicide risk, especially in temporal lobe epilepsy
	Haste et al. (1998)	No association with epilepsy for male or females
	Motta et al. (1998)	Suicide in epileptics five times the general population.
Spinal Cord Injury	Harris et al. (1996)	Case reports on incidence
Chronic Abdominal Pain	Magni et al. (1998)	4,964 patients. Rates for thoughts about death, wishing to die, suicidal ideation and suicide attempts were 2-to 3-times more frequent in those with chronic abdominal pain compared with those without
Diabetes	Kyvik et al. (1994)	Young men with IDDM found to have a higher risk of suicide than expected
	Suris et al. (1996)	See data combined with cancer above
Parkinsons Disease	Stenager et al. (1994)	No difference between expected and observed suicide rates (men or women) in a Danish cohort
Multiple Sclerosis	Stenager et al. (1992)	53 suicides in a cohort of 5,525 compared to expected rate of 29
	Stenager et al. (1996)	Association found, particularly in younger males (40-49 years) often linked to deterioration, previous attempts and mental health
	Koch-Henriksen et al. (1998)	An increased risk of death from suicide indirectly attributed to MS (3.8% of 6,068 register cases)
	Flachenecker & Hartung (1996)	Suicide more common than in controls, especially in first 5 years after diagnosis (review study)

Table 1 (cont'd). Studies of suicide and euthanasia associated with terminal illness

Condition	Studies	Findings
Bone Marrow Transplanta- tion	Molassiotis & Morris (1997)	Case reports of three bone marrow recipients, two of whom committed suicide and one with suicidal ideation after treatment
End-Stage Renal Disease	Nagel et al. (1997)	Retrospective study of 24 patients; high suicide burden at end-stage renal disease can be ameliorated by kidney transplant procedures
Neurol. Diseases Overall	Stenager & Stenager (1992)	Review of all studies on Neurological diseases. Suicide associated with multiple sclerosis and spinal cord lesions and some epilepsy. Other disease unclear if there is an association (Huntington's chorea, spinal cord lesions, cranial trauma, brain tumours, Parkinson's disease, vascular disorders, and migraine)
Asthma	Suris et al. (1996)	See data combined with cancer above

A few conditions, notably pregnancy, are protective of suicide (Marzuk et al. 1997). Understanding why this is so, the links with current knowledge and pathways for addressing this phenomenon, and providing potential services need to be explored. This can only be done if a clear understanding of the relevant factors and linking pathways is gathered from the literature to inform such provision.

STUDYING EUTHANASIA AND SUICIDE IN RELATION TO TERMINAL ILLNESS

Given legal considerations, euthanasia in any form is difficult to study. Despite good evidence that the practice is prevalent globally, the majority of in-depth and informative studies emerge from the Netherlands where there are legal directives which allow for systematic reporting of euthanasia of any form and overall a diminished likelihood of prosecution, although the legal right to prosecute still remained in the Dutch law for the duration of the twentieth century with change only recently emerging. (Van der Maas et al. 1992; Van der Wal et al. 1996).

Suicide data is more readily available, but under-reporting may well be a factor given the social stigma associated with the act. This under-reporting may be a particular problem in cases where terminal illness is present. Data sources invariably utilize hospital medical records, general practitioner records, death certificates, or coroner reports. All may provide limited detail of the co-existence of terminal illness with suicidal acts and they may also under-report deliberate end of life acts.

END-OF-LIFE ACTS

The end-of-life by deliberate act(s) has been sub-divided into a number of different concepts.

Suicide refers to a situation where a person ends his/her own life.

Assisted suicide relates to aid in the act, with an articulated request.

Physician Assisted Suicide (PAS) occurs when a doctor provides medication that the patient can subsequently use to achieve a suicide, although the doctor does not necessarily directly assist in this act.

Euthanasia refers to all acts that result in a managed death, invariably seen as painless and at the patient's request.

Active euthanasia is a result of treatments that hasten death.

Passive euthanasia occurs when death is hastened but as a result of not treating, or withholding a treatment, rather than initiating an action.

Non-voluntary euthanasia occurs in situations where death is hastened but the patient is not able to or does not provide full consent.

Pain management with double effect can also be seen under the euthanasia umbrella, whereby doctors prescribe pain medication that has the side effect of hastening death.

AIDS PATIENTS, SUICIDE, AND EUTHANASIA

A range of end-of-life behavior has been well catalogued for HIV and AIDS patients (Starace and Sherr 1998), and this may form an interesting perspective from which studies can generalize to other terminal illness. End-of-life decisions can include suicidal thoughts or ideation, suicidal acts either noted as deliberate self-harm or para-suicide, death by suicide or death by the numerous forms of euthanasia. The suicidal burden has been associated with the trauma of HIV seropositivity and bleak future prospects (Sherr 1995), the social stigma concomitant with diagnosis (Green 1995), isolation (Miller 1995), withdrawal of family support (Bor et al. 1992), strained sexual relationships (Sherr 1995b), loss of hope (Sherr 1995c), and a shortened future (Sherr 1995d) as well as the effects of the various opportunistic infections on the quality of life of the diagnosed individual (Lamping and Satchdev 1995).

From the start of the HIV epidemic suicidal thoughts and acts have been documented (Perry et al. 1990), although the effects of new treatments on these have not been explored to date (Kravcik et al. 1997). Studies have shown bimodal time peaks where clusters of suicidal thoughts and acts occur at or around HIV diagnosis and again at end-stage illness (Sherr 1995; Perry 1993). This would point to two key foci, which may well generalize to other terminal illnesses. The first concerns suicide and self-harm associated with crisis, trauma, and adjustment at the time of diagnosis and the second relates more to the exploration of quality of death, end-of-life decision-making and control at end-stage illness. The euthanasia debate is pertinent to this latter phase as well (Starace and Sherr 1998; Van den Boom 1995; Laane 1995). Thus, the extensive study of the phenomena in HIV disease may form a model of understanding for all terminal illnesses where life-ending questions may arise in these two phases. Other terminal illnesses may have similar patterns. Early studies in HIV disease noted a higher rate of suicidal behaviors in the presence of this virus (Marzuk 1988; Kizer 1988; Piott 1989; Marzuk 1991; Coté 1992; Pugh

1993; Copeland 1993; Mancoske 1995). These findings have been consistently upheld (Fugelstad 1992; Barbieri et al. 1989; Pueschel & Heineman 1995), with variation in the degree rather than in the finding. Not a single study has looked at the effect of the new therapeutic optimism (Deeks 1997) and availability on suicidal behavior, and one could postulate that combination therapies may have reduced the bleak outlook (Mocroft 1998) and may contribute to a reduction in suicidal behaviors. However, the onerus nature of the therapeutic regimen may be a complicating factor and these speculations need urgent attention to understand their effect on end-of-life intentions and mediating factors. This would be particularly helpful in guiding other diseases where new therapeutic breakthroughs may emerge and change the mental health as well as the physical health burden on the individual. Within the general literature the increase in suicides by young men (Hawton 1998) may have invoked the emergence of AIDS and HIV as an explanation for this trend. This seems inadequate now, given that the increase in suicide in young men persists (and grows) despite the new combination treatments for HIV.

PREDICTORS

Care and prevention can be informed by a detailed understanding of predictors or covariates of suicidal behavior. A past history of suicidal acts, pre-dating and post-dating HIV diagnosis, have been found to correlate with subsequent suicidal acts (Sherr 1995; Catalan & Pugh 1995). Starace & Sherr (1998) emphasize additional factors such as social isolation, utilization of avoidant or denial coping strategies, drug use, physical illness (Rabkin et al. 1993), hopelessness (Beck 1985; Catalan 1992), and bereavement (Rabkin et al. 1993; Sherr 1993; 1995). The findings are not universal and others have not found similar correlates (Rosengard and Folkman 1997; Riccio 1993). Gender, lifestyle factors, and sexuality are also found to be related to suicidal behavior (Gala et al. 1992; Catalan et al. 1995b). HIV seems to compound the suicidal burden (Catalan 1991).

The literature from HIV disease would highlight the need for ongoing vigilance in focusing attention on traditional risk factors that may increase the burden, such as previous attempts, mental illness, gender, and lifestyle. The bimodal distribution of strain may target services to initiate at the time of diagnosis as well as at the onset of burdens associated with end-stage illness. Counseling provision has been the cornerstone of the HIV and AIDS epidemic, but is often sorely lacking in other diseases. Self-help groups have also played a role for this population, and their support should be harnessed for other terminal illness.

For euthanasia the data is less helpful. The concentration of studies within a few geographic centers is the first limitation, but this is confounded by the fact that these studies tend to describe detailed demographic and epidemiological variables but few mental health or psychological factors. There are a few exceptions to this, notably the study by Van den Boom (1995) who reported on the emotional adjustment in those planning euthanasia as well as complicated grief among survivors. Of interest was the fact that planning euthanasia, feeling in control, and having an opportunity to discuss death anxiety and management of death was seen as adaptive. This planning tended to occur during early stage HIV disease. Those who planned euthanasia did not automatically invoke it as their means of death; indeed only half of those planning euthanasia carried it out. Euthanasia in the Dutch context cannot be generalized given the experience of the profession with approaching death by euthanasia and backed up by the general provision of medical care and the absence of financial burden. Furthermore, survivors of a loved one who died by euthanasia displayed no elevated levels of complicated grief reactions in this study.

Valente (1998) noted that patients with terminal illness are likely to contemplate ending their life but notes that prompt pain management can dissipate these thoughts.

CHANGING VIEWS ON EUTHANASIA

As the topic of euthanasia becomes easier to study, a growing rate of euthanasia requests can be monitored in the literature (Emanuel et al. 1998; Vidalis et al. 1998; Di Mola et al. 1996; Kirschner and Eikeles 1998). However, it is not easy to establish whether this reflects an increase in the phenomenon, a greater willingness to report the phenomenon, an increased interest to study euthanasia requests, or a true increase in requests. The answer may well be a combination effect.

In general, the studies show a growing dialogue on the issue of Euthanasia among professionals and an increasing willingness to consider, endorse, assist, and carry out acts of euthanasia in a variety of forms. The issues raise enormous ethical considerations and the debate in the international literature is in its formative stages, benefiting from multidisciplinary perspectives. A selection of studies and their findings are summarized in table 2 below.

Table 2. Euthanasia: The changing mood

Study	Sample	Views on Euthanasia	Willingness to consider any form of Euthanasia
Emanuel et al. (1998)	n=355 (Oncology)	53% comfort with helping 24% regretted 40% fear prosecution	15.8% participated in some way
Vidalis et al. (1998)	n=417	44.3% against life extension	Not reported
Kitchener (1998)	n=1,218	69% wanted change in euthanasia laws	
Di Mola et al. (1996)	n=685 doctors	35% thought euthanasia was wrong	39% received requests 4% complied with requests
Kirschner & Elkeles (1998)	n=473 doctors	Half endorse discussion Half have seen patients who express wishes for EU	4% comply with request
Rutecki et al. (1997)	n=125 nephrologist	48% would never initiate even if it was legal	

Table 2 (cont'd). Euthanasia: The changing mood

Study	Sample	Views on Euthanasia	Willingness to consider any form of Euthanasia
Siaw & Tan (1996)	n=1,028 Doctors and medical students		15.6% will to assist
			9.8% would comply with active EU request
			97.6% would withhold life support on request
			78.6% would withdraw life support on request
			88% would consider double effect
Hermandez et al. (1997)	n=114 doctors and students	92% believe life sacred; 69% believe brain dead patients should not receive futile treatment	32% agree with passive EU 18% agreed with active EU 19.3% felt hey had exceeded therapy ever
Meier et al. (1998)	n=1,902	11% noted occasions when they could consider hastening death	18.3% received EU request 3.3% written prescription 4.7% administered at least one lethal injection
Verhoef & Kinsella (1996)	n=866	42% felt it right on occasion to be able to practice EU	
Asch & DeKay (1997)	n=1,1399 nurses	30% believed EU unethical	19% engaged in EU
Kowalski (1997)	n=539 nurses	agreement: 92% withdrawal of life support 85% double effect 53% PAS 44% active EU	46% willing to participate if it was legal
Forde et al. (1997)	n=974 doctors	17% believed in active end of life	6% reported actions to hasten death
Suarez-Almazor et al. (1997)	n=197 doctors	60-80% opposed EU	
Howard et al. (1997)	n=355 Oncologists	48.1% could imagine EU situations	85.8% found EU acceptable for their patients as an issue
Slome et al. (1992)	N=68 doctors		24-27% noted an intention to assist
Slome et al. (1996)	n=114		48-51% intention to assist 53% at least one participation in the past
Hogg et al. (1977)	n=1,153 doctors		60% expressed intention to assist
Leiser et al. (1996)	n=215 nurses		38% indirect requests 54% had direct requests 59% intention to assist 15% participated at least once in the past

Note: This table does not include studies from the Netherlands given the different legal background

Studies of professionals show a varied impact of euthanasia related issues. In all studies there was a consistent minority who showed a willingness to consider the issue of euthanasia and another group who performed a variety of situations in which euthanasia behavior was recalled. Studies that refer to patient or general population views show a greater endorsement of euthanasia-associated beliefs by patients (Sherr & Starace 1999). Of the nine patient studies located since 1993 the majority were related to those diagnosed with a terminal illness. Four studies reported on HIV patients (Breitbart et al. 1996; Fleerackers et al. 1996; Ogden 1994; Tindall 1993), two on the elderly (Tijmstra et al. 1997; Cicirelli 1997), two on cancer (Kuuppeiomaki 1997; Suarez Almazor et al. 1997), and the final one looked at general practitioner patients (Graber et al. 1996).

CONCLUSION

The link between different forms of terminal illness and suicide/euthanasia seems to be clear. Some terminal illness have benefited from detailed study, whereas others seem somewhat neglected. A clearer understanding would be helpful, given that there are some illness specific factors that are associated with a patient's response as well as societal response in their willingness to consider, tolerate, or facilitate various options on end-of-life. For many people, the advent of terminal illness may either create a situation or add to a situation where life becomes too difficult to face and death becomes the only option. There is a complex set of factors surrounding the psychology of adjusting to and accommodating mental health reactions to terminal illness.

No studies have been possible in the area of combined suicide and euthanasia to see if the availability of the latter affects the need for the former. It is unclear whether those seeking suicide as a solution to their crisis are similar to those who explore the euthanasia routes. Much of the debate centers on the life shattering impact of terminal illness, quality of life, and quality of death. The complexity of euthanasia linked decisions and their legal situation confounds the area of study. The role of palliative care is vital, if quality and control is to be a reality for the dying patient. There is a clear place for mental health understanding at all stages of illness, from diagnosis to end-stage disease. Some illnesses are studied in-depth, while others seem neglected. Protocols for handling all terminal illness should articulate elements of end-of-life as well as address its management.

Lorraine Sherr, Reader in Clinical and Health Psychology, Royal Free & University College Medical School, London, United Kingdom.

Fabrizio Starace, Director, Consultation Psychiatry & Behavioral Epidemiology Service, Cotugno Hospital, Naples, Italy.

NOTES

This chapter is dedicated to Stanley Reiss, taken too soon but whose memory is large as life.

REFERENCES

Abramson, N., J. Stokes, N. Weinreab, and W. Clark. 1998. Euthanasia and doctor assisted suicide responses by oncologists and non oncologists. *Southern Medical Journal* 91(7): 637-42.

Allebeck, P., and C. Bolund. 1991. Suicides and suicide attempts in cancer patients. *Psychological Medicine*. 21(4): 979-84.

American Medical Association. 1994. Euthanasia/physician-assisted suicide: lessons in the Dutch experience. *Issues in Law & Medicine* 10: 81-90.

Anderws, K. 1996. Euthanasia in chronic severe disablement. *British Medical Bulletin* 52(2): 280-8.

Aro, A.R., M. Henriksson, P. Leinikki, and J. Lonnqvist. 1995. Fear of AIDS and suicide in Finland: a review. *AIDS Care* 7 (Suppl.2): S187-97.

Asch, D., and M. DeKay. 1997. Euthanasia among US critical care nurses. Practices attitudes and social and professional correlates. *Medical Care* 35(9): 890-900.

Asch D., J. Shea, M. Jedriezwski, and C. Bosk. 1997. The limits of suffering critical care nurses views of hospital care at the end of life. *Social Science and Medicine* 45: 1661-8.

Back, A.L., J.I. Wallace, H.E. Starks, R.A. Peariman. 1996. Physician-assisted suicide and euthanasia in Washington State. Patient requests and physician responses. *JAMA* 275: 919-25.

Banazak, D.A. 1996. Late-life depression in primary care. How well are we doing? *Journal of General and Internal Medicine* 11: 163-7.

Barraclough, B.M. 1987. The suicide rate of epilepsy. *Acta Psychiatrica Scandinavica* 76(4): 339-45.

Baume P., and E. O'Malley. 1994. Euthanasia: attitudes and practices of medical practitioners. *Medical Journal of Australia* 161: 137-44.

Beck, A., R. Steer, M. Kovacs, and B. Garison. 1985. Hopelessness and eventual suicide. *American Journal of Psychiatry* 145: 559-63.

Block, J., and J. Billings. 1995. Patient requests for euthanasia and assisted suicide in terminal illness. The role of the Psychiatrist. *Psychosomatics* 36(5): 445-57

Bor, R., R. Milier, and E. Goldman. 1992. *Theory and practice of HIV counselling: A systemic approach.* London: Cassell.

Breitbart, W., E. Bruera, H. Chochinov, and M. Lynch. 1995. Neuropsychiatric syndromes and psychological symptoms in patients with advanced cancer. *Journal of Pain and Symptom Management* 10: 131-41.

Breitbart, W., B. Rosenfeld, and S. Passik. 1996. Interest in physician assisted suicide among ambulatory HIV infected patients. *American Journal of Psychiatry* 153(2): 238-42.

Brown, J.H., P. Henteleff, S. Barakat, and C.J. Rowe. 1986. Is it normal for terminally ill patients to desire death? *American Journal of Psychiatry* 143: 208-11.

Cantor, N.L. 1996. Can health care providers obtain judicial intervention against surrogates who demand "medically inappropriate" life support for incompetent patients? *Critical Care Medicine* 24: 883-7.

Charlesworth, M. 1993. *Bioethics in a liberal society.* London: Cambridge University Press.

Catalan, J. 1991. Deliberate self harm and HIV disease. In *HIV and AIDS Related Suicidal Behaviour,* eds. J. Beskow, M. Bellini, J. Faria, and A. Kerllliof, 47-54. Bologna: Monduzzi Editore.

Catalan, J., D. Seijas, T. Lief, A. Pergami, and A Burgess. 1995b. Suicidal behaviour in 111V infection a case control study of deliberate self harm in people with HIV infection. *Archives of Suicide Research* 1: 85-96.

Catalan, J. I. Klimes, A. Day, A. Garrod, A. Bond, and J Gallway. 1992a. The psychosocial impact of HIV infection on gay men. *Journal of Psychiatry* 11: 774-9.

Catalan, J., I. Klimes, A. Bond, A. Day, A. Garrod, and C. Rizza. 1992b. The Psychosocial impact of HIV infection in men with haemophilia. *Journal of Psychosomatic Research* 36: 409-16.

Catalan, J., and K. Pugh. 1995. Suicidal behaviour and HIV infection: is there a link? *AIDS Care* 7 (Suppl.2): S117-21.

Carvajal, M.J.C. Vicioso, J.M. Santamaria, and A. Bosco. 1995. AIDS and suicide issues in Spain. *AIDS Care* 7(Suppl.2): S135-8.

Cicirelli, V. 1997. Relationship of psychosocial and background variables to older adults end of life decisions. *Psychology and Aging* 12(1): 72-83.

Cooke, M., L. Gourlay, L. Collette, A. Boccellari, M. Chesney, and S. Folkman. 1997. Informal caregivers and the intention to hasten AIDS related death. *Annals of Internal Medicine* 158(1): 69-75.

Cote, T.R., R.I. Biggar, and A.L. Dannenberg. 1992. Risk of suicide among persons with AIDS. A national assessment *JAMA* 268(15): 2066-8.

Di Mola, G., P. Borsellino, C. Brunelli, M. Gallucci, A. Gamba, M. Lusignani, C. Regazzo, A Santosuosso, M. Tamburini, and F. Toscani. 1996. Attitudes toward euthanasia of physician members of the Italian Society for Palliative Care. *Annals of Oncology* 7(9): 907-12.

Doukas, D.J., D. Waterhouse, D.W. Gorenflo, and S. Seid. 1995. Attitudes and behaviors on physician assisted death: a study of Michigan oncologists. *Journal of Clinical Oncology* 13: 1055-61.

Emanuel, E., D.L. Fairclough, E.R. Daniels, and B.R. Ciarridge. 1996. Euthanasia and physician-assisted suicide: attitudes and experiences of oncology patients, oncologists, and the public. *Lancet* 347: 1805-10.

Fainsinger, R., E. Bruera, M.J. Miller, J. Hanson, and T. MacEachern. 1991. Symptom control during the last week of life on a palliative care unit. *Journal of Palliative Care* 7: 5-11.

Fishbain, D. 1995. Chronic pain and suicide *Psychotherapy and Psychosomatics* 63: 154-5.

———. 1996. Current research on chronic pain and suicide. *American Journal of Public Health* 86(9): 1320-1.

Flachenecker, P., and H.P. Hartung. 1996. Course of illness and prognosis of multiple sclerosis. The natural illness course. *Nervenarzt* 67(6): 435-43.

Flavin, D., J. Franiclin, and R. Frances. 1986. AIDS and suicidal behavior in alcohol dependent homosexual men. *American Journal of Psychiatry* 143: 1440-2.

Folker, A.P., H. Holtug, A.B. Jensen, K. Kappel, J.K. Nielsen, and M. Norup. 1996. Experiences and attitudes towards end-of-life decisions amongst Danish physicians. *Bioethics* 10: 233-49.

Forde, R., O.G. Aasland, and E. Falkum. 1997. The ethics of euthanasia: attitudes and practice among Norwegian physicians. *Social Science and Medicine* 45: 887-92.

Fried, T.R., M.D. Stein, P.S. O'Sullivan, D.W. Brock, and D.H. Novack. 1993. Limits of patient autonomy. Physician attitudes and practices regarding life-sustaining treatments and euthanasia. *Archives of Internal Medicine* 153: 722-8.

Emanuel, E., E. Daniels, D. Fiarciough, and B. Clarridge. 1998. The practice of euthanasia and physician assisted suicide in the US adherence to proposed safeguards and effects on physicians. *JAMA* 280(6): 507-13.

Gala, C., A. Pergarni, J. Catalan, M. Riccio, F. Durbano, and M. Musicco. 1992. Risk of deliberate self harm and factors associated with suicidal behavior among asymptomatic individuals with HIV infection. *Acta Psychiatrica Scanadavica* 86: 70-5.

Ganzini, L., D. Fenn, M. Lee, R. Heintz, and J. Bloom. 1996. Attitudes of Oregon psychiatrists toward physician assisted suicide. *American Journal of Psychiatry* 153(11): 1469-75.

Ganzini, L., M.A. Lee, R.T. Heintz, J.D. Bloom, and D.S. Fenn. 1994. The effect of depression treatment on elderly patients' preferences for life-sustaining medical therapy. *American Journal of Psychiatry* 151: 1631-6.

Garel, M., S. Gosme, M. Kaminski, and M. Cuffini. 1997. Ethical decision making in neonatal intensive care. Survey among nursing staff in 2 French centres. *Archives de Pediatrie* 4(7): 662-70.

Ginsburg, M.L., C. Quirt, A.D. Ginsburg, and W.J. MacKiliop. 1995. Psychiatric illness and psychosocial concerns of patients with newly diagnosed lung cancer *Canadian Medical Association Journal* 152(5): 701-8.

Grabbe, L., A. Demi, M. Camann, and L. Potter. 1997. The health status of elderly persons in the last year of life a comparison of deaths by suicide, injury, and natural causes *American Journal of Public Health* 87(3): 434-7.

Graber, M., B. Levy, R. Weir, and R. Oppliger. 1996. Patients views about physician participation in assisted suicide and euthanasia. *Journal of General Internal Medicine* 11(2) 71-6.

Green, G. 1995. AIDS and euthanasia. *AIDS Care* 7(Suppl.2): S169-73.

Groenewoud, S., Van der Maas P., G. Van der Wal, M.W. Hengeveld, A.J. Tholen, W.J. Schdel, A. van der Heide. 1997. Physician-assisted death in psychiatry practice in the Netherlands. *New England Journal of Medicine* 336: 1795-1801.

Haghbin, Z., J. Streitzer, G.P. Danko.1998. Assisted suicide and AIDS patients. A survey of physicians' attitudes. *Psychosomatics* 39(1): 18-23.

Haghbin, Z., B. Streitzer, and G.P. Danko. 1996. Assisted suicide in AIDS: physicians' attitudes. 149th Annual Meeting of the American Psychiatric Association, New York, May [abstract Tu. 5.45].

Hanson, L.C., M. Danis, E. Mutran, and N.L. Keenan. 1994. Impact of patient incompetence on decisions to use or withhold life-sustaining treatment. *American Journal of Medicine* 97: 235-41.

Haste, F., J. Chariton, and R. Jenkins. 1988. Potential for suicide prevention in primary care? An analysis of factors associated with suicide. *British Journal of General Practice* 48: 1759-63.

Hendin, H. 1994. Seduced by death: Doctors, patients and the Dutch cure. *Issues in Law and Medicine* 20: 123-68.

Hendin H., and G. Kierman. 1993. Physician-assisted suicide: the dangers of legalization. *American Journal of Psvchiatrv* 150: 143-5.

Harris, E.C., and B.M. Barraclough. 1994. Suicide as an outcome for medical disorders. *Medicine* 73(6): 281-96.

Harris, E.C., B.M. Barraclough, D.J. Grundy, E.S. Bamford, and H.M. Inskip. 1996. Attempted suicide and completed suicide in traumatic spinal cord injury. Case reports. *Spinal Cord* 34(12):752-3.

Hawton, K. and J. Catalan. 1987. Attempted suicide. Oxford: Oxford University Press.

Hernandez Arriaga, J., A. Morales-Estrada, and G. Cortes Gallo. 1997. Survey of physicians' attitudes to terminal patients. *Revista de Investigacion Clinica* 49(6): 497-500.

Hogg, R., K. Heath, G. Bally, P. Cornelisse, B. Yip, and M. O'Shaughnessy. 1997. Attitudes of Canadian Physicians toward legalizing physician assisted suicide for persons with HIV disease. Abstract, III International AIDS Impact Conference, Melbourne.

Horte, L., R. Stensman, U. Suddqvist-Stensman. 1996. Physical disease among 21 suicide cases interviews of relatives and friends. *Scandinavian Journal of Social Medicine* 24: 53-8.

Howard, O., D. Fairclough, E. Daniels, and E. Emanuel. 1997. Physician desire for euthanasia and assisted suicide would physicians practice. *Journal of Clinical Oncology* 15: 428-32.

Huyse, F., and W. van Tiiburg. 1993. Euthanasia policy in the Netherlands the role of consultationliaison psychiatrists. *Hospital and Community Psychiatry* 44: 733-8.

Kinsella, T.D., and M. Verhoef. 1993. Alberta euthanasia survey: 1. Physicians' opinions about the morality and legalization of active euthanasia. *Canadian Medical Association Journal* 148: 1921-6.

Kirschner, R., and T. Elkeles. 1998. Physician practice patterns and attitudes to euthanasia in Germany: A Representative survey of physicians. *Gesundheitswesen* 60: 247-53.

Kitchener, B. 1998. Nurse characteristics and attitudes to active voluntary euthanasia a survey in the Australian Capital Territory. *Journal of Advanced Nursing* 28: 70-6.

Kravcik, S., N. Hawley Foss, G. Victor, L.J. Angel, G. Garber, S. Page, N. Denorme, L. O'Reilly, and D. Cameron. 1997. Causes of death of HIV infected persons in Ottawa Ontario 1984-95 *Archives of Internal Medicine* 157: 2069-73.

Kowalski, S. 1997. Nevada nurses attitudes regarding physician assisted suicide. *Clinical Nurse Specialist* 11: 109-15.

Kuhse, H. and P. Singer. 1988. Doctors' practices and attitudes regarding voluntary euthanasia. *Medical Journal of Australia* 148: 623-7.

Kuhse, H., P. Singer, P. Baume, M. Clark, and M. Rickard. 1997. End of life decisions in Australian Medical practice. *Medical Journal of Australia* 166: 191-6.

Kuuppeiomaki M. 1997. Attitude of cancer patients their relatives nurses and doctors to active euthanasia. *Hoitotiede* 9: 186-93.

Lee, M.A., H.D. Nelson, V.P. Tilden, L. Ganzini, T.A. Schmidt, and S.W. Tolie. 1996. Legalizing assisted suicide. Views of physicians in Oregon. *New England Journal of Medicine* 334: 310-5.

Leiser, R.I., T.F. Mitchell, S. Hahn, L. Siome, and D.I. Abrams. 1998. Nurses' attitudes and beliefs toward assisted suicide in AIDS. *Journal of the Association of Nurses in AIDS Care.* 9: 26-33.

Macdonald, W. 1998. Situational factors and attitudes toward voluntary euthanasia. *Social Science and Medicine* 46: 73-81.

Magnani, K., M. Ercolani, and L. Grassi. 1997. Atteggiamento del medico nei confronti della richiesta di eutanasia e di suicidio assistito. Abs. XVI Congresso Nazionale Società Italiana di Medicina Psicosomatica, Parma, 1-4 maggio.

Magni, G., S. Rigatti-Luchini, F. Fracca, and L. Merskey. 1998. Suicidality in chronic abdominal pain: an analysis of the Hispanic Health and Nutrition Examination Survey (HHANES). *Pain* 76: 137-44.

Mancoske, R.I., C.M. Wadsworth, D.S. Dugas, and J.A. Hasney. 1995. Suicide risk among people living with AIDS. *Social Work* 40: 783-7.

Marzuk, P. 1991. Suicidal behaviour and HIV illness. *International Review of Psychiatry* 3: 365-71.

Marzuk, P., H. Tierney, K. Tardiff, E. Gross, E. Morgan, M. Hsu. 1988. Increased risk of suicide in persons with AIDS. *JAMA* 259: 1333-7.

Marzuk P., K. Tardiff, A. Leon, C. Hirsch, L. Portera, N. Hartweil, and M. Iqbal. 1997. Lower risk of suicide during pregnancy. *American Journal of Psychiatry* 154: 122-3.

Matsukawa, Y., S. Sawada, T. Hayama, H. Usui, and T. Horie. 1994. Suicide in patients with systemic lupus erythematosus: a clinical analysis of seven suicidal patients. *Lupus* 3: 31-5.

Meier, D., C. Emmons, S. Wallenstein, T. Quili, R. Morrison, C. Cassel. 1998. A national survey of physician assisted suicide and euthanasia in the United States. *NEJM* 338: 1193-201.

Miller, R. 1995. Suicide and AIDS problem identification during counseling. *AIDS Care* 7: S199-205.

Moiassiotis, A., and P.J. Morris. 1997. Suicide and suicidal ideation after marrow transplantation. *Bone Marrow Transplantation* 19: 87-90.

Motta, E., K. Miller, D. Rosciszewska, and E. Klosinska. 1998. Depression in epileptic patients with and without history of suicidal attempts: preliminary report. *Psychiatria Polska* 32: 199-208.

Muller, M.T., G. van der Wal, J.T. van Eijk, and M.W. Ribbe. 1994. Voluntary active euthanasia and physician-assisted suicide in Dutch nursing homes: are the requirements for prudent practice properly met? *Journal of the American Geriatrics Society* 42: 624-9.

Muller, M., B. Onwuteaka-Philipsen, G. van der Wal, B. van Eijk, and M. Ribbe. 1996. The role of the social network in active euthanasia and physician-assisted suicide. *Public Health* 110: 271-5.

Norup, M. 1998. Limits of neonatal treatment a survey of attitudes in the Danish Population. *Journal of Medical Ethics* 24: 200-6.

Ogden, R. 1994. Euthanasia Assisted suicide and AIDS. Toronto: Peroglyphics Publishing.

Onwuteaka Philipsen, B., and G. van der Wal. 1998. Cases of euthanasia and physician assisted suicide among AIDS patients reported to the Public Prosecutor in North Holland. *Public Health* 112: 53-6.

Sherr, L. 1995b. Coping with Psychosexual Problems in the Context of HIV Infection. *Sexual and Marital Therapy* 10: 307-19.

Sherr L., and B. Hedge. 1995c. Psychological Needs and HIV/AIDS. *British Journal of Clinical Psychology* 12: 203-9.

Sherr, L. 1995d. The experience of grief: Psychological aspects of grief. In *Grief and AIDS*, ed. L. Sherr, 1-28. Chichester: Wiley and Son.

Slome, L., B. Moulton, C. Huffine, R. Gorter, D. Abrams. 1992. Physicians' attitudes toward assisted suicide in AIDS. *JAIDS* 5: 712-8.

Slome, L., T. Mitchell, E. Charlebois, B. Moulton, D. Abrams. 1997. Physician assisted suicide and patients with HIV. *NEJM* 336: 417-21.

Starace, F. Epidemiology of suicide among persons with AIDS. 1995. *AIDS Care* 7: S123-8.

Starace, F., and L. Sherr. 1998. Suicidal behaviours euthanasia and AIDS. *AIDS* 12: 6281-6

Steinberg, M., J. Najman, C. Cartwright, S. MacDonald, and G. Williams. 1997. End of life decision making. Community and medical practitioners perspectives. *Medical Journal of Australia* 1666: 131-5.

Stenager, E., and F. Stenager. 1992. Suicide in patients with neurological diseases. Methodological problems. *Archives of Neurology* 49: 1296-303.

Stenager, E., C. Madsen, E. Stenager, and S. Boidsen. 1998. Suicide in patients with stroke: epidemiological study. *British Medical Journal* 316: 1206-8.

Suarez Almazor, M., M. Beizile, and E. Bruera. 1997. Euthanasia and physician assisted suicide: a comparative survey of physicians, terminally ill cancer patients and the general population. *Journal of Clinical Oncology* 15: 418-27.

Sullivan, M., S. Rapp, D. Fitzgibbon, and C. Chapman. 1997. Pain and the choice to hasten death in patients with painful metastatic cancer. *Journal of Palliative Care* 13: 18-28.

Sullivan, M.D. and S. Youngner. 1994. Depression, competence, and the right to refuse lifesaving medical treatment. *American Journal of Psychiatry* 151: 971-8.

Suris, J.C., N. Parera, and C. Puig. 1996. Chronic illness and emotional distress in adolescence. *Journal of Adolescent Health* 19: 153-6.

Stevens, C.A., and R. Hassan. 1994. Management of death, dying and euthanasia: attitudes and practices of medical practitioners in South Australia. *Journal of Medical Ethics* 20: 41-46.

Thompson, S.C., A. Manjikian, A. Ambrose, L.A. Ireland, and E.M. Stevenson. 1998. HIV positive tests at Coronal Services in Victoria 1989-1996: lessons for HIV surveillance. *Australian & New Zealand Journal of Public Health* 22: 532-5.

Tijmstra, T., G. Kempen, and J. Ormei. 1997. End of life and termination of life opinions of elderly persons with health problems. *Nederlands Tijdschrift voor geneeskunde* 141: 2444-8.

Tindali, B., S. Forde, A. Carr, S. Barker, and D. Cooper. 1993. Attitudes to euthanasia and assisted suicide in a group of homosexual men with advanced HIV disease. JAIDS 6: 1069-70.

Uchitomi, Y., H. Okamura, H. Minagawa, A. Kugaya, M. Fukue, A. Kagaya, N. Oomori, and S. Yamawaki. 1995. A survey of Japanese physicians' attitudes and practice in caring for terminally ill cancer patients. *Psychiatry and Clinical Neurosciences* 49: 53-57.

Valente, S.M., and D. Trainor. 1998. Rational suicide among patients who are terminally ill. *AORN Journal* 68: 252-8.

Van den Boom, F. 1995. AIDS, euthanasia and grief. *AIDS Care* 7: S5175-85.

Van der Heide, A., P. van der Maas, G. van der Wal, C. Graaff, S. Kester, L. Koilee, R. de Leeuw, and R. Holl. 1997. Medical end of life decisions made for neonates and infants in the Netherlands. *Lancet* 350: 251-5.

Van der Maas, P.J., L. van Deleden Pijnenborg, and C. Loman. 1991. Euthanasia and other medical decisions concerning the ending of life. *Lancet* 338: 669-74.

Van der Maas, P.J., G. van der Wal, I. Haverkate, C.L.M. De Graaf, S. Kester, B.D. Onwuteaka-Philipsen, A. van der Heide, S.M. Bosma, and D.L. Willems. 1996. Euthanasia, physician-assisted suicide, and other medical practices involving the end of life in the Netherlands, 1990-1995. *NEJM* 335: 1699-1705.

Van der Maas, P., J. van Delders, and L. Pijnenborg. 1992. Euthanasia and other medical decisions concerning the end of life. Health Policy Monograph. New York: Elsevier.

Van der Wal, E., H. van Leenen, and C. Spreeuwenberg. 1992. Euthanasia and assisted suicide: how often is it practiced by family doctors in the Netherlands. *Family Practice* 19: 130-4.

Van der Wal G., J.T. van Eijk, H.S. Leenen, and C. Spreeuwenberg. 1992b. The use of drugs for euthanasia and assisted suicide in family practice. *Nederlands Tijdschrift voor Geneeskunde* 136: 1299-1305.

Van der Wal, G., J.T. van Eijk, H.S. Leenen, and C. Spreeuwenberg. 1992c. Euthanasia and assisted suicide. Do Dutch family doctors act prudently? *Family Practice* 9: 135-40.

Van der Wal, G., and R. Dilimann. 1994. Euthanasia in the Netherlands. *BMJ* 308: 1346-9.

Van der Wal, G., and B. Onwuteaka Philipsen. 1996. Cases of euthanasia and assisted suicide reported to the public prosecutor in North Holland over 10 years. *BMJ* 312: 612-3.

Van der Wal, G., J.T. van Eijk, H.S. Leenen, and C. Spreeuwenberg. 1991a. Euthanasia and assisted suicide by physicians in the home situation. I. Diagnoses, age and sex of patients. *Nederlands Tijdschrift voor Geneeskunde* 135: 1593-1598.

————. 1991b. Euthanasia and assisted suicide by physicians in the home situation. Suffering of the patients. *Nederlands Tijdschrift voor Geneeskunde* 135: 1599-1603.

Van der Wal G., M.T. Muller, L.M. Christ, M.W. Ribbe, T. van Eijk. 1994. Voluntary active euthanasia and physician-assisted suicide in Dutch nursing homes: requests and administration. *Journal of the American Geriatrics Society* 42: 620-3.

Van der Wal, G., P.J. van der Maas, S.M. Bosma, B.D. Onwuteaka-Philipsen, D.L. Willems, I. Haverkate, and P.J. Kostense. 1996. Evaluation of the notification procedure for physician-assisted death in the Netherlands. *NEJM* 335: 1706-11.

van Thiel,. G., K. van Delden de Haan, and A. Huibers. 1997. Retrospective study of doctors end of life decisions in caring for mentally handicapped people in institutions in the Netherlands. *BMJ* 315: 88-91.

Ventafridda, V., C. Ripamonti, F. De Conno, M. Tamburini, and B.R. Cassileth. 1990. Symptom prevalence and control during cancer patients' last days of life. *Journal of Palliative Care* 6: 7-11.

Verhoef, M., and G. van der Wal. 1997. Euthanasia in family practice in the Netherlands. Toward a better understanding. *Canadian Family Physician* 43: 231-7.

Verhoef, M., and T.D. Kinsella. 1993. Alberta euthanasia survey: 2. Physicians' opinions about the acceptance of active euthanasia as a medical act and the reporting of such practice. *Canadian Medical Association Journal* 148: 1929-33.

————. 1996. Alberta Euthanasia survey. 3 year follow up. *Canadian Medical Association Journal* 155: 885-90.

Vidalis, A., T. Dardavessis, and G. Kaprinis. 1998. Euthanasia in Greece: moral and ethical dilemmas. *Aging* 10: 93-101.

Wachter, R.M., J.M. Luce, N. Hearst, and B. Lo. 1989. Decisions about resuscitation: inequities among patients with different diseases but similar prognoses. *Annals of Internal Medicine* 1: 525-32.

Ward, B., and P.A. Tate. 1994. Attitudes among NHS doctors to requests for euthanasia. *BMJ* 308: 1332-4.

CHAPTER NINE

NAOKI WATANABE, MANABU TAGUCHI &
KAZUO HASEGAWA

SUICIDE

Implications for an Aging Society

Japan is a country with a population of 125,197,000 and is situated in Far East Asia. The land is small for its occupants at 377,727sq/km per person, compared to the United States at 9,809,386sq/km per person and a population of 263,034,000 (Time Atlas 1999). After World War II, Japan experienced rapid technical and economic growth that led to abrupt changes in lifestyle, prolongation of life, and shifts in life values.

Figure 1 shows the annual change of suicide rates by sex and age groups. For both males and females, young adult suicide rates have decreased from 1950 to 1997. Just after World War II there was a general feeling of disappointment with the defeat of the war and disorientation of young adults (who are now elderly) with their life goals. Also, it was a time of epidemic expansion of tuberculosis and low expectations from treatment. The average life expectancy in 1949 was 56.2 for males and 59.8 for females. (The proportion of elderly in 1950 was 4.9%.)

In spite of the fact that there are many suicides committed today by young students with the identified problems of bullying and violence in schools, the suicide rates of young adults is decreasing. The recent increase in the suicide rate of middle-aged males could be associated with the recent economic depression since 1989. The total number of suicides rose to 23,494 with the suicide rate at 18.8 per 100,000 persons in 1998. As for the elderly, suicide rates (Figure 1) declined from 1950 to 1997. Nevertheless, suicide rates of the elderly are much higher than that of any other age group. Compared with other countries, suicide rates in Japan increase as people age for both males and females.

Rapid growth is occurring in the proportion of elderly in Japan. In the 50 years following World War II, the average life expectancy has increased about 30 years from the age of 50 to 80. Such a markedly higher average life expectancy has brought about several problems for the elderly, such as dementia, an increase in diseases, burden to families, and medical costs. Until the year 2050, it is estimated

D.N. Weisstub, D.C. Thomasma, S. Gauthier & G.F. Tomossy (eds.), Aging: Decisions at the End of Life,
155-169.
© *2001 Kluwer Academic Publishers. Printed in the Netherlands.*

that the proportion of elderly will rise to over 30% of the total population (Table 1). In 1998, the average life expectancy for males was 77.16 and 84.01 for females; furthermore, Japan still maintains the highest longevity rate in the world. While the life expectancy for females still continues to climb, a slight decline in that of males has occurred (down from 78).

This research focuses upon regional differences in suicide rates and upon deepening the knowledge for such differences in suicide rates of the elderly. Interview protocols were used to gain information.

Figure 1. Annual Change of Suicide Rates by Sex and Age in Japan.

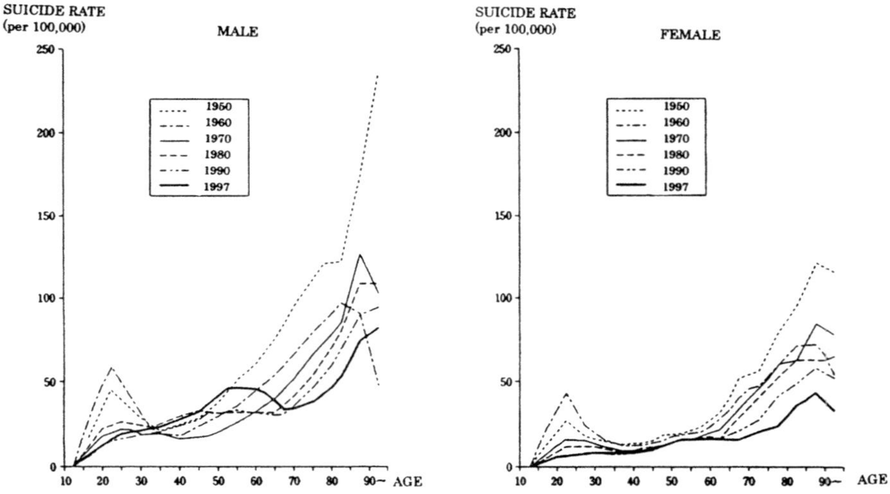

Source: Ministry of Health and Welfare "Suicide Statistics" 1999

Table 1. Estimated Future Population and Elderly Proportion in Japan

Year	Population (1000)	65+ (1000)	Proportion (%)
2000	126,892	21,870	17.2
2010	127,623	28,126	22.0
2020	124,133	33,335	26.9
2025	120,913	33,116	27.4
2050	100,496	32,454	32.3

Source: National Institute for Social Security and Population problems (January, 1997).

As seen in Figure 2, Japan is a country with high suicide rates. When comparing national suicide rates, Japan falls in the mid-spectrum akin to Germany (Takahashi et al. 1995). However as for the elderly, Figure 2 shows that Japan is one of the countries with a higher than average elderly suicide rate for both males and females. Japanese males have similar trends to the U.S.A. or Sweden wherein suicide increases with age. In the U.S.A., women across age groups have much lower suicide rates than men. In Japan, in general, males have slightly higher suicide rates than women in all age groups. However, there is little data from which to conclude reasons for such differences.

Figure 2. International Comparison of Suicide Rates

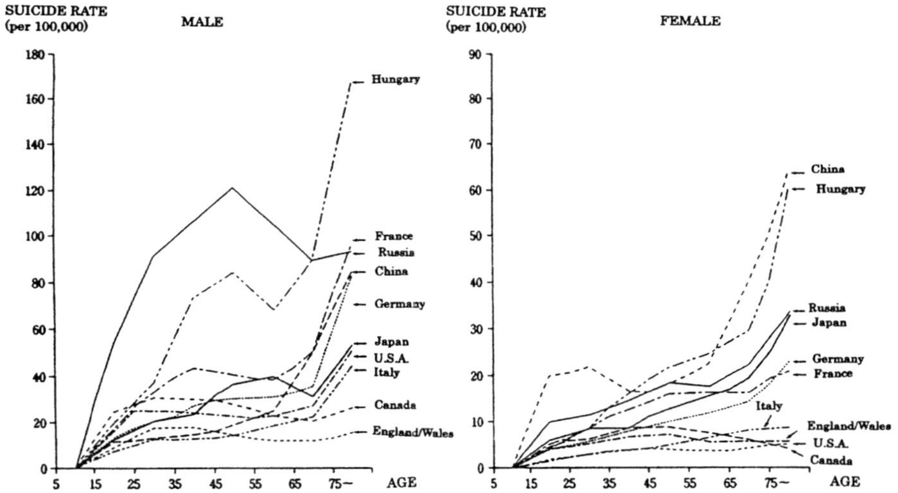

Sources: Ministry of Health and Welfare "Suicide Statistics" 1999

WHO, World Health Statistics Annual 1994,1995,1996

In order to understand suicide patterns, this research focuses upon regional differences in Japan. Two prefectures in Japan, Akita and Niigata, annually place first and second among other prefectures in regard to suicide rates of the elderly (65+). In contrast, Okinawa Prefecture has the lowest elderly suicide rate. Akita Prefecture is situated in the northern part of Japan with a population of 1,204,000 in 1997. This area is famous for rice production and as an agricultural area, and in winter is snow-laden. Okinawa Prefecture is a southern island with a population of 1,285,000 in 1997. It is warm all-year round there. Akita Prefecture has a suicide rate four times higher among the elderly than Okinawa Prefecture. Table 2 shows the suicide rates of the elderly for the year 1995 in Akita Prefecture and Okinawa

Prefecture. In Akita Prefecture, the rate was 66.62 suicides per 100,000 persons whereas Okinawa had 20.72 per 100,000 persons. An interesting finding is that elderly women in Okinawa Prefecture commit suicide less often than elderly men. In Akita Prefecture, suicide rates for males are 79.04 and those for females are 57.86 per 100,000 persons, while in Okinawa Prefecture suicide rates for males are 42.42 and those for females 6.84 per 100,000 persons. Besides, Akita Prefecture is also well known for the highest prevalence of cerebro-vascular disease (181.6 per 100,000 persons in 1997) and also of vascular dementia.

Table 2. Comparison of Elderly Suicide Rates (65+, per 100,000 Population, 1995)

	Akita Prefecture	Okinawa Prefecture	Japan as Total
Male	79.04	42.42	38.2
Female	57.86	6.84	25.9
Total	66.22	20.72	30.9

Our intent is to investigate why regional differences concerning suicide exist. Our aim is to find contributing factors of suicide and suicide prevention in each prefecture. The previous suicide research targeted suicide completers and their families. However based on the assumptions of John McIntosh (McIntosh 1997) we changed our focus and tried to determine how the normal elderly would spend and enjoy their life in districts where they have different suicide rates. From this viewpoint we chose two districts, namely Akita Prefecture with a high elderly suicide rate and Okinawa Prefecture with a low elderly suicide rate, and investigated what types of differences might exist between the two districts.

SUBJECTS AND METHODS

The subjects, as shown in Table 3 (p. 159), are 110 elder persons (65+, 29 males and 81 females, with the mean age of 70.7) from Yuri Town in Akita Prefecture and 105 elder persons (65+, 26 males and 79 females with the mean age of 74.0) from Motobu Town in Okinawa Prefecture.

We visited Yuri Town, an agricultural district with a population of about 6,000 inhabitants with approximately 20% elderly in July 1997. The population has shown a gradual annual decrease as the young productive population moves into big cities to find better jobs with higher incomes. The town is situated in the southern part of Akita Prefecture. In the summer of 1997, a primary research team of four stayed in Yuri Town for four days and visited eight villages with the assistance of four official health care professionals from the prefecture. Prior to the team's visit, the public health officials called for elderly volunteers to attend interviews at the public halls of each village. For the older persons who had difficulty filling out the questionnaire, public nurses and our research staff assisted them. We kept in mind that the research is not only for the research itself but also for the process in which the community members would become aware of health promotion. The same procedure was followed in Motobu Town in Okinawa Prefecture in September 1998. Using the same questionnaire and research protocol in Motobu Town with 150,000 inhabitants, six community centers were visited and the elderly were asked to voluntarily fill out the questionnaire. In Motobu Town, the suicide rate of the elderly has not yet been reported. Comparisons were made between answers from the elderly of Yuri Town in Akita Prefecture and those of Motobu Town in Okinawa Prefecture. Researchers attempted to identify recurrent themes related to choices in lifestyle among the non-suicidal. Furthermore, researchers observed the religious practices of elderly women who worshiped a god-representative; it was hypothesized that this worldview may contribute to the living status of the elderly women that in turn could prevent suicide. It should be noted that during the course of this research, the public health officials restricted use of the term "suicide" by researchers for fear this might stimulate suicidal tendencies.

Table 3. Subjects (Mean age ±S.D.)

	Yuri-town		Motobu-town	
	Male	*Female*	*Male*	*Female*
Number	29 (69. ± 5.7)	81 (73. ± 5.7)	26 (73.4 ± 6.2)	79 (76.1 ± 5.5)
Total	110(70.7 ± 5.9)		105(74.0 ± 6.1)	

The questionnaire items were: 1) personal variables (age, sex, academic career, marital status, etc.); 2) family variables (number of generations and their members, etc.); 3) health variables (use of alcohol, tobacco, illnesses, and the use of medical services); 4) geriatric depression scale (short version). The full score is 15 points (Sheik et al. 1986); 5) coping inventory for stressful situations (CISS). The characteristic coping behaviors of the individuals might be classified as a) task-oriented coping, b) emotion-oriented coping, and c) avoidance-oriented coping. The full score for each is 40 points (Ender et al. 1990); 6) QOL (Quality of Life)-Assessment Scale mixed with PGC (Philadelphia Geriatric Center) Morale Scale and Life-Satisfaction Index (LSI-A). The scale is divided into three categories: a) present life satisfaction, b) psychological stability and c) energy for living. Each category has a full score of eight points (Ishihara et al. 1992); 7) social support inquiry: from spouses, other family members living together, children or relatives outside the family and friends or neighbors. It is divided into three categories, namely a) positive support, b) physical support, and c) negative support; 8) suicide ideation questionnaire: Beck's Suicide Intent scale was shortened into the following five questions (Beck et al. 1979), a) within one month did you have an intention to die? b) within one month, did you want to kill yourself? c) for how long did you wish to kill yourself last time? d) have you ever written a suicide note? and e) do you ever think of suicide and its method now? 9) loss experiences within one year: bereavement and separation from spouse, brothers and sisters, children, friends, work, economic loss, loss of social role, and bereavement due to pets. The answers of the elderly from both districts were compared using the chi square test or unpaired t-test.

RESULTS

Regional Characteristics

As seen in the Table 4, the older person living alone was significantly higher in Motobu Town than in Yuri Town. The person who lived with six people in Yuri Town (5.0±1.9) was higher than in Motobu Town, while the person who lived with a spouse was higher in Motobu Town (3.1 ±1.9) than in Yuri Town. In Yuri Town more three-generation families existed and in Motobu Town living with a spouse was most common.

Concerning occupation, as shown in Table 4, in Yuri Town most elderly were engaged in some farm work (77.1%), whereas in Motobu Town 39.6% of the elderly worked regularly.

We found no significant differences concerning hobbies or activities. However in Yuri Town it was characteristic that many elderly answered that farming was their hobby.

Health

In regard to physical illnesses, as shown in Table 4, the number of the elderly concerned with heart disease (16 persons 19.1% > 8 persons 7.8%) and gastrointestinal disease (16 persons 19.1% > 1 person 1.0%) was significantly higher in Yuri Town than in Motobu Town. Nevertheless, most elderly in both towns were under treatment and taking medication. As for the use of alcohol, the elderly in Yuri Town drank significantly more than those in Motobu Town.

Geriatric Depression Scale (GDS)

The GDS score for older persons was 4.0±2.7 in Yuri Town and 3.9±2.3 in Motobu Town with no significant differences noted.

Table 4. Regional Characteristics (n(%) ***p<.001).

	Yuri-town	Motobu-town
Spouses		
Healthy	86(78.2)	70(70.0)
Divorce	1(0.9)	0(0.0)
Bereavement	23(20.9)	29(29.0)
Living alone ***	2(1.8)	16(15.4)
Working situations ***		
Working	84(77.1)	40(39.6)
Not working	25(22.9)	58(57.4)
Hobby activities		
Yes	77(71.3)	78(76.5)
No	31(28.7)	24(23.5)
Illnesses		
Yes	84(76.4)	80(77.7)
No	26(23.6)	23(22.3)
Under treatment		
Yes	82(74.6)	69(4.9)
No	28(25.5)	34(33.0)
Medication		
Yes	82(74.6)	69(67.0)
No	28(25.5)	34(33.0)
Smoking		
Yes	8(7.4)	11(10.9)
No	100(92.6)	90(89.1)
Drinking situations		
Every day	31(28.4)	5(4.9)
Some times	6(5.5)	10(9.7)
none	72(66.1)	88(85.4)
Psychiatric history		
None	103(95.4)	101(97.1)
Previous visit	5(4.6)	3(2.9)
In one year did you have following experiences?		
Bereavement or separation		
Spouses	3(2.8)	2(2.1)
Brothers and sisters	13(12.0)	10(4.9)
Children	5(4.6)	2(2.1)
Friends	18(16.7)	18(18.8)
Pets	4(3.7)	5(5.2)
Retirement	5(4.6)	5(5.3)
Economical loss	4(3.7)	5(5.2)
Loss of social role	2(1.9)	4(4.2)
Within one month did you have a wish to die?		
Sometimes yes, Yes	8(7.4)	3(3.0)
Within one month did you want to kill yourself?		
Sometimes yes, Yes	2(1.9)	3(2.9)

Subjective Quality of Life

As shown in Figure 3, the subscales of present satisfaction, psychological stability and energy for living and, also, the total score of the QOL scale showed no significant differences between the two towns.

Figure 3. Subjective Quality of Life.

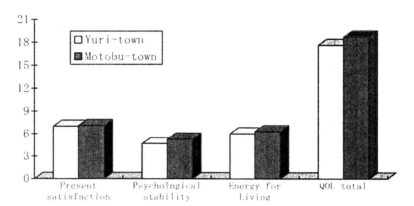

Coping Inventory in Stressful Situations

As we can see in Figure 4 (p. 163) no significant differences in task-oriented behavior and avoidance-oriented behavior was found. However the score of the emotion-oriented behavior was significantly higher in Yuri Town than in Motobu Town.

Social Support

A further comparison was made of social supports in Yuri Town and Motobu Town concerning friends and neighbors. In the SSI (Social Support Inquiry), the categories of positive and physical support were similar in both Yuri Town and Motobu Town, with the exception of one item, "people would be ready to lend money," which was noted in Motobu town. We found that there were significant differences in the items of negative support between these two groups. The items of negative support were 1) "bother me with over-engagement," 2) "complaining," 3) "make me angry and irritated," and 4) "make trouble for me." The scores of these items in Yuri Town were significantly higher than in Motobu Town. These results were also applicable for the family members who lived together or separately. As for spouses the scores of two items, namely "bother me with over-engagement" and "make trouble for me," were higher in Yuri Town than in Motobu Town.

Figure 4. Coping Inventory in Stressful Situations.

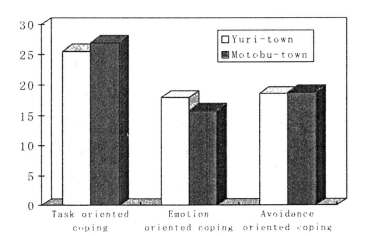

Loss Experiences and Suicidal Ideation

In Table 4 (p. 161), no significant differences between the two groups concerning loss experiences within one year such as bereavement, separation, retirement, economic loss and loss of social role were found. Furthermore, no significant differences existed in the two groups concerning answers to, " Did you want to kill yourself within one month?" and "Did you have an intention to die within one month?"

DISCUSSION

Physical Illnesses

More than 80% of the elderly in both areas had some illnesses, although this was a survey on the healthy elderly. We found no specific correlation between physical illnesses and suicide. The majority of the elderly was also under treatment. As long as the life value and the physical activities of the elderly matched each other, there might be no problem at all. However, once the physical function was limited, then the life value itself would be in danger. Accordingly, as long as the elderly could move by themselves they could hold their life value as "working hard in the field."

However, once their physical function was limited and the elderly were unable to go to work in the field, then their life value might experience a tremendous loss.

Differences in Family Composition and Life Style

There were differences in the family composition where the elderly belong in Yuri Town and Motobu Town. In Yuri Town, the area with a high suicide rate, the elderly were living with more than three generations. These elderly were dealing with farming and tried to instill their working skills to their sons and grandchildren. The women also accumulated their wisdom for life by working in the rice field, dealing with household chores and child rearing, and how to live during the hard winters with heavy snowfall. The elderly were respected by the younger people also because they could teach them the skills of farming and the wisdom of life. However, after World War II the international economy was influenced by technological and industrial change and agricultural politics to reduce the amount of farmland, and the younger generation started to move from farmland to the cities where they became the working power of the industries. The elderly in Yuri Town might feel that to wake up early in the morning and to farm is a specific life value. The greetings in the community were "Are you still working?" Many elderly say to each other, "Are you working?" Farming is also seen as their hobby. The confirmation of the elderly that they could still work might be the value of their existence, which they could show to other family members and their community. In many families with more than three generations in Yuri Town, the accepted life style is that the younger generation goes into the city to work at a company and the elderly go into the field to farm. The elderly no longer need to teach the younger generation about farming or life. In addition, the younger generation earns much more money than they could do farming. Based upon this economic reality, there is a horizontal information gap between the two generations even if they are living together. The elderly seem to feel lonely even if they live in more than three-generation families. Once the older person, who believes that farming is his hobby and simultaneously his life value, gets ill and losses the ability to move, then it could easily lead to the loss of life value which is a critical situation.

On the other hand, the elderly in Motobu Town live mostly with their spouses. The weather is warm and the structure of the houses is built to openly face the outside so that anyone can easily see inside the house and talk to the inhabitants. There is a strong community consciousness and the elderly in Motobu Town did not feel lonely even if they lived alone because they meet each other very often and know each other well. Their personality is generally open-minded so that they tend to dance at any time and at any moment.

QOL and Depressive Feelings

No significant differences concerning QOL and depressive feelings in the two groups of healthy samples were found. Before beginning the survey, our hypothesis was that "the elderly who live in the area with higher suicide rates might have a lower QOL than those in the area with lower suicide rates." This hypothesis could not be proven however. A research team from the Department of Psychiatry of Niigata University researched suicide in Higashikubiki County in Niigata Prefecture and began a campaign of psychological education for the elderly and the inhabitants to inform them about depression, as early detection and early treatment could reduce suicide. The study group showed a statistical decrease in the number of elderly suicides since the ten years from its inception. Our survey, however, targeted the normal elderly and we could find no significant data that depression is stronger in Yuri Town than in Motobu Town. We need to specifically survey for the prevalence rate of depression and its correlation with suicide. In the United States, it is estimated that half to two thirds of completed suicides might have been in a state of depression before committing suicide. However we could find no evidence that the differences of the suicide rates in the two towns might correlate with the depressive feelings of the inhabitants. We suppose that other mediating factors besides depression might play a role in the process leading to suicide.

Differences in Social Support

Interesting results emerged from the Social Support Inquiry. Compared to Yuri Town, the older adults in Motobu Town showed significantly lower scores of negative support. The elderly in Yuri Town were very sensitive concerning negative support from their spouse, other family members, and family members living outside of the immediate family, and friends and neighbors. In Motobu Town there is an organization of elderly called "Yuimahru" which is a traditional self-help organization that may have contributed to differences in expectations from outsiders. A major difference was found in the category of negative support. In Motobu Town older adults showed significantly fewer answers for negative support questions. Whereas in Yuri Town, most of the elderly admitted that they were overly involved and bothered by others including spouses, family members living together, family members living outside the immediate family, and friends and neighbors.

This research began as an inquiry to better understand resilience and risk factors in the elderly in Japan regarding death by suicide. The study was developed following the observation that there are strong differences in suicide rates between rural and urban areas in Japan (Watanabe 1995). This research focused upon factors that contributed to an elderly person's decision to commit suicide despite their older age (the highest suicide age in Niigata Prefecture was reported at 103 years of age). There was a curiosity about the fact that those who chose death by suicide seemed to have had some type of social relationship – such as many victims had lived with three generations, resided in beautiful surroundings, lived together with their spouses, had daily contact with other family members, and maintained social roles in the community.

Results from our interviews and literature review indicated that the elderly who committed suicide were especially sensitive about their physical conditions; once they developed an illness, they interpreted it as a great loss experience for them and a burden to others. The older people in this district were regarded by community members as honest and hard working and were highly respected. However, once they experienced some stressful events, such as loss experiences and abandonment experiences, their suicide risk increased and they were reported by family members to have shown a state of lower quality of life and behaviors that demonstrated emotional coping. Results from this study may be useful for establishing an effective intervention program for those at risk for suicide.

Findings showed that there were significant differences between the urban and rural settings regarding age, number of family members, and emotion-oriented coping behavior. Also, there were differences in work and use of alcohol habits. Intensity of depressed feelings was correlated only with hobby activities. There were no significant differences in subgroups of the GDS-scores between the Akita Group and Okinawa Group.

Results from this study do not support the prevailing research regarding suicide in the elderly. For instance, Tamura (1979) suggests that elderly at risk from the region with highest suicide rates have extended family support and have a tendency to drink alcohol every day; their hobby activities are minimal and they set higher value on their work as farmers. (In our study, depressed feelings in the elderly in both groups (as reported by cohorts) were not correlated with suicidal ideation and suicidal thought. Most interestingly, other studies suggest that illness could not be declared as one of the leading factors for suicide. Healthy elderly also had a higher percentage of illnesses and there was no correlation of illnesses with the intensity of depressed feelings (Dorpat et.al. 1968; Whitlock 1986; Conwell et.al. 1991). However, in our study suicide was inversely related to hobby activities and subsequent intensity of depressed feelings as reported by survivors. For instance, the subgroup with depressed feelings had fewer hobbies. This subgroup was dominant in Yuri Town. Therefore, the encouragement to develop hobbies for the elderly might be one of the ways to reduce suicide.

In this study the comparison of the data between Akita and Okinawa made clear that the elderly in Yuri Town in Akita Prefecture are very sensitive to their surroundings and social environment. They are mostly afraid to "lose face" toward others which was named as "masochism" by Nakakuki (1994). They want to be regarded by others as "diligent and self-sacrificing." As long as the elderly devote themselves to their family and work hard, they feel respected and safe in their family and also in their community. However, given changes to their financial status, social role and identity as farmers – they are at risk due to loss factors (Tamura 1979). When comparing communities, the elderly in Akita Prefecture seemed more concerned not to lose face. As long as they were healthy and hard working they felt safe, secure, and accepted in the community. However, once their activities were disrupted due to severe illnesses, they tended to withdraw from the community and suffer symptoms of depression which increased their suicidal thoughts and behaviors. Most of the elderly seemed to be afraid to lose their mental and physical abilities due to severe illness; they made frequent visits to their medical centers. This could be a strong indication of anxiety in this population (perhaps as much as depression). The tragic story called "Narayama Bushikou" is a masterpiece of Shichiro Fukazawa that is based upon a legend of abandoning the elderly in the mountains. Similar legends were abundant in many areas of Japan when the country was poor. Orin, the heroine, was 69 years old. She ordered her son to carry her on his back to the mountain "Narayama" and abandon her as a way to decrease the burden on the family so that other family members could have more to eat. Orin was a very hard working old woman and it was astounding that she declared to end her life by herself, although she was loved by her son, his spouse, and children. Such self-sacrificial thoughts and behavior hit the mind of many Japanese people and it was regarded with empathy as a tragic however irresistible and understandable deed (Ohi 1995).

In Okinawa prefecture the female elderly were especially stable and seemed to feel secure and safe in their community, even though they lived alone or together with their spouses. The female elderly carry an important role in their community and participate in religious ceremonies by taking the role as a mediator between God and the people. They are respected and also able to provide leadership in various ceremonies throughout the cycles of the year. They have a strong emotional tie with other community members. As it is relatively warm throughout the year, their houses are also open to the outside community. Everybody can look in each others' houses and start talking with each other from outside and from inside the houses. There is no strict barrier between individuality and community.

Perhaps the so-called "masochistic thinking" by the elderly in Akita Prefecture is a rather traditional way of thinking in Japan. It seems to be found less in Okinawa where they have nourished their own culture as an island. In big cities other values for development and meaningful activities are evident. The elderly tend to live separately from their children. However they have a lot of opportunity to meet them, go to meet other friends, and enjoy theaters, and so on. There is a good transportation system that allows them to keep good communication despite the fact that they are living alone.

CONCLUSION

1. We could extract characteristic features in the region with the highest suicide rate where older people are living with many family members and have a tendency to drink alcohol every day. Their hobby activities scarcely exist and they set higher value on their work as a farmer (Tamura 1979).
2. Depressed feelings in the elderly in both groups were not correlated with suicidal ideation and suicidal thought.
3. Illness could not be declared as one of the leading factors for suicide. The healthy elderly also had a higher percentage of illnesses and there was no correlation of illnesses with the intensity of depressed feelings (Dorpat et.al. 1968; Whitlock 1986; Conwell et.al. 1991).

4. We did find a correlation between hobby activities and the intensity of depressed feelings. The higher subgroup of depressed feelings had less hobby activities. This subgroup was dominant in Yuri Town. To have a hobby and enjoy it might be one of the ways to prevent suicide.

5. The comparison of the data between Akita and Okinawa made clear that the elderly in Yuri Town in Akita Prefecture are very sensitive to the people in their community much more so than those in Motobu Town in Okinawa Prefecture. Personality characteristics could also play a role for their viewpoint of living, which is to be investigated further.

Naoki Watanabe, Associate Professor, School of Medicine, St. Marianna University Tokyo, Japan.

Manabu Taguchi, Research Scientist, School of Medicine, St. Marianna University Tokyo, Japan.

Kazuo Hasegawa, Emeritus Professor, School of Medicine, St. Marianna University Tokyo, Japan.

REFERENCES

Beck, A.T., M. Kovacs, and A. Weissman. 1979. Assessment of suicidal intention. The scale of suicidal ideation. *Journal of Counseling and Clinical Psychology* 47: 343-52.

Conwell, Y., K. Olsen, E.D. Caine, and C.J. Flanney. 1991. Suicide in later life: Psychological autopsy findings. *International Psychogeriatrics* 3: 59-66.

Dorpat, T.L., W.F. Anderson, and H.S. Ripley. 1968. The relationship of physical illness to suicide. In *Suicidal behaviors: Diagnosis and management*, ed. H.P.L. Resnik, 209-19. Boston: Little Brown

Endler, N.S., and J.D.A. Parker. 1990. Multidimensional assessment of coping. A critical evaluation. *Journal of Personality and Social Psychology* 58(5),: 844-54.

Ishihara, O., K. Naitou, and K. Nagashima. 1992. Trial for making the QOL evaluation scale based on psychological traits based on a subjective scale. *Social Science on Elderly (Pohnen Shakai Kagaku)* 14: 43-51. (Japanese)

Journal of Health and Welfare Statistics. 2000. Tokyo: Health and Welfare Statistics.

McIntosh, J., M. Eikens, L. Talcott, B. Quick, and R.W. Hubbard. 1997. Reasons for living among healthy, community-dwelling elders. Paper presented at the 30th Annual Conference of the American Association of Suicidology, Memphis Tennessee.

Nakakuki, M. 1994. Normal and developmental aspects of masochism: Transcultural and clinical implications. *Psychiatry* 57: 244-57.

Ohi, G. 1995. Ethical orientations and dignified death. *Psychiatry and Clinical Neuroscience* 49(Suppl.1): S155-9.

Sheik, J.I., and J.A. Yesavage. 1986. Geriatric Depression Scale(GDS). Recent evidence and development of a shorter version. *Clinical Gerontologist* 5(1/2): 165-73.

Suicide Statistics. 1999. Tokyo: Ministry of Health and Welfare.

Takahashi,Y. 1998. *Senile depression.* Tokyo: Nihon Hyouronsha. (Japanese)

Takahashi, Y., H. Hirasawa,K. Koyama, O. Asakawa, M. Kido, H. Onose, M. Udagawa, Y. Ishikawa, and M. Uno. 1995. Suicide and aging in Japan: An examination of treated elderly suicide attempters. *International Psychogeriatrics* 7(2): 239-52.

Tamura,K. 1979. Life value and death of the Japanese – From case study on the elderly suicide in the area of population decrease in rural area. In *Suicide prevention and death and life concept*, Study Group on Suicide Prevention and Life Saving in Japan, 159-64, Tokyo: Seiwa Shoten. (Japanese)

Time Atlas. 1999. *Atlas of the World (Concise Edition)*. New York: Time.

Watanabe, N., K. Hasegawa, and Y. Yoshinaga. 1995. Suicide in later life in Japan. *International Psychogeriatrics* 7(2): 253-61.

Whitlock, F.A. 1986. Suicide and physical illness. In *Suicide,* ed. A. Roy, 151-70. Baltimore: Williams & Wilkins.

CHAPTER TEN

ARTHUR SCHAFER

RESEARCH ON ELDERLY SUBJECTS

Striking the Right Balance

THE CENTRAL DILEMMA

Many chronic and disabling diseases disproportionately or exclusively affect the elderly. Alzheimer's dementia would be an obvious illustrative example. Unless medical scientists are able to conduct research upon elderly human subjects, prospects for the development of effective therapies will be poor. In the search for reliable diagnostic tests and for effective therapies, animal models are helpful but, ultimately, not adequate (Schafer 1981). Scientific progress in the diagnosis and treatment of such disease requires that procedures and drugs be tested on those who have the disease or who are at high risk of developing it.

Unfortunately, the very factors that make it imperative to conduct research upon the elderly also pose serious ethical obstacles to recruiting them as research subjects. Stereotypes of the elderly as necessarily possessing diminished mental capacity ought to be rejected, of course. There is as much heterogeneity among the population of elderly people, as there is among the population of young or middle-aged people. Thus, for purposes of recruiting volunteers to become research subjects there is no reason automatically to discriminate between the old and others. It would, nevertheless, be naive to deny that, among the population of those with special vulnerabilities, one would expect to find that old people are present in disproportionate numbers. Old people who are, for example, disabled by stroke or ill with Parkinson's disease tend to be physically and mentally vulnerable. They are not just physically frail; they are, often, emotionally vulnerable and/or intellectually impaired. Frequently they are institutionalized, which circumstance poses its own special ethical challenge.

Thus, research on the elderly is both ethically required and ethically suspect. Scientific progress toward the effective diagnosis and treatment of such diseases as senile dementia, which many would regard as an ethical imperative, requires

D.N. Weisstub, D.C. Thomasma, S. Gauthier & G.F. Tomossy (eds.), Aging: Decisions at the End of Life, 171-205.
© 2001 Kluwer Academic Publishers. Printed in the Netherlands.

experimentation that uses demented subjects. But it is a fundamental requirement of research ethics that subjects should only be recruited from that class of volunteers who are competent to give valid consent. Almost by definition, those who are demented are not competent to give such consent – a genuine ethical dilemma.

THE CASE AGAINST USING THE VULNERABLE ELDERLY AS RESEARCH SUBJECTS

The *locus classicus* for the case against experimentation on frail patients would have to be the highly influential and much-cited article by Hans Jonas (1969): "Philosophical reflections on experimenting with human subjects."

In this article, Jonas focuses attention upon the vulnerability of very sick patients to exploitation, especially when those patients are dying or suffering from serious chronic illnesses. He does not specifically single out the frail elderly, but they would constitute a paradigm case of the kind of patient whose vulnerability to exploitation concerns him. "The afflicted should not be called upon to bear additional burden and risk [involved in clinical trials], ...they are society's special trust and the physician's particular trust – these are elementary responses of our moral sense" (Jonas 1969, 238).

Jonas is a leading advocate of the principle that averting a disastrous outcome always carries greater moral weight than promoting a good one (ibid., 229). It follows from this principle that one ought always to assign a higher priority to the value of minimizing the risk of harm to research subjects than one assigns to the value of promoting their best interests or promoting the best interests of others. Jonas recognizes, of course, that without medical experimentation on human subjects the prospects for advancing scientific knowledge of disease will be seriously hampered, if not utterly stifled. Nor would he deny that the burden of individual suffering and disease would increase if such scientific progress does not occur, or occurs at a much reduced pace. It seems clear that for a community to eschew the benefits of modern medical research, whenever such research poses risks or hazards to research subjects, would impose a heavy burden of avoidable disease on its members.

Consider the implications for any society that would consider adopting a rigorous Jonas-type policy of prohibiting all risky experimentation on research subjects. Virtually every clinical trial exposes its research subjects to some risk. Even a seemingly innocuous experiment to determine the efficacy of mouth rinse in preventing tooth decay for seniors, such as the current T.E.E.T.H. multi-centered trial,[1] exposes subjects to the risk that the rinse may, for a few people, irritate the lining of the mouth. Moreover, this clinical trial requires that each subject undergo a panoramic X-ray both at the beginning and at the end of the trial, which exposes every volunteer to a low dose of radiation. There may be no known adverse side effects from such a low dose of radiation, but the potential for harm cannot be said to be non-existent. What is true of a comparatively innocuous mouth wash experiment on elderly volunteers will, *a fortiori*, be true for trials involving powerful new drugs or innovative surgical techniques. The point to note here is that there will be very few experiments on human subjects about which it can be said confidently: "This experiment is entirely without risk of harm to any of its human research subjects."

Those who adopt a Jonas-type ethical stance to medical experimentation would, presumably, decry even such an apparently benign clinical trial as T.E.E.T.H. on the grounds that it is unethical to invite any person, and especially any vulnerable person, to volunteer for any experiment that involves any degree of risk of any level of harm. Moreover, for those clinical trials in which the likelihood of harm is high and from which the harm itself, should it occur, would be grave, even the "superlative good" of public health would not, according to Jonas, provide adequate justification for carrying out vitally important research (Jonas 1969, 226-7). By contrast, proponents of experimentation upon human subjects will argue that the prevention of disease or its cure for victims, actual or potential, will sometimes provide a sufficient warrant for carrying out research involving risks to human subjects. The value of the end that is to be promoted, relief of suffering, say, will sometimes justify the risk of harm to research subjects entailed by the means. Against such consequentialist reasoning, Jonas insists that the essentially melioristic goals of medical research must be subordinated to the sacrosanctity of the individual research subject. On Jonas' view, even when the potential benefits from research would be significant, would accrue to a large number of present and future sufferers, and would far outweigh the likely harms, the principle of absolute protection for the "dignity" of the individual remains inviolable.

The ethical position taken by Hans Jonas is not, however, quite as rigorously absolutist as it may at first sight appear to be. Even the ethically abstemious Jonas accepts that when a community is faced with the sort of "clear and present danger," which is posed by a raging epidemic, it may be morally legitimate to seek and accept the submission of human subjects to research which is hazardous. Thus, given the urgency and vast social dimensions of the current AIDS pandemic, a randomized clinical trial intended to test an AIDS vaccine might offer the kind of "transcendent social sanction" which Jonas requires in order to justify human experimentation. Research on patients suffering from senile dementia, however, would likely not meet even Jonas' somewhat modified but still strict ethical standards if the research posed any risks whatsoever to cognitively impaired patients.

THE CASE AGAINST HANS JONAS – THE RIGHT TO BE TREATED AS A PERSON

There is an inescapable aura of paternalism about the approach to experimentation advocated by Jonas. Individuals must be protected against the hazards of medical experimentation, he insists, even when they voluntarily consent to participate as research subjects, in full knowledge of whatever hazards (and benefits) may flow from their decision. Thus, Jonas might be seen as targeting elderly potential research subjects for special protection on dubiously legitimate paternalistic grounds simply because they are more likely to be cognitively impaired (Halper 1980).

Against this kind of paternalistic approach, Benjamin Freedman has argued that "there does seem to exist a positive right of informed consent, which exists in both therapeutic and experimental settings" (Freedman 1975). In other words, to deny to a competent individual, of whatever age, the opportunity to give (or withhold) informed consent to participation in a clinical trial is to undermine that person's right to be treated as an autonomous person.

When old people are capable of giving valid consent to becoming research subjects, (that is, consent which is both informed and voluntary), to deny them the right to make the decision to participate in medical experimentation is to attack their dignity as persons. It is to treat them as objects rather than as rational agents. When well-meaning health care professionals or family members refuse permission for an elderly person to become a research subject, they typically believe that they are acting benevolently. Indeed, their refusal even to consult grandma or grandpa about becoming a research subject derives, most often, from their conviction that they are acting to promote the old person's long-term best interests. The danger here is that well-meaning family may be guilty of underestimating the older person's capacities "to act with as much enlightened self-interest as any other adult citizen in deciding whether a particular research project is something which is desirable for him or her" (Leader and Neuwirth 1978).

Benevolent caregivers will often be correct in their calculations, on behalf of older patients, of the comparative benefits and harms likely to arise from participation. Nevertheless, the usurpation of the old person's right to make this decision for herself will often be experienced as a diminution of autonomy. This is a point of some moral importance. To act in such a way that one reduces further the already questionable autonomy of an elderly person carries its own considerable risks of harm (Goldstein 1978). It has, perhaps, been insufficiently noticed that old persons may suffer more inadvertent harm from the paternalistic interventions of those who care for them than they suffer deliberate harm from those with malice in their hearts (Schafer 1988).

COMPETENCE AND THE ELDERLY SUBJECT

When competent older patients have their competence unrecognized or inappropriately denied they may be harmed in several important ways. The most important of these potential harms is, perhaps, the undermining of their dignity as autonomous persons when they are treated "as children." Of scarcely less importance is the harm which arises from losing what may, for them, count as a life-enhancing opportunity to exercise their generous impulses – to behave as an altruistic member of the community in a project which promises important benefits to others, as well as to oneself. It must be admitted, of course, that the decision to allow an elderly person of limited competence to participate, altruistically, as a volunteer subject for a research project is morally troubling. When there is a high risk of grave harm, when the research is not intended to be even partly therapeutic for the volunteer, and when the competence of the elderly person is seriously compromised, few would defend the "right" of the elderly to participate. By contrast, when the risk is low and the harm trivial, when the research is intended to be at least partly therapeutic for the volunteer, and when there are periods of clear-minded lucidity during which the elderly person expresses a wish to volunteer, then few would raise strong moral objections to participation as a research subject. Sharpest disagreement is likely to occur within the gray intermediate zone, when risk and stake are neither frightening high nor inconsequentially low, and when competence is difficult to ascertain or fluctuates wildly.

Even when an investigator fulfils her responsibility to disclose all materially relevant facts about the experiment to a potential research subject, and even when the subject gives his consent voluntarily, let us suppose with no element whatsoever of coercion, the consent given by the subject will not be valid consent if the subject lacks competence. When that is the case, then either the incompetent individual should be excluded from participation in the experiment or (an alternative to be discussed later) a surrogate should be identified who can legitimately make this decision on behalf of the incompetent patient.

The questions of how competency is to be defined, how it is to be measured, and how much of it is required for a patient to be considered eligible for recruitment to a clinical trial, are all of them difficult and complex. Attempts to provide adequate answers have proliferated and now constitute a substantial body of literature in their own right. Despite this large body of literature, or because of it, there is still much controversy about what constitutes the most satisfactory answer to each of the questions posed above. The issues may appear so purely theoretical as to be of interest exclusively to professional philosophers but, in reality, much of practical importance hinges on the answers that are accepted. If, for example, excessively stringent criteria are adopted in the process of ascertaining whether a potential research subject is competent, many patients who might benefit from participation will be excluded, to their disadvantage and to the disadvantage of society and future generations. Alternatively, if inappropriately lax standards of competency are put into place, then highly vulnerable patients may be exploited and harmed in ways that cannot be morally justified by appeal to their (the patients') valid consent. Errors in either direction expose an investigator to moral criticism.

So, how competent is competent enough? James Drane offers a promising approach in our quest for a defensible answer to this difficult question. In the course of his discussion of what constitutes patient competency to give informed consent Drane proposes that: "[A]s the consequences flowing from the patient's decisions become more serious, competency standards for valid consent or refusal of consent [ought to] become more stringent" (Drane 1985, 18). Drane does not direct his attention specifically towards the issue of the competency of elderly prospective research subjects to give consent to participation in a clinical trial. It seems clear, however, that when such research poses serious risks to the subject, especially when these risks are not counter-balanced by proportionately high expected benefits, he would insist upon the presence of a high level of competency – "an understanding that is both technical and personal, intellectual and emotional" (ibid., 20). Where the risk and stake for the research subject are low, advocates of variable competency, such as Drane, would accept a proportionately low level of comprehension from the prospective subject. Even patients/subjects who are mentally or emotionally disturbed, suffering from anxiety, pain, or reactive depression might qualify as competent to volunteer for research which threatens them with no more than minor inconvenience and which holds out hope of producing some benefits to them personally or major benefits to society.

Once it is recognized that competence (or "decisional-capacity," as some prefer to label it) is not a matter of all or nothing-at-all but is, rather, a sliding scale of capacities with many shades and degrees in between the poles at each end of the spectrum, one is also more likely to recognize that from the fact that an elderly person is incompetent to live independently, it should not be inferred automatically that she is incompetent to perform such other tasks as giving informed consent to participation in a clinical trial. A patient in the early stages of dementia, for example, might be judged incompetent to decide whether to participate in a therapeutic trial which presented high risks of grave harm, but this same patient might nevertheless be judged to possess the modest level of competence required to decide whether to participate in a trial that posed only slight risks of negligible harm. The former decision requires of a prospective research subject, who has a lot to lose, that she possess a clear understanding of the risks and benefits involved, the ability to weigh and balance these risks and benefits against the risks and benefits of the available alternatives, and a clearly thought out understanding of her own attitudes and values to these risks and benefits. A patient with senile dementia, even in the early stages of the disease process, is unlikely to possess this kind of decisional capacity. The latter decision, by contrast, requires an altogether more modest level of decision-making capacity, such that even a moderately demented person might be found competent to choose whether or not to participate.

In sum, with respect to the issue of "experimentation on elderly prospective subjects of less-than-full mental competence" it seems more appropriate, morally-speaking, to adopt a nuanced and qualified view rather than to adopt a position of either total rejection or total acceptance. It is too simple to ask the question "Is this patient competent to give valid consent to become a research subject?" The ethically appropriate question would be, rather, "Is this patient competent to give valid consent to become a research subject for this particular experiment, given its risks and benefits and given the patient's capacities relative to the complexities of the decision involved?" A patient in the early or middle stages of senile dementia might, for example, be judged competent to participate in a low-risk clinical trial but incompetent to decide to participate in a trial which posed high risks of serious harm. When investigators attempt to balance the competing concerns of (a) protecting subjects from the harms that could befall them as a result of their impaired decision-making capacity and (b) permitting prospective subjects autonomously to determine the shape of their own lives, the adoption of a variable standard of competence seems likely to contribute to the achievement of a morally defensible balance.

If an excessively restrictive standard of decisional competency were to be adopted and if, as a result, some frail elderly patients were wrongly denied the opportunity to become research volunteers, the mistake would not be a trivial one. For, as Freedman (1975, 33) forcefully reminds us, the physician-experimenter faces ethical risks not just from one direction but, simultaneously, from two opposing directions:

> When the consent [of the potential research subject] is of doubtful validity...the physician experiences a conflict between two duties. He will not be ethically well-protected by choosing not to experiment, for there exists the possibility...that he is violating a duty in so choosing.

This reminder is important because, to this day, many scientists and regulators of scientific research appear to believe that there is what might be called a "fail-safe" position, viz., the position that when consent is of doubtful validity one is morally obliged not to recruit the patient as a subject for research. In other words, the preponderant view among the scientific and ethical communities is that the burden of proof rests on those investigators who seek to accrue subjects with questionable decision-making capacity. By contrast, the gravamen of Freedman's argument is that there is no such burden of proof on the investigator, because researchers have an equally important duty to potential subjects not to deny them, wrongfully, the opportunity to participate in what may for them be a meaningful life project. If Freedman is correct, and I believe that he is, then those who follow a Jonas-like ultra-strict standard for the recruitment of patients as research subjects are vulnerable to serious moral criticisms, viz., the criticism that they are slighting the autonomy of patients and the additional criticism that they are neglecting the very real benefits which many patients derive from their participation in clinical trials.

In one of the earliest modern studies of the ethical issues raised by medical research on human subjects, Beecher (1970) cites a survey of prison inmates who unashamedly admit that altruism rather than money was their primary motivation for participating in a study involving malaria. I say "unashamedly" because in our cynical culture many people are reluctant to admit that they were moved to act by altruistic motives – perhaps for fear of being pejoratively labeled a "do-good-er" – perhaps for fear of being disbelieved by cynical listeners. Indeed, one would expect that prison inmates would be more reluctant than most to admit (except, perhaps, during a parole board hearing) to harboring generous impulses. This lends added credibility to Beecher's findings. Other studies (e.g., Annas, Glantz, and Katz 1977) confirm that in addition to altruism, prisoner volunteers for research participation are motivated by hope for a variety of benefits, including self-respect, curiosity, and relief from boredom. A nursing home differs in some important respects from a prison, but many of these same benefits would seem plausibly to accrue to residents who choose to become research subjects. One could argue, of course, that just because life in such institutions as prisons and nursing homes is so often bleak, boring, and depressing, their inmates are vulnerable to "illegitimate" inducements, for example, the "excitement" of participation in medical experimentation. Their consent could, in consequence, be viewed as coerced rather than voluntary. Since the effect of such an argument would be effectively to disallow any choices which either prisoners or the institutionalized elderly might make in an effort to improve their lives, it smacks of unacceptable paternalism.

The claim, here defended, that researchers have an equal obligation to investigate both the potential harms and the potential benefits of participation in any particular research trial, and to communicate their findings of harm and benefit to prospective research subjects, would not be universally accepted. Wicclair (1993, 168), for example, contends that "the ethical obligation to make a conscientious effort to uncover unknown potential harms and burdens to [potential] subjects" is weightier than "the similar obligation with respect to potential benefits to [potential] subjects." He offers two main arguments in support of this conclusion.

The first argument appeals to the best interests of the prospective research subject. Wicclair claims that, from the point of view of the research subject, to experience unexpected harms and burdens is worse than to miss the experience of correspondingly significant benefits. Those who are risk aversive will likely agree with this conclusion. But not every elderly person will be risk aversive, and if large numbers of the elderly would prefer to risk some degree of harm as a result of becoming research subjects, in return for gaining the opportunity for some degree of benefit, would it not be presumptuous of Wicclair or any other well-meaning protector of the elderly to foreclose that decision on their behalf?

Wicclair's (1993, 168) second argument is that "when there was insufficient disclosure of risks by investigators, subjects can claim to have been inappropriately used as a means to benefit others," whereas "a similar charge cannot be made when subjects fail to volunteer because there was inadequate disclosure of potential benefits by investigators." This point must, I think, be conceded. But does the charge of having "been inappropriately used as a means to benefit others" carry greater moral weight than the charge of having been denied the opportunity to benefit from participation as a research subject in this or that experiment? This is the key question, and Wicclair does not offer us any good reason to prefer his conclusion rather than its opposite.

There are, in addition, two ancillary arguments which Wicclair (1993, 169) offers in support of his contention that it is better to err on the side of wrongful exclusion rather than on the side of wrongful inclusion. He asserts that when elderly persons claim the right to refuse to serve as research subjects this rights-claim has a weight and legitimacy not enjoyed by the opposite claim, viz., that they have a positive right to participate in research studies of their choosing. Wicclair's contention seems plausible, since there would be near-universal agreement that no one has a positive right to participate in a study of her choosing, while equally strong agreement could be expected for the claim that everyone has a right not to participate.

If Wicclair's point is reformulated, however, it loses much of its plausibility. He is clearly correct to suggest that no one has an automatic right to participate in a study of her choosing. Resources for research are limited, after all, and only a small minority of those who might wish to be accommodated as subjects for certain popular studies – such as those studies which are testing some new anti-AIDS drugs, with a high expected benefit to participants – can realistically be enrolled. But if such a study is being undertaken, would there not be a strong moral consensus that every person whose particular circumstances fit the study's protocol has a right not to be illegitimately excluded from the opportunity to participate as a subject in the study? Would it not be a case of wrongful discrimination if someone were so excluded on discriminatory grounds of their race, their sex, or their presumed (but not demonstrated) incompetence? This is not to deny, of course, that race or sex or age may sometimes be legitimate criteria for inclusion or exclusion. But when such considerations are medically and morally irrelevant to the clinical trial, anyone who was denied the opportunity to become a research subject on untenable grounds could surely complain of wrongful discrimination.

Wicclair's (1993, 169) second ancillary argument is that "since expected benefits provide an important inducement to prospective subjects, it is to be expected that investigators will conscientiously identify and disclose potential benefits." His point is that no special encouragement is required to induce investigators to disclose potential benefits since they would do so enthusiastically in any event. This generalization about the motivation of clinical researchers has a certain common sense appeal. Why should it be necessary to reinforce the already strong inclination of researchers to communicate possible benefits to patients? It is, after all, the potential harms that they are most likely to feel tempted to conceal or down-play, and it is therefore their duty to disclose such harms that needs to be emphasized.

The duty to disclose potential harms does need to be stressed. That seems incontestable. But I would argue that the duty to disclose potential benefits also requires to be stressed. In a litigious society, when a certain class of people has been identified as less-eligible to participate in an important social activity, such as clinical research trials, a prudent researcher might well feel that it would be better to discourage such elderly volunteers by minimizing or simply not disclosing potential benefits. In this way, she could avoid "the hassle" of ethics committees, suspicion from over-protective relatives, and possible litigation if things go badly for volunteers whose frailty puts their decision-making competence into question. Such an attitude on the part of researchers would be understandable, but not morally defensible. Every prospective research subject is entitled to decide for herself, after full disclosure of both potential benefits and potential burdens, whether or not she wishes to participate. The right of informed consent requires that the prospective subject make her decision in the light of all materially relevant information.

In sum, once we recognize that researchers have a moral duty to recognize valid as well as invalid consent, and that these moral duties are equally weighty, we are forced also to recognize that there is no fail-safe position. Whether any particular patient, however elderly or infirm, is capable of giving a valid consent to experimentation, is a question that can only be answered after careful examination of the circumstances of the case. Pace Jonas, it cannot be answered *a priori*.

THE CASE AGAINST HANS JONAS: SOCIAL UTILITY

Setting aside, for the moment, the anti-Jonas argument which supports recruitment of elderly patients for risky (to some degree) research by appealing either to the autonomy or to the overall best interests of the potential subject, we turn now to a different argument, one which appeals to the social utility of permitting at least some medical experimentation on elderly persons.

There is virtually unanimous consensus among demographers that an increasingly large number of people, especially in advanced industrial or post-industrial societies, will be surviving to very old age (Zimmer et al. 1985). More of us will be living into old age, and more old people will live into old, old age. As a concomitant to this phenomenon, popularly labeled "the aging population," more of us will live long enough to suffer from those disabling and chronic diseases that primarily or uniquely affect the aged. In order for scientists to improve their understanding of such diseases sufficiently to develop effective therapies or, better still, preventive strategies, it is necessary for society to promote and pursue a vigorous program of geriatric research.

Even with respect to those diseases (cancer and heart disease, for example, and stroke), which affect some younger as well as many elderly people, if elderly patients are to receive proper medical treatment it is necessary that researchers acquire adequate knowledge concerning the safety and efficacy of drugs and of surgical treatments when used in the treatment of the old. As Bell, May, and Stewart (1987) observe:

> Results of well-controlled clinical studies using healthy young or middle-aged adults who are not taking medications that might interfere with the research project cannot be generalized to persons in whom the aging process is coupled with multiple chronic illnesses and several concomitant medications.

The important point to note here is that the aging process itself, together with some disease, states more prevalent among the elderly make it unwise to extrapolate data acquired from research on younger study populations to older populations.

The need for extensive research on the elderly becomes even more exigent when one takes into account the great diversity to be found within the category of "the elderly," some of whom enjoy robust good health and mental acuity while others are acutely ill and mental compromised. Many elderly people are dependent upon a great range of drugs for a variety of serious conditions, but some elderly people live medication-free lives. Stereotypes of the elderly as physically frail and mentally incompetent disguise the heterogeneity to be found within this group. In consequence, research on the elderly will have to be equally sensitive to diversity and will need to focus on many sub-populations within the general category of "the old". Urinary incontinence, for example, is a problem from which many old people suffer. But because there are several types of urinary incontinence, which differentially affect particular sub-groups of the elderly and, because this condition will often co-exist in the elderly with such other conditions as diabetes mellitus, congestive heart failure, or symptomatic arthritis, clinical trials which seek to ascertain the efficacy of experimental drug therapy will face special difficulties not usually encountered with drug trials on younger populations (Williams and Pannill 1982).

Research in geriatric medicine thus poses some unique problems. For example, "scientists must consider the importance of cohort effects, the need to separate age-related effects from the effects of disease and psycho-social factors" (Zimmer et al. 1985) Norms and clinical practices developed by research on general populations will not necessarily serve older patients well (Avorn and Gurwitz 1990).

Moreover, this requirement for research on elderly patients themselves arises not only within the realm of geriatric medicine but, much more broadly, within the realm of all gerontological research and gerontological policy-formation. The elderly, it should be noted, require services from a range of professionals other than physicians and nurses. The same empirical argument that establishes that geriatric research is necessary if we are to develop effective medical treatments for the elderly, also establishes that gerontological research on elderly subjects is necessary if we are to develop effective social and psychological programs and sound gerontological social policy. If we desire to live in a society in which elderly patients and clients are well-served by social workers, clinical psychologists, and policy planners, as well as by doctors and nurses, we have no choice, *pace* Jonas, but to accept the necessity of research on the elderly.

Jonas' argument affirms that society has no right to demand of its members that they make significant personal sacrifices on behalf of the public good. Fair enough. Nevertheless, it is widely accepted in many western societies that society also has an obligation to provide at least minimally decent medical care for everyone. This obligation could plausibly be interpreted to entail that society has an obligation to promote the development of effective treatments. To forego medical experimentation involving some non-trivial level of risk would be functionally equivalent to abandoning most medical research. To forego medical research in this dramatic fashion would be to deprive ourselves and future generations of the benefits of new diagnostic tests, new treatments, and new potential cures for dreadful diseases and crippling disabilities. Such moral reasoning, when offered in criticism of Jonas' position, leads directly to the conclusion that every member of society has an assignable *prima facie* obligation to promote health through a willingness to volunteer for research. There would not appear to be any obvious reason why the elderly, considered as a group, would be exempt from such a basic social obligation. Indeed, for any individual, young or old, to accept the benefits of others' volunteer efforts of this sort, without being willing to make similar efforts herself, smacks of what we might label "moral parasitism." We are here dealing with that familiar game theoretic issue: the problem of the free rider. Old people are no more exempt from moral strictures, in particular the stricture to contribute to the collective good, from which they in turn benefit, than any other group in society.

Thus, one might argue, against Jonas, that there is a significant ethical cost attached to not doing research on vulnerable potential research subjects. As one physician (James 1978, 409) expresses the moral point:

In the discussion of ethical considerations relating to clinical research, the rights of the
unborn generations to benefit from the fruits of research musts also be weighed. It can
be debated that no man today has the free and moral right to condemn his grandchildren
to the same perils of disease to which he is exposed by virtue of the present lack of
effective scientific information, and his failure to participate in a search for it.

Without extensive scientific research, employing elderly human volunteers as
subjects, we not only forego the opportunity to acquire new and more effective
treatments for the present generation of old people and for future generations, but we
risk allowing actual harm to older people as a result of their treatment with
insufficiently tested therapies. Moreover, it is worth keeping in mind that even when
inadequately tested therapies turn out not to be actively harmful they may be
inefficacious. Given the fact of scarce medical resources, one person's provision is,
so to say, another person's deprivation. Thus, the issue of cost-effectiveness carries
with it a moral dimension which, when combined with the imperative of promoting
health, tends to add further support to the conclusion that we would be ethically
remiss if we were to abstain from medical research which is likely to yield beneficial
knowledge.

Suppose, however, for the sake of argument, that one were to concede to Jonas
that self-sacrifice, in non-emergency situations, is supererogatory, and that "our
descendants have the right to be left an unplundered planet; they do not have a right
to new miracle cures" (Jonas 1969, 230). Let us also concede that, after all, Jonas
was right when he insists that neither present social welfare nor concern for future
social welfare obligates individuals to participate in clinical trials as research
subjects. Accept that neither of these important goals could justify society in
demanding of individuals that they accept the personal sacrifices associated with
becoming a volunteer for hazardous experimentation. Having made all these
concessions, however, one might nevertheless question why it would be wrong for
society to accept the voluntary informed enrollment as research subjects of those
who are motivated to participate, whether by altruism, by a sense of moral
obligation, or by perceived self-interest.

One could also question the a-historical and asocial conception of the self that is
presupposed by Jonas' extreme individualism, with its concomitant near-exclusive
focus on individual rights. As McIntyre (1981) has reminded us, the process by
which individuals construct their identity and achieve the status of autonomous
beings is one that necessarily takes place through community membership. Once the
reciprocal ties of the individual to his or her community have been understood and
their importance recognized, one may feel that what is needed is less an exclusive
stress on protecting the rights of the individual and more a stress on finding the
correct balance between the rights of individuals and the needs of the community
from which individuals derive their identities and draw their strength. A patient-
centered ethos, or a research-subject-centered ethos requires to be counterbalanced
by the morally legitimate claims of the wider society (Hardwig 1990). Rights need
to be counterbalanced by responsibilities. At the very least, community needs and
the individual's responsibility to promote the satisfaction of those needs should not

be slighted or ignored. The communitarian perspective is one that deserves to be considered and to be assigned some moral weight in policy deliberations (Nelson 1992).

MORALLY COERCIVE APPEALS FOR VOLUNTARY SELF-SACRIFICE

Jonas argues against the moral permissibility of any appeal to members of society that they volunteer to become research subjects. He opposes such appeals on the grounds that the mere issuing of the appeal, the calling for volunteers, inevitably exerts moral and social pressures on potential subjects (Nelson 1992, 233). This argument has special force when the appeal is directed toward frail elderly patients by their personal physicians, nurses, or nursing home caregivers. Many of the factors that make such patients highly desirable as research subjects, for example, their continuous presence in an institution where their health can be easily monitored and their progress tracked, also render them highly vulnerable to exploitation. This is especially true for patients with cognitive deficits who, according to some estimates, constitute almost one half the population of most nursing homes (Zimmer et al. 1985, 277). Moreover, even for those frail elderly patients who are cognitively intact, serious moral problems arise when attempts are made to recruit them as research subjects, some of which are discussed below.

Of course, it would be quite wrong to suggest that all or even most elderly persons are "frail," either physically or intellectually. Rather, the suggestion is merely that various kinds of frailty are increasingly present with advancing age. Aging patients are typically in a weakened physical state and often experience emotional upset. Many are likely to feel both dependent upon and submissive toward those who are responsible for their care and treatment. It is well established by empirical research that patients who are ill tend to suffer a concomitant diminution of their autonomy (Ackerman 1982). The problem becomes more serious still when the patient is gravely ill and/or institutionalized. To be removed from one's familiar surroundings and one's familiar routines is inevitably a disorienting experience, and when this deracinating process is accompanied by a significant loss of privacy and the substitution of unfamiliar care-givers for friends and family, the effects on autonomy can be marked. It seems unreasonable, therefore, to expect that an elderly patient (or any patient, for that matter) who is gravely ill and/or suffering significantly from pain, discomfort, emotional fatigue or distress, perhaps living in an institution and heavily dependent upon its staff, will nevertheless be able to attend to the subtleties of a complex research project, much less possess the capacity freely and rationally to assess potential risks and benefits.

This line of reasoning has led to the proposal that patients who are in a seriously weakened or otherwise vulnerable position be assigned (or, at an earlier stage, be encouraged to designate) an "advocate," someone whose obligation it would be to represent their wishes or needs. To some authors, the family seems the appropriate place to look for such an advocate. Others, suspicious of family dynamics, argue that when a patient requires either an advocate or even a surrogate decision-maker the person chosen should be independent of the family system (Hardwig 1990).

PHYSICIAN-RESEARCHER CONFLICTS OF INTEREST

Many commentators have observed that patients have traditionally tended to think and behave toward their doctors in a submissive manner – admiring and obedient in the manner of good children responding to parental commands (Katz 1984). This observation is much less true today, at least in most Western societies, than it would have been prior to the advent of the consumers' and women's movements of the 1960s. Even so, it is still likely to be true for many geriatric patients. Given that many elderly patients grew to maturity in an era marked by a higher degree of deference to authority than that which prevails in contemporary society, it will often be the case that they manifest a submissive deference toward their care-givers which can be easily exploited (consciously or unconsciously) by physician-researchers who solicit their own patients to participate in a clinical trial.

The role of personal physician diverges significantly from that of scientific investigator, though the same person often takes both roles. Thus, it seems potentially exploitative, and therefore ethically dubious, to use the trust accumulated by physicians in their traditional role as healers for purposes related to their distinct role as scientific investigators (Schafer 1982). The performance of dual roles raises the specter of potential conflict-of-interest, viz., conflict of interest between the duty of undivided loyalty to an individual patient, which inheres in the physician role, and the duty of advancing medical knowledge, which is the primary goal of medical research.

Since each of these roles – that of scientific investigator, on the one hand, and personal physician, on the other – defines itself by reference to a different primary purpose, the possibility of conflict of interest is an ever-present danger. Because physicians have a plurality of responsibilities, they must attempt to reconcile and balance a plurality of value commitments. The Hippocratic Oath, in all of its many version may insist that for a physician "the life and health of my patient shall be my first consideration," but most modern physicians would acknowledge that they have such additional moral responsibilities as an obligation to safeguard public health, to protect the interests of future generations, to promote the well-being of society, to obey the law. These multiple responsibilities and obligations will occasionally come into conflict with each other. Even when there is no outright conflict, there may be tensions between competing values which are difficult to reconcile.

This worrying problem is not confined, of course, exclusively to geriatric research. All patients, and certainly all acutely ill patients, are vulnerable to implicit coercion when their personal physician invites them to participate in clinical research. As Berkowitz (1978, 241) rightly observes, "[i]t is difficult to imagine that an institutionalized subject or a subject dependent upon his or her physician for treatment and/or relief from suffering, would not be influenced by the care-giver's desire for patient participation in any research proposed." Wherever possible, therefore, the invitation to participate in clinical research should come not from the patient's personal physician or caregiver but from some entirely independent person who exercises no power over the patient. In this way, those whose primary role is to serve as physician caregivers would be distinct from those whose primary role is to serve as researchers. Explicit assurances to patients – that no untoward consequences whatsoever will follow from a refusal to volunteer for research – should also be mandatory. Where credible evidence exists that patients nevertheless feel intimidated, it may be necessary to withdraw the invitation, unless the element of coercion can somehow be successfully counter-acted.

To re-iterate: the primary purpose of research is to contribute to knowledge; the primary purpose of medical treatment is to benefit the individual patient undergoing the treatment. In the case of what is often labeled "non-therapeutic research," the fact that it may be carried out by a medical doctor upon a research subject who happens to be, simultaneously, his/her patient, is coincidental and should be treated as morally irrelevant, for the doctor deals with the subject exclusively in her (the doctor's) role as a scientist.

Unfortunately, in some situations it is exceedingly difficult or even impossible to eliminate role conflicts between the "physician-as-healer" and the "physician-as-researcher." The difficulty is generated partly by the fact that the distinction between therapy and research is often in practice quite blurred. The distinction is less clear-cut than one might wish because of the large number of gradations that exist between the experimental and the therapeutic ends of the medical spectrum. It is sometimes not easy to say confidently whether a particular clinical trial is more accurately to be labeled as "therapy" or as "research." For example, much medical research is explicitly intended to be of direct benefit to the patient, as when a promising new experimental drug is given to cancer patients under controlled conditions with the aim of discovering whether it is more beneficial, overall, than some older drug. Thus, elderly patients who are in the early stages of a dementing process may gain access to a new but promising drug by agreeing to become subjects in a clinical trial of a new therapy designed to slow mental deterioration. Alternatively, or additionally, the elderly patient may benefit indirectly by receiving especially careful medical attention from highly trained specialists.

This activity sounds as if it could properly be labeled "therapeutic." However, one needs to keep in mind that one of the essential purposes underlying so-called therapeutic experimentation is to contribute to medical knowledge. Procedures may be undertaken, in pursuit of this objective, which are not strictly necessary for the treatment of the patient. When treatment is combined with research, systems of treatment are chosen partly with a view to curing the patient but partly, also with a view to testing new procedures or comparing the efficacy of various established procedures. The patient who is also a research subject may thereby be exposed to added hazards, discomforts, or inconveniences. Despite these negative aspects of research, it will sometimes be to the direct and immediate benefit of a patient to become a research subject. (As outlined in the previous paragraph.) The key point to note, however, is that patients who are potential research subjects, especially when they are physically or mentally frail, will typically need assistance in weighing and balancing the competing values involved in any decision to volunteer as a subject. Wherever possible, that assistance should be given by persons whose professional orientation does not put them into a conflict of interest situation.

THE INSTITUTIONALIZED ELDERLY

When elderly patients become institutionalized, perhaps because of chronic illness or physical frailty, the mere fact of institutionalization should be considered as a liberty-limiting factor: "For example, an individual [who has been institutionalized] may not believe that he is free to refuse participation or withdraw because of fear of subtle discriminatory practices or attitudes among the care-givers" (Cassel 1988). If some elderly patients do harbor such fears and if, in consequence, their consent to become research subjects is more reflective of a desire to avoid possible untoward consequences than it is reflective of a genuine wish to participate, one would have to say that their consent was coerced rather than voluntary.

There is also a legitimate worry, on the part of those concerned to protect the elderly from exploitation, that the experience of institutional life is likely, for many nursing home residents, to have reduced still further their already compromised ability to engage in autonomous decision-making. Once individuals become acculturated to allowing nurses or attendants to make personal decisions on their behalf, there is a marked deterioration in their sense of efficacy and independence. Thus it comes about that those old people whose physical or mentality frailty has led to their being institutionalized in the first place then suffer from a further reduction in their capacity to function autonomously as a result of paternalistic control from their institutional care-givers. The experience of dependency (on institutional care-givers or family) together with the experience of interdependency (via the peer pressure towards conformity from fellow institutional residents) combine to generate serious doubts about whether they are capable of giving voluntary consent to become research subjects (Ratzan 1980).

On the other hand, participation as a subject for scientific research can be a highly beneficial experience for nursing home residents, even when the research is not intended to be therapeutic for its volunteers. Those who live within such institutions not uncommonly find their lives marked by loneliness, with few opportunities for significant social interaction. Most people, young or old, require such social interactions to thrive, and this is no less true for the institutionalized elderly than for others. Indeed, because the richness of their lives has inevitably suffered as a result of constrained living circumstances, it is likely to be more true for them than for others. People seem to need involvement with some larger social project or purpose in order to give meaning to their lives. As noted above, the willing participation in a scientific project the goal of which is to advance medical knowledge can be a source of great emotional satisfaction to those who make the choice to become research subjects. For reasons such as these, there is some weight to the claim that it would be wrong to deny such a potentially worthwhile activity to elderly people solely on the grounds that their lives are lived within a care institution.

Thus, if we keep in mind Freedman's warning that there is a moral cost to over-protectiveness which counter-balances the more commonly noted moral cost attaching to under-protection of vulnerable subjects, we will not automatically conclude that nursing home residents ought to be entirely excluded from the opportunity to decide whether or not to become research subjects. Once the opposing dangers are both recognized, one is compelled to admit that experimentation on elderly nursing home residents cannot be accepted or rejected *uberhaupt.* Some kind of discretionary balancing in each individual case of the likely benefits and harms appears to be unavoidable. But, of course, if it is decided to permit institutionalized elderly prospective subjects to participate in research as volunteers, it might be morally required to set careful limitations to such research.

That is, one might propose that, since cognitive impairment and/or institutionalization increase(s) vulnerability to exploitation, it would be morally desirable, perhaps even morally mandatory, that special restrictions be applied to research carried out using individuals who already suffer with such significant burdens. Hans Jonas, for example, proposes that if it is decided to carry out research using vulnerable people as subjects – something of which he disapproves in any event – then the research carried out must be specific to the disease or disability from which the volunteer is herself suffering (Jonas 1969, 241). This requirement – that research upon vulnerable subjects be "patient-specific" – has now been incorporated into the National Commission's (1978) Report and Recommendations in Research Involving Those Institutionalized as Mentally Infirm.

Jonas also proposes a sliding scale of eligibility for recruitment as a research subject. Those who are most knowledgeable about the research, viz., those with membership in the research community itself, should be regarded as the first and most eligible for accrual to clinical trials. Of the entire pool of possible research subjects, physicians are typically the best informed, healthiest, and least vulnerable to coercion from fellow researchers. Thus, they ought to be viewed as the prime group from whom subjects should be drawn. As Jonas rightly notes, there is a long and honorable history of self-experimentation among physicians (Jonas 1969, 243). If members of the medical community should properly be regarded as the first and the most highly preferred candidates for recruitment as research subjects, the last to be considered for recruitment should be those who are physically or intellectually frail. Those who are highly vulnerable – owing to such factors as their illness, their diminished comprehension or their institutional captivity – are, in consequence, least able to give valid consent and should, therefore, be regarded as the least preferred candidates for recruitment. As Jonas (1969, 237) puts the point:

> The poorer in knowledge, motivation, and freedom of decision (and that, alas, means the more readily available in terms of numbers and possible manipulation), the more sparingly and indeed reluctantly should the reservoir be used, and the more compelling must therefore become the countervailing justification.

The research ethics policy to which this argument points might be formulated as follows: recruit first among those who are least vulnerable to exploitation; recruit last, and only when absolutely necessary, among those who are most vulnerable.

Moreover, as a matter of general policy, one's reluctance to allow participation by the frail elderly in a clinical trial should increase as the risks to research subjects increase. When research involves high risks of serious harm then one would want to insist with special vigor that the most vulnerable prospective research subjects be placed at the very back of the queue rather than at the front. There is a clear and obvious moral distinction to be drawn between research which involves no more than minimal risks of no more than trivial harms, on the one hand, and research which is likely to be outright hazardous to many or all of those who participate, on the other.[2] One would also view the accrual of vulnerable patients to clinical research trials with greater sympathy when the research is intended, to a significant degree, to be therapeutic for the patient, and when the patient's non-participation as a research subject carries risks of harm (from the disease) that are of the same order of magnitude as those posed by the research itself. To put the matter crudely, a patient suffering from advanced senile dementia or rapidly metastasizing cancer has "very little to lose" when standard therapy offers virtually no hope of remission or decent quality of life. This is not to suggest that patients suffering from otherwise "hopeless" medical conditions have nothing to lose because of their severe dementia or their imminent death. Participation as a research subject may expose them to such additional burdens, for example, as the discomfort of being poked, prodded, and tested. Even the least invasive medical experiments are likely to involve some degree of discomfort or inconvenience, if not the risk or actuality of harm.

Nevertheless, while recognizing the virtue of the sort of carefully restrictive policy advocated by, among others, Jonas and The National Commission, it must be acknowledged, in practice, that for certain kinds of medical research the only suitable or the only available prospective subjects may be individuals who rank high on the vulnerability scale. That is, for reasons explained earlier, there may be no way to achieve some supremely important advances in scientific knowledge without conducting clinical trials on those whose lives are already heavily burdened by illness and disability. If society is concerned, for example, about the plight of patients whose lives are blighted by senile dementia, or concerned about the fate of long-term nursing home residents, it must find ways of facilitating research into both dementia and the quality of care delivered in such institutions. When the research is intended to be therapeutic, then it also holds out the possibility of real benefits to those who voluntarily participate, as well as to future generations of patients. But even when the research has little or no therapeutic potential for its subjects, it may nevertheless produce a significantly favorable surplus of good over bad consequences. Those attracted to the consequentialist approach to ethical decision-making will find such considerations to be of considerable moral significance.

INFORMED CONSENT

Individuals who are considering whether to volunteer as subjects for medical research have to confront what for many is a difficult and complex task: evaluation of the comparative risks and benefits associated with each option. Each person must decide what she stands to gain and what she stands to lose by agreeing or refusing to participate. Physicians are also required to make such case-by-case calculations as a basis for the recommendations they make to patients who are also potential research subjects. But those patients who contemplate becoming research subjects are entitled, legally as well as morally, to undertake their own evaluation of the risks and benefits, and to bring their own attitudes and values to bear in reaching a decision. Individuals who choose to become research subjects ought never to be viewed simply as experimental "raw material." Rather, they are entitled to view themselves and to be viewed by the experimenter as "joint adventurers" or "partners" in the enterprise. Consent makes this relationship possible, and represents the duty of fidelity and loyalty between researcher and subject (Ramsey 1970). The doctrine of informed consent has been accepted widely as a, and perhaps as the, fundamentally important moral requirement for every kind of research on every kind of human subject.

Consent is important, but it must be consent based upon adequate information, communicated in a form which patients can understand. If people agree to become research subjects without having been given adequate information in a form which they can understand, then they have not really had an opportunity to decide their own fate.

Is there any reason to think that the generally accepted moral standards or principles governing experimentation on human subjects – including the principle requiring that informed consent be obtained from potential research subjects – should be any different for older than for younger people? Young (1978, 68-9) answers this question negatively: "While it is true that the aged have special needs, it is not clear that special ethical principles are needed to respond to the moral dilemmas raised by health care and biomedical research in the aged." Sachs and Cassel (1990, 235) agree with this view, claiming that: [T]here are no *a priori* reasons to treat differently the research participation of older people." They argue that since the vast majority of older people are cognitively unimpaired and live outside controlling institutions, the elderly should not be considered as a special category for purposes of research ethics. The fear is, of course, that if the elderly are classified as a special group in this context, the process of so classifying them may perpetuate, in an ethically undesirable manner, their segregation from general society.

Even if Young, Sachs, and Cassel are correct, that is, even if the same moral standards and principles which obtain for research on people of middle age also apply to those who are old, might it nevertheless be true that special regulations are necessary when applying these standards and principles to the elderly? Given that special regulations exist to protect the institutionalized mentally disabled, prisoners, pregnant women, fetuses and children, it would not be a surprise to discover that we also need special regulations to protect elderly research subjects, as a group.

It is acknowledged, of course, that people of whatever age who are cognitively impaired or who are institutionalized or both are especially vulnerable to exploitation. Accordingly, members of these groups require special regulatory protection and special moral consideration. The same is true for such populations as children, or the mentally ill. But, since the category of "elderly people" encompasses a full spectrum of the human condition – from the robustly healthy to the physically frail, from the mentally acute to the mentally confused, and from the living-independently to the institutionalized-dependent – individuals who fall within this category of "elderly" are sufficiently heterogenous that regulatory bodies are justified in deciding not to treat them as a separate population (Sachs and Cassell 1990), or so it is argued.

The appeal of this position derives some of its force from the fact that no one, in these enlightened times, wants to be guilty of the sin of discriminatory stereotyping. Thus, since the majority of older people are cognitively intact and live independently, age may be a factor but it will generally not be the most important factor when, for example, discussing issues of capacity to give informed consent. We could label this "the assimilationist" position, since it insists that no special criteria are needed for the protection of experimental subjects who happen to be old.

A very different answer to our question – "Do we need special ethical guidelines for research on the elderly?" – is offered by Ratzan (1980, 37), who argues in favor of the conclusion that "being old makes you different." The cluster of physiological and psychological changes that are a normal concomitant of the aging process may well be relevant to the process of obtaining informed consent from elderly people. We could label this view the "special category" position. Ratzan claims that the elderly have different values and weaknesses from the young (and perhaps different strengths, as well). He further claims that when the elderly are invited to participate as research subjects, it is morally wrong not to take special precautions to safeguard their dignity and their autonomy. Empirical evidence is cited in support of the claim that the cognitive abilities of elderly research subjects decline with normal aging and are often different from those of younger subjects. Comprehension by any elderly subject of the proposed research ought to be viewed as problematic. Memory declines, as does learning ability, when compared to that of younger subjects (ibid.). Impaired hearing and vision are also much more common among the elderly than among the young, which can easily result in communication difficulties if researchers do not take special measures, such as large-type print and increased voice volume, when seeking informed consent from potential research subjects who are elderly. Since informed consent forms are notoriously convoluted and legalistic in their language, not to mention esoteric in the vocabulary they employ and, since elderly people have generally had fewer opportunities to undertake higher education and may, in consequence, have a less well-developed ability to comprehend such obfuscatory forms, specially designed forms may be required (Tymchuk 1988).

This makes the problem-solving process of informed consent a slower and more arduous process for many elderly subjects than it would be for their younger counterparts. Ratzan concedes that although the problem-solving abilities of elderly people often involve deterioration, at least some of these differences may be regarded as adaptations rather than as impairments. Even so, if researchers are to obtain valid informed consent from apparently competent elderly research volunteers, the researchers must attend carefully to aging-induced differences and must ensure that the process by which consent is obtained takes proper account of the strengths but also of the weaknesses which old age typically produces in the quality of decision-making.

Strategies which might be employed to facilitate properly informed consent among prospective subjects who are elderly include, for example, having family present at the interview, presenting relevant materials in both written and oral form, and perhaps on video as well, so that it can be replayed later at leisure, simplifying information so that it can be easily comprehended by those with lower educational attainments, ensuring that printed materials employ large print and that voice communications are loud enough to be audible to the hearing impaired, where these are necessary for hearing or sight-impaired individuals and, perhaps most important of all, ensuring, by means of follow-up questions or questionnaires, that the relevant information has been successfully comprehended (Tymchuk 1986). One researcher has proposed that it may be necessary, in order effectively to protect the rights of elderly individuals, to employ comprehension tests before participation in research investigations (Taub 1980, 686). Of course, such special consideration should be given to all prospective research subjects who suffer from special disabilities but, since the prevalence of such disabilities increases with age, it seems only prudent to emphasize their importance when dealing with an ageing population.

In addition to such potentially important physiological differences between elderly and middle aged populations, there is some reason to think that the old may have an age-distinctive point of view as to what counts as a potentially significant harm, a harm worth taking into consideration when deciding whether to become a research subject. Berkowitz (1978) warns, for example, that since concerns about vision and mobility loss typically loom much larger in the minds of most older people than they do in the minds of others, for the obvious reasons, "even a minuscule [sic] risk of vision or mobility impairment may become a material consideration affecting consent and therefore requiring disclosure to this particular population." In other words, when deciding which risks are materially irrelevant and, therefore, need not be communicated, investigators must take care not to project their own sense of priorities upon elderly potential research subjects. A risk which seems trivial to younger persons may seem significant to the elderly.

In an earlier section of this paper, "Physician-Researcher Conflicts of Interest," the potentially coercive effect of physicians recruiting their own geriatric patients was discussed. It was noted that valid consent to become a research subject must not only be informed, it must also be voluntary, that is, uncoerced. This entails that those doing geriatric research should pay careful heed to the fact that the lives of elderly people are not infrequently lived within a "pervasive web of dependence" (Berkowitz 1978, 35). The dependence of many elderly patients – those living independently but, even more so, those who have been institutionalized – on their families, friends, neighbors, and government agencies to provide them with assistance in living, can easily lead to an attitude on their part of passive acceptance, resignation, and compliance (Brody 1977). If such an attitude of complaisant passivity is a normal concomitant of the aging process, then it may often be the case that the consent given by older people to participate in medical research is not truly voluntary consent. Moreover, as Strain and Chappell (1982, 528) note, when compared to younger prospective subjects, the elderly may not be as aware of what is entailed in volunteering to become a research subject, and they may not appreciate fully that they have a right to refuse.

Without falling into the trap of discriminatory stereotyping of old people as a group, it should be possible to recognize that when the old, and especially the very old, are to be recruited for scientific experimentation, it behooves those inviting such participation to take those steps necessary to ensure that the volunteers are truly volunteers and not simply human raw material for the progress of science. If the research community is properly sensitized to those processes of aging which are most likely to affect an elderly patient's capacity to give valid consent, then those elderly patients who require special protection are most likely to receive it. It is not a bad thing that scientists engaged in geriatric research should receive periodic reminders of the danger that some of the elderly who "volunteer" to participate in medical research may, for one or another of the reasons discussed above, be merely acquiescing rather than properly volunteering to participate. Unless special safeguards are in place, the quality of the consent given by elderly patients must be regarded with some skepticism. This line of argument reinforces the position of those who, like Hans Jonas, wish to eschew the recruitment of old people as research subjects, or at the least accept their participation only after rigorously ascertaining that their consent is fully informed and properly voluntary. On the other hand, it is equally important to stress that those who view the participation of elderly subjects in research as a potential boon to subjects, society, and future generations will not view these objections as insuperable.

When the risks to research subjects are low, when the harm, should it occur, is trivial, and when the possible benefits to elderly potential research subjects are considerable, many will feel inclined to accept a weak standard of decision-making capacity as adequate to the situation (Wicclair 1993, 166). Thus, an elderly person whose cognitive impairment nevertheless permits some minimal awareness of what is transpiring might nevertheless be judged to possess sufficient decision-making capacity to enable her to assent to participation in a low-risk/high benefit clinical trial, even when the direct benefits accrue to others. (On this point, see also the earlier discussion, in the section titled: "Competence and the Elderly Subject.)

Some may feel that, when proper consent is beyond an impaired patient's capacity, the acceptance of mere "assent" puts us on a slippery slope to ethical danger and degradation. However, the possibility of therapeutic benefit to the individual subject should go at least some distance to offset this moral concern, as would evidence (from family members, say) that the individual, when competent, expressed such values or desires as might count as authorization for participation in low-risk research. The availability of an advance directive giving explicit instructions would, of course, carry more weight still. How much additional weight one ought to assign to such advance directives, with respect to the recruitment of individuals as research subjects, is discussed below.

INFORMED CONSENT BY ADVANCE AUTHORIZATION

If the topic of research on elderly subjects occasions much controversy, the subtopic of research on incompetent elderly subject generates greater controversy still. The arguments of Hans Jonas, discussed at the outset of this paper, are directed generally against the recruitment for research purposes of patients who are vulnerable to abuse or exploitation. They weigh most heavily, however, against the enrollment of patients who are totally incompetent to give valid consent, patients such as the seriously demented elderly. It seems clear that Jonas would favor a virtually absolute ban on the use of incompetent elderly patients as research subjects. That is to say, on his view, if an individual, as a result of Alzheimer's disease, stroke or multi-infarct dementia, becomes cognitively impaired to such a degree that she loses the capacity to give or withhold informed consent, then it becomes ethically impermissible to use her as a research subject. Impermissible, full stop. Similar arguments are offered by Paul Ramsey against the use of children for non-therapeutic research (Ramsey 1970), and apply with equal force to patients of any age who are incompetent to give informed consent.

There is no denying the moral appeal of such a blanket restriction. As a result of illness, the incompetent may have lost their autonomy, but this loss, so far from rendering them prime material for exploitation by the rest of society, ought to lead us to respect absolutely such dignity as remains to them. Because the incompetent elderly are typically institutionalized, at least in most advanced western industrial societies, it is often both convenient and efficient to enroll them in clinical trials. But convenience and efficiency, however much such considerations may weigh with harried researchers scrambling to enlist statistically adequate numbers of research subjects, ought to be subordinated to the moral requirement of offering maximal protection to vulnerable populations.

On reflection, however, a blanket prohibition of the sort defended by Jonas and Ramsey seems difficult to defend. As discussed earlier, participation in a clinical trial sometimes offers the possibility of significant health (or other) benefits to research subjects. If we were to prohibit, tout court, the participation of both young children and the incompetent elderly from clinical research because they lack competence to give valid consent, we would sometimes be denying to them the opportunity to gain important health benefits. It is not impossible, as discussed previously, that participation in a clinical trial could provide subjects with access to some new and highly effective drug treatment – one with the potential to restore a significant level of cognitive functioning to seriously demented patients.

For this reason, even philosophers who agree with a Jonas-like approach to human experimentation should be willing to permit the enrollment of incompetent patients in those clinical trials which reasonably offer to the patient/subject substantially more benefit than harm. With respect to research having therapeutic potential, it would be unfair to incompetent patients never to allow them to participate. The objective of protecting such patients from harm or exploitation can be achieved by other means, such as allowing a surrogate decision-maker to give or withhold consent on their behalf. Even if society were to adopt an outright prohibition against the participation of incompetent elderly patients as experimental subjects, concern for the interests and well-being of such incompetent patients would require that an exception to be made, at least for those clinical trials whose prospective benefits significantly outweigh their likely harms, compared to other alternatives available to the patient. In short, research with therapeutic potential ought not to be ruled out *a priori*.

There is a second reason, closely related to the first, why one should hesitate before endorsing any absolute prohibition against using the incompetent elderly as research subjects. Imagine this situation. A physician-investigator who is treating a patient with early-stage Alzheimer's dementia asks the patient whether she would be willing to give advance authorization to be enrolled later in a clinical trial of a new drug therapy with potential to reverse the dementing process from which she is suffering. That is to say, she is being invited now, while still competent, to give valid consent to her participation at a later time, at which later time her dementia will have progressed to the point that she will then have become mentally incompetent to give valid consent. If the patient refuses to give such advance consent then, of course, it would be a clear violation of research ethics subsequently to enroll her as a subject in such a trial. But if the patient agrees to give such advance authorization, then it would seem to violate both the principle of benevolence and the principle of patient autonomy to refuse to honor such a directive.[3]

The most obvious motive for an elderly patient to give advance authorization to becoming a research subject would be a present desire to gain future access to potentially effective treatment. But some patients suffering with degenerative diseases which cause progressive cognitive impairment may wish to give advance authorization to their becoming research subjects even when there is little or no prospect of future therapeutic benefits for themselves. They may do so from other motive, such as a desire to benefit future generations or to advance scientific knowledge. If the elderly patient is presently competent to give voluntary informed consent to becoming a research subject, if the patient understands the potential benefits of the research for others and the possible risks of harm to herself, and if the patient wishes now to authorize her future participation in a clinical research trial from such motives as altruism or a sense of social obligation, would it not be a case of unwarranted paternalism to deny the patient this right?

Dworkin (1993) has argued, persuasively in my view, that respect for the personhood of currently competent patients requires that we continue to respect their autonomously chosen values even after they have lost their rational decision-making capacity. If, therefore, at a time when I am still capable of formulating a plan for my life as a whole, I give advance authorization for my future participation as a research subject in a clinical trial, others have a moral obligation to respect my current values as an autonomous person by acting upon those desires, even after I have, through cognitive deterioration, lost both my autonomy and my decision-making capacity. In other words, what happens to me after I have lost competence should be governed by the attitudes and values that I have explicitly expressed before losing competence.

However, not everyone accepts the moral legitimacy of this position. Dresser, for example (1986), argues that one becomes a quite different person as a result of severe cognitive deterioration. If the demented person is viewed as having become a different person from the person she was when competent, then caregivers of the demented person are bound to consider only her current needs and wishes. The patient has ceased to be the person with the values she once held. She has, instead, become a different person who, although incompetent, has a quite different set of needs and desires from those previously felt and expressed (by the person she used to be). On this view, care givers and researchers are morally bound to assign decisive priority to the fulfillment of these current needs when they are inconsistent with those previously expressed.

Any attempted adjudication of the complex metaphysical issues of personal identity – when does a person cease to be "the same" person? – is beyond the scope of this paper. Those who feel the force of Dworkin's appeal to the weight of autonomously-chosen values, will insist that it is obligatory for society to adhere to the demented patient's earlier wishes, at least when the patient's current (demented) preferences do not explicitly conflict with those earlier expressed. The most difficult situations, however, and the ones posing the greatest ethical tension, will be those in which a competent patient has expressed a wish to be treated in a certain manner which that "same" patient, now demented, actively opposes.

Here is one possible example of such a situation. A person, when still competent, exercises her right to autonomous decision-making by directing that she should be enrolled (or continued) in a clinical trial even after her dementia progresses. That person, now demented and no longer capable of rational decision-making, protests strongly against some features of participation in the trial (the need for a daily injection, let us say). This discordance between previous (competently expressed) and current (incompetent) wishes puts the researchers into an extremely awkward dilemma. If they regard the advance directive of the patient as a kind of Ulysses-contract, instructing them to disregard any later (non-autonomous) changes-of-mind in favor of respecting her current (autonomous) values, then they will feel justified in disregarding the demented research subject's later protests.

The knowledge that one's advance directive will be later respected is likely to provide some degree of comfort to patients, which comfort provides a consequentialist argument in favor of privileging advance directives. Contrariwise, if they regard the current interests of demented patients as morally decisive and, if they construe those interests as accurately expressed by the patient's protest against participation in the clinical trial, then they will feel obliged to ignore the patient's advance directive in favor of respecting the patient's current experiential wishes. There is, however, an additional factor of some moral relevance, one that militates against privileging an advance directive over a currently expressed wish. The non-therapeutic treatment of non-co-operative or even actively resisting demented patients, as part of a research protocol, may have difficult-to-measure effects on the morale and well-being of health-care personnel, not to mention possible effects on the family and friends of the incompetent patient who witness or learn of what is happening. If the practice of allowing an advance directive to over-ride currently expressed wishes were to produce distress or lowered-morale for care-givers, family, friends, researchers, then this would provide a consequentialist argument against following the instructions of the advance directive.

To recapitulate, there are several competing ways of conceptualizing this moral conflict. It could be viewed as a conflict between the values of two distinct persons: the autonomous desires expressed by the competent patient while she was still competent, and the current desires expressed by the patient now that she has entirely lost decision-making competence. Conceived in this way, it might seem sensible to ignore the values and wishes of the autonomous person since that person no longer exists and, instead, to respect the current wishes of the demented person.

Alternatively, the moral conflict could be viewed as a clash between the sometimes competing values of autonomy and benevolence. Conceived in this way, one would have to chose between respecting the (previous) autonomous values of the patient, even when following those wishes would cause suffering or distress to the incompetent person she has become, or assigning a higher value to the minimization of present pain and distress, thereby sacrificing her interest in having her autonomous wishes respected even after she has lost autonomy. For a person who cares deeply about making something morally significant out of her progressive dementia and who believes that the best way to do this is to bind her future self, Ulysses-like, to participate in a non-therapeutic clinical trial of a new anti-Alzheimer's drug, the inability to bind herself in this way could seem disrespectful of her autonomy. Moreover, if she now suffers from or feels distressed by the knowledge that her wishes may not be respected after she has ceased to be competent, this suffering/distress would itself count as a consequentialist reason to accept an obligation to respect her autonomy. In other words, if the principle of benevolence is allowed invariably to trump that of autonomy, one consequence could be that additional suffering would be caused to those who feel strongly committed to controlling what happens to them in their post-competent life. Thus, such a policy, although aiming to minimize patient suffering, could be counter-productive.

In the light of such considerations, we might wish to re-conceptualize the conflict as one between two competing assessments of how best to minimize suffering and distress. The problem for caregivers and researchers would then be a practical one: how to assess, consequentially, the comparative benefits and harms of adopting one policy rather than another. It seems likely, however, that there exist wide divergences among individual patients concerning the value they would assign, respectively, to controlling the overall shape of their post-competent lives via advance directives and the value they would assign to promoting their comfort and experiential interest once they have lost competence. This is a seriously complicating factor, since any blanket policy of always assigning a higher priority to one or the other would be guaranteed to run afoul of the value priorities of some individuals. On the other hand, a policy of assigning weight to one or other value on a case-by-case basis, which would take individual differences of value priority into account, would face the possibly insuperable difficulty of discovering for each patient just what they value and how much they value it in various actual and hypothetical situations. This would require eliciting from them information both about the respective weight they assign to the sometimes competing values of autonomy and beneficence and the respective weight they assign to minimizing different kinds of distress.

Jaworska (1999) attempts to navigate a passage out of this difficult-to-resolve clash of values (between autonomy and beneficence) by arguing, against Dworkin, that we ought to respect the current interests of demented patients, not because (as Dresser claims) they are different persons from the persons they once were but, rather, because "many of these patients may still be capable of autonomy to a significant degree and that they may still have authority concerning their well-being". For Jaworska, in contrast to Dworkin, the salient question for those wishing to show respect for an Alzheimer's patient becomes neither, "Can this patient reason thoroughly and come to a rational decision?" nor "Does he grasp his life as a whole?" but, rather, "Does this patient still value?"(ibid.)

By "the capacity to value," she means "the capacity to originate the appropriate bases for one's decisions" (id., 134). Consider, for example, how an Alzheimer's patient might value listening to music "as a way of holding on, as a way to still lead a recognizably human existence despite his disease" (id., 120). A demented patient is said to be capable of valuing (a critical interest), and not simply of desiring (an experiential interest), if she can give some rationale for choosing the activities she in fact chooses (ibid.). If, as Jaworska argues, the capacity to value is not completely lost in dementia, then to the extent that it is retained, "respect for the immediate interests of a demented person is contrary neither to his well-being nor to the respect for his autonomy" (ibid.).

Although Jaworska seems to be advancing a position much different from Dworkin's, the differences in the end may amount to little more than nuance and subtle shading. Dworkin is concerned primarily with patients who are suffering from late-stage Alzheimer's disease, patients who invariably lack not only a sense of their own lives as a whole, but lack also what Jaworska calls the capacity to value. Moreover, Dworkin and Jaworska both agree that in the early stages of dementia, a person will still be capable of autonomous decision-making, which capacity imposes on their caregivers an obligation to respect their wishes. Such differences as do exist between their approaches manifest themselves, if at all, only in the middle-stages of dementia, when a patient might possess what Jaworska calls the "capacity to value," but lack the capacity to formulate a plan for her life taken as a whole, which latter capacity Dworkin associates closely with being an autonomous agent. Even in this middle-zone of dementia, however, the differences between them are, on close inspection, less striking than one might at first expect. Dworkin argues, in the name of patient autonomy, that we should be willing to override the patient's current preferences when those current preferences conflict with the patient's previously expressed autonomous life-plan. But, perhaps surprisingly, Jaworska also appears to be willing, where appropriate, to allow the demented person's earlier autonomous wishes to override her current preferences: "The caregiver must learn to pay attention to the person's values rather than to her concrete, yet perhaps ill-informed selection of options" (id., 134).

Where they differ, Dworkin and Jaworska, is in this: that Dworkin wishes to privilege the values of a person at that point in time when she possesses sufficient autonomy to lead her life according to her own life-plan, whereas Jaworska wishes to privilege the values of a person at that point in time (presumably much later in the course of the dementing process) when she is capable of self-governance only in the rather more limited sense of possessing some sort of rationale for her preferences. We may, then, she concedes, set aside the expressed preferences in favor of those that more closely fit the underlying value that the patient is seeking to promote.

As between these two conceptions of autonomous decision-making, the traditional one (Dworkin's) and the revisionist one (Jaworska's), which is the more defensible? On Dworkin's conception, we are entitled to ignore Ulysses' plea to be unbound from the mast, safe in the knowledge that by so doing we are nevertheless respecting his autonomy. We should ignore Ulysses' currently expressed desire, despite the force and passion with which he both feels and expresses it, and despite the present suffering to him which our refusal entails, because we know that his overall life-plan, formulated in a reflective moment, includes the desire both to hear the Sirens sing and to survive the experience without being lured by siren-song into the whirlpool of Scylla or smashed on the rock of Charybdis. It would appear, however, that on Jaworska's rather "thin" conception of autonomous decision-making the plea of Ulysses to be unbound would count as autonomous and deserving of our respect so long as he can give us some rationale for his wish to be released.

Well, there is no doubt that Ulysses bound to the mast of his ship could provide us with reasons or a rationale for his wishes, so Jaworska would, presumably, advocate that we release him to pursue his fatal passion, despite his stringent earlier instructions to do no such thing. For many, this will seem a telling objection to Jaworska's position. The judgment of Ulysses' men – who chose to ignore his current wishes in favor of what they took to his authentic/real/objective/critical/long-term wishes – somehow seems more genuinely respectful of his autonomy than would be the alternative of acceding to his present suicidal pleadings. Conceding, however, that there is some appeal to Jaworska's thin conception of autonomy, at least when one is confronted by an elderly demented patient, one could adopt a policy of allowing individual patients, while they are still competent, to choose for themselves the circumstances in which they would prefer to have their contemporaneous interests set aside in favor of their more reflective (fat-autonomy) interests. That is, those who have begun to suffer from cognitive deterioration but who still qualify as autonomous in the full or strong sense could decide to instruct others whether or not to privilege their subsequent weakly autonomous wishes. Ulysses clearly instructed his men to ignore his subsequent wishes, at least during the period of time when he was under the potentially fatal attraction of the Sirens. But others might wish to instruct their caregivers that they desire to have their choices respected so long as they are capable of giving any sort of reason or rationale for those choices at the time they are made.

Keeping in mind the ethical requirement that research subjects retain the right to withdraw from a clinical trial at any time, each approach would offer a somewhat different answer to the question: at what point does a dementing research subject lose her right to change her mind? For Dworkin, the die will be cast once the subject is no longer capable of formulating a life-plan; for Jaworska it would be when the subject is no longer capable of giving some sort of rationale for her wishes. Deciding when either of these points has been reached in the life of any dementing patient will be a difficult challenge.

Which still leaves unaddressed the question: how should caregivers proceed, with respect to enrolling a currently demented patient as a research subject in a clinical trial when the patient has not given explicit advance instructions, while competent, as to their wishes?

SURROGATE INFORMED CONSENT

People who are suffering from a progressive diminution of their cognitive abilities frequently designate a member of their family or a close friend to become their surrogate decision-maker at that point in time when they have lost competence. They may do so formally by vesting durable power of attorney in this person, or informally by indicating to the treating physicians and nurses that so-and-so is the person to whom the medical team should refer when a decision must be made. When a patient's cognitive impairment has advanced to such a point that she no longer seems competent to make important decisions and, when an important health care decision has to be made, it has now become customary for the medical caregivers to seek instructions from a designated surrogate, when one exists. It does not seem a very big stretch to assign to a proxy decision-maker the further role of deciding whether to consent to the patient's participation in research with therapeutic potential, for this continues to fall under the rubric of "health care decision-making." When, however, the question to be decided is whether the incompetent patient should participate in research with hazards for the patient, but with little or no therapeutic potential, reluctance to accept a proxy decision is likely to be much increased.

Regardless of whether the decision to be made by a surrogate involves consent to therapy, consent to participation in research with therapeutic potential, or consent to become a research subject for a trial which is not intended to be therapeutic for its participants, the key question to be answered is: to what standard should the surrogate appeal in making his decision? The most widely canvassed decision-making standard is one labeled "substituted judgment." On this standard, the role of the surrogate is first to determine what the incompetent patient would herself have wanted and then to instruct the health care providers or researchers accordingly. Suppose, for example, that a surrogate must decide, on behalf of an incompetent elderly patient, whether or not to consent to her participation in a non-therapeutic clinical trial. According to the substituted judgment standard the surrogate should ask himself: What would she (the patient), given her attitudes and values before she became incompetent, have wanted to do in this kind of situation? In this way, the decision which issues from the proxy decision-maker, whether to approve participation or to withhold consent, will be one which seeks to respect the autonomous values of the patient, even though disease has now robbed the patient of the ability to inform us herself about what these values were.

A surrogate decision-maker will sometimes have enjoyed opportunities to discuss with the patient, before her impairment, such issues as the patient's attitude towards participating in clinical trials, with or without therapeutic potential. Such explicit advance discussions are, alas, comparatively rare but, when they occur, they give a surrogate decision-maker confidence that he will be able successfully to represent the wishes of the patient at a time when the patient can no longer represent herself. Much more often, however, there will have been no antecedent discussion between the proxy and the patient about the matter upon which a decision must now be made. The patient herself may not have given much thought to the general question of becoming a research subject, and it is even less likely that she will have given serious consideration to the question of whether she wishes to be enrolled in a clinical trial at a time in her life when she is suffering from advanced dementia. If she has thought about the matter, she may not have reached any clear-cut decision. And even if she has reached such a decision, she is unlikely, as mentioned above, to have communicated it clearly to those close to her.

Thus, although substituted judgment may be widely accepted as the ideal to which health care proxies should aspire, as a practical matter it seems unlikely that proxies will very often possess sufficient evidence to be confident that their decisions on behalf of the patient do indeed approximate closely the ones which the patient would have made for herself. Empirical research tends to confirm the worrying extent to which surrogate decision-making fails closely to track the decisions that the incompetent elderly would have made for themselves (Uhlmann, Pearlman, and Cain 1988). For example, when a series of hypothetical clinical scenarios concerned with life-sustaining treatment decisions is presented to older adults and their potential proxies, the lack of correspondence between the choices made by the old people and the choices made on their behalf by their potential proxies was significant (Zweibel and Cassel 1989; Uhlmann, Pearlman, and Cain 1988).

Part of the difficulty, though only part, appears to rest with the kind of question that is, all too often, posed by the physician to the surrogate. The wrong questions to ask are questions such as "What do you want us to do for the patient?" or "What is your recommendation?" Such vague questions focus the surrogates' attention on their own preferences rather than on the incompetent patients' previously expressed attitudes and values. They tend, thereby, to elicit answers that diverge significantly from what the patient would have wanted. By contrast, recent empirical research indicates that when proxy decision-makers are explicitly asked to make the choice which they believe the elderly person would have wanted for herself, their answers are not only more accurate, they also have the ancillary benefit of reducing "animosity and discord between family members and physicians" (Tomlinson et al. 1990, 60).

Because of the difficulties facing surrogate decision-makers in their efforts to ascertain what the patient would have wanted had the patient been competent to decide for herself, the American College of Physicians [ACP] (1989) has concluded that the only proper role for the substituted judgment standard is a negative one: surrogates ought not to consent to the enlistment of an incompetent person as a subject for experimentation if there is evidence that the subject herself would have refused.[4] Absent such evidence, surrogates should base their decision to give or withhold consent for participation in research solely upon what they understand to be in the best interests of the patient. Thus, if a proxy decision-maker knows from prior discussions with the patient that she would have wanted to participate in research, including non-therapeutic research, but the patient is now incompetent and the proxy believes that participation would not be in the patient's best interests, then the proxy ought to ignore the patient's prior wishes and should refuse permission. At least, this is the case according to the ACP.

The strictures of Hans Jonas, discussed at the outset of this paper, exhort us to view incompetent as the least eligible candidates for recruitment as research subjects, especially when a research project is not intended to be directly therapeutic for its subjects. When the subjects to be recruited are incompetent elderly patients, the moral objection to such recruitment is, indeed, weighty. Wherever and whenever feasible, non-therapeutic research, especially that which involves more than minimal risks, should recruit its subjects from within the group of those who are competent to give valid consent to their own participation. The American College of Physicians, in its position paper cited above, adopts a stance which seems on the face to be only marginally more permissive than that advocated by Jonas. It declares flatly that surrogates should not consent to non-therapeutic research "which presents more than a minimal risk of harm or discomfort" (ACP 1989, 844-5). But the ACP also advocates that a national review body be established "to evaluate and make a final determination on research protocols involving incompetent persons that may not otherwise be allowed under the guidelines set forth here, such as non-therapeutic research which poses more than a minimal risk of harm or discomfort to cognitively impaired subjects (ibid., 846).

Critics might want to object that the ACP position on non-therapeutic research involving incompetent elderly patients manages, simultaneously, to be too restrictive and too permissive. It is too restrictive because it prevents a surrogate decision-maker from following the advance instructions of a competent elderly patient who clearly expresses the wish to participate in future non-therapeutic research, notwithstanding the risks that such participation may carry. Since there are significant potential benefits [as discussed above] accruing to individuals who agree altruistically to participate as research subjects, even when the research is not intended to be therapeutic, the exclusively negative role which the ACP assigns to the substituted judgment standard could be viewed as a diminution of patient autonomy. The ACP guideline seems less than adequate because the purely negative role it allocates to substituted judgment results in a difficult to defend slighting of the earlier attitudes and values of the now demented patient. In this regard, the guidelines for non-therapeutic research, as proposed by Wicclair (1993, 182), seem ethically more satisfactory: "For research without therapeutic potential, surrogates should be instructed that they may consent if ...considering the nature of the proposed study, the potential benefits and harms to subjects, and what is known about the cognitively impaired person, they can reasonably conclude that the person would have consented...." As a safeguard, Wicclair sensibly proposes that "the reasoning of surrogates should be scrutinized by persons who do not have a vested interest in recruiting subjects."

A further ethical doubt about the ACP proposed guidelines for non-therapeutic research on incompetent patients arises when one considers the scope for abuse inherent in their proposal that a national review body be established which would have very broad decision-making powers. It is more than a little worrying, for example, that the national review body would have the power to approve the recruitment of incompetent elderly patients for highly risky clinical trials absent any strong evidence that the patients, while still competent, gave explicit authorization for such participation. Where the research is not intended to provide health benefits to the patients who become subjects and, where the risks of harm or discomfort to vulnerable subjects are more-than-minimal, it would seem ethically mandatory to eschew the recruitment of such incompetent patients unless one had solid evidence that they had given advance authorization for such participation. Guidelines that do not absolutely require such advance authorization run the risk of undermining the social contract between citizens and society, and raise the possible specter of vulnerable patients being unjustly exploited as "guinea pigs" solely for the benefit of others.

CONCLUSION

As this *tour d'horizon* has made clear, it is no easy matter to chart an ethically defensible course through the minefield of problems which fit under the label "the ethics of research on elderly subjects." Striking the right balance between such competing objectives as promoting socially important medical research, on the one hand, and protecting the dignity and the interests of vulnerable elderly patients, on the other, requires that each god be given his or her proper due. If our sole objective were to encourage the most rapid possible development of effective therapies for the diseases that afflict an aging population, we would view restrictive guidelines as "bureaucratic red tape," and would favor maximal de-regulation. We would favor a policy of loosening or eliminating governmental "fetters" so that the medical/scientific research community could "get on with the job." If, on the other hand, our sole objective were to provide maximal protection against harm to elderly patients, especially those who are vulnerable because cognitively impaired, we would favor total exclusion of such patients from the pool of those eligible for non-therapeutic research.

Simplicity has its attractions, but both of these positions, when stated in such an over-simplified way, become self-defeating. A society which stressed the importance of research but which neglected to stress the rights and interests of potential research subjects would be one in which the research enterprise itself would fall into such disrepute that the stream of both volunteers and financial support on which it crucially relies would dwindle markedly. However much the public might cheer new research advances, public confidence in the moral integrity of the research process is a frail plant that could easily sustain damage if the research process came to be seen as exploitative of the most vulnerable members of society. On the other hand, the protection of potential research subjects, especially those who are most vulnerable to harm because of their cognitive impairments, could easily mutate into the sort of paternalistic over-protectiveness which destroys the very autonomy it seeks to respect and thereby undermines the best interest of the elderly it purports to defend.

The task of negotiating the right balance between such competing foundational values is almost certain to be an ongoing process. Solutions that elicit broad public support at one historical juncture may require to be modified in the light of changing public attitudes and values. Shifts and modest changes of emphasis are to be expected and should be welcomed as part of healthy social evolution. What can be confidently concluded is that any social policy that pursued either objective – promotion of research or protection of research subjects – without serious regard to the other, would be self-defeating. A society that eschewed medical advance in the fight against debilitating disease would be as unappealing as one that set no moral bounds on the pursuit of such advance.

Arthur Schafer, Professor of Philosophy and Director, Centre for Professional and Applied Ethics, University of Manitoba, Canada.

NOTES

[1] The "Teeth" Trial (Trial to Enhance Elders' Tooth and Oral Health) is an ongoing double-blind multi-center randomized controlled trial designed to evaluate the effect of chlorhexidine rinses on tooth mortality in the elderly, financially supported by the U.S. National Institutes of Health.

[2] The National Commission (1978) proposes regulations that define "minimal risk" as "the risk that is normally encountered in the daily lives, or in the routine medical or psychological examination of normal persons", though Ratzan (1980, 33-4) questions whether their characterization of "minimal risk" is as appropriate for elderly as for younger subjects.

[3] As Wicclair (1993, 180) points out, advance informed consent will often not be possible and even when possible it is unlikely, for various reasons, to become a common phenomenon.

[4] The policy advocated in the ACP (1989) position paper would cover such cases as those reported by Warren et al. (1986, 315) in which proxies gave consent for their relatives to participate in research despite their belief that the relatives would not themselves have consented when competent.

REFERENCES

Ackerman, T.F. 1982. Why doctors should intervene. *Hastings Center Report* 12 (4): 14-17.

ACP (American College of Physicians). 1989. Cognitively impaired subjects. *Annals of Internal Medicine* 111(10): 843-8.

Annas, G. J., L. H. Glantz, and B. F. Katz. 1977. *The law of informed consent to human experimentation: The subject's dilemma*. Cambridge: Ballinger.

Avorn, J., and J. Gurwitz. 1990. Principles of pharmacology. In *Geriatric medicine*, 2d ed., eds, C.K. Cassel, et al., 66-77. New York: Springer-Verlag.

Beecher, H. K. 1970. *Research and the individual: Human studies*. Boston: Little Brown.

Bell, J.A., F.E. May, and R.B. Stewart. 1987. Clinical research in the elderly: Ethical and methodological considerations. *Drug Intelligence and Clinical Pharmacy* 21: 1002-7.

Berkowitz, S. 1978. Informed consent, research, and the elderly. *The Gerontologist* 18(3): 237-243.

Brody, E.M. 1977. Environmental factors in dependency. In *Care of the elderly: Meeting the challenge of dependency* eds. A.N. Exton-Smith, and J.G.Evans, 81-95. New York: Grune and Stratton.

Cassel, C. K. 1988. Ethical issues in the conduct of research in long term care. *The Gerontologist* 28(3 Supp): 90-6.

Drane, J. 1985. The many faces of competency. *Hastings Center Report* 15(2): 17-21.

Dresser, R. 1986. Life, death, and incompetent patients: Conceptual infirmities and hidden values in the law. *Arizona Law Review* 28(3): 373-405.

Dworkin, R. 1993. *Life's dominion*. New York: Alfred A. Knopf.

Freedman, B. 1975. A moral theory of informed consent. *Hastings Center Report* 5(4): 32-9.

Goldstein, J. 1978. On the right of the institutionalized mentally infirm. Paper prepared for the National Commission for the Protection of Human Subjects of Bio-medical and Behavioral Research, *Report and recommendations: Research involving those institutionalized as mentally infirm*. DHEW Publication No. (OS) 78-0006, Washington.

Halper, T. 1980. The double-edged sword: Paternalism as a policy in the problems of aging. *Millbank Memorial Fund Quarterly / Health and Society* 58(3): 472-99.

Hardwig, J. 1990. What about the family? *Hastings Center Report* 10(2): 5-10.

James, G. 1978. "Clinical research in achieving the right to health. *Annals of the New York Academy of Sciences* 169: 301. Cited in R. Q. Marston. Medical science, the clinical trial and society. In *Contemporary issues in bioethics*, eds. T. L. Beauchamp, and L. Walters, 407-410, (Encino, Calif.: Dickenson Publishing), 409.

Jaworska, A. 1999. Respecting the margins of agency: Alzheimer's [sic] patients and the capacity to value. *Philosophy and Public Affairs* 28(2): 105-138.

Jonas, H. 1969. Philosophical reflections on experimenting with human subjects. *Daedalus* 98(2): 219-47.

Katz, J. 1984. *The silent world of doctor and patient*. New York: The Free Press.

Leader, M.A., and E. Neuwirth. 1978. Clinical research and the noninstitutionalized elderly: A model for subject recruitment. *Journal of the American Geriatric Society* 26(1): 27-31.

MacIntyre, A. 1981. *After Virtue*. Notre Dame, Ind.: University of Notre Dame Press.

National Commission for the Protection of Human Subjects of Biomedical and Behavioral Research. 1978. *Report and recommendations in research involving those institutionalized as mentally infirm*. DHEW Publication No. (OS) 78-0006, Washington.

Nelson, J. Lindemann. 1992. Taking families seriously. *Hastings Center Report* 22(4): 6-12.

Ratzan, R.M. 1980. "Being old makes you different": The ethics of research with elderly subjects. *Hastings Center Report* 10(5): 32-42.

Ramsey, P. 1970. *The patient as person*. New Haven: Yale University Press.

Sachs, G.A., and Christine K. Cassel. 1990. Biomedical research involving older human subjects. *Law, Medicine & Health Care* 18: 234-43.

Schafer, A. 1981. The ethics of research on human beings: A critical review of the issues and arguments. In Research advances in alcohol and drug problems, eds. Y. Israel, F.B. Glaser, H. Kalant, R.E. Popham, W. Schmidt, R.G. Smart, 475-511. New York: Plenum Press, 1981.

————. 1982. The ethics of the randomized clinical trial. *New England Journal of Medicine* 307(12): 719-720.

————. 1988. Civil liberties and the elderly patient. In *Ethics and aging: The right to live, the right to die*, eds. J.E. Thornton, and E.R. Winkler, 208-14. Vancouver: University of British Columbia Press.

Strain, L.A., and N.L. Chappell. 1982. Problems and strategies: Ethical concerns in survey research with the elderly. *The Gerontologist* 22(6): 526-31.

Taub, H.A. 1980. Informed consent, memory and age. *The Gerontologist* 20(6): 686-90.

Tomlinson, T., K. Howe, M. Notman, and D. Rossmiller. 1990. An empirical study of proxy consent for elderly persons. *The Gerontologist* 30(1): 54-64.

Tymchuk, A., J.G. Ouslander, and N. Rader. 1986. Informing the elderly: A comparison of four methods. *Journal of the American Geriatric Society* 34(11): 818-822.

Tymchuk, A.J., J.G. Ouslander, B. Bahbar, and J. Fitten. 1988. Medical decision-making among elderly people in long term care. *The Gerontologist* 28(3 Suppl): 59-63.

Uhlmann, R. F., R. A. Pearlman, and K. Cain. 1988. Physicians' and spouses' predictions of elderly patients' resuscitation preferences. *Journal of Gerontology* 43: M115-21.

Warren, J. W., J. Sobal, J.H. Tenney, J.M. Hoopes, D. Damron, S. Levenson, B.R. DeForge, H.L. Muncie Jr. 1986. Informed consent by proxy: An issue in research with elderly patients. *New England Journal of Medicine* 315(18):1124-8.

Wicclair, M. R. 1993. *Ethics and the elderly*. New York: Oxford University Press.

Williams, M.E., and F.C. Pannill. 1982. Urinary incontinence in the elderly. *Annals of Internal Medicine* 97: 851-7.

Young, E.W.D. 1978. Aging and the aged: Health care and research in the aged. In *Encyclopedia of bioethics*, ed. Warren T. Reich, 65-9. New York: Free Press.

Zimmer, A. W., E. Calkins, E. Hadley, A. Ostfield, J. Kaye, and D. Kaye. 1985. Conducting clinical research in geriatric populations. *Annals of Internal Medicine* 103(2):276-83.

Zweibel, N.R., and C.K. Cassel. 1989. Treatment choices at the end of life: A comparison of decisions by older patients and their physician-selected proxies. *The Gerontologist* 29(5): 615-21.

CHAPTER ELEVEN

DAVID C. THOMASMA

COMMUNITY CONSENT FOR RESEARCH ON THE IMPAIRED ELDERLY

It has been noted often that the elderly do not participate enough in research protocols, though they require a substantial proportion of health care interventions that benefit from research. Less than one percent of elderly persons sign on to research projects, but persons over 65 years of age require twice the health care dollars as younger generations, and those over 85 expend three times as much on health care (Callahan 1990). It seems important from the standpoint of justice to raise the number of elderly who could and should participate in biomedical research. One of the major impediments to this increase is the fact that a large number of projects that might benefit the elderly regard research on neurological diseases that impair their ability to consent, not the least of which are projects on Alzheimer's disease and on strokes (Whitehouse 1998). There are also other considerations as explored by Professor Schafer earlier in the preceding chapter.

The bedrock of research ethics since the Nuremberg Code and the subsequent Helsinki Accords has been the doctrine of informed consent. This paper, complementing two others I have written on neurologically impaired persons as possible subjects for research (Thomasma 1996; 2000a), highlights a strategy for entering partially-incompetent and neurologically impaired subjects on research protocols through a process of community, rather than personal, consent. It represents a movement away from an individualistic view of consent and, to some extent, modifies traditional norms in biomedical and behavioral research. The bulk of the paper is formed by supportive arguments in favor of this position as well as by caveats and necessary conditions for implementing the idea.

INFORMED CONSENT

Protecting the vulnerable from harm is a virtue essential for modern scientific and medical research. The Nazi experience was the first time in human history that a powerful state marshaled the forces of medicine to its own ends. Those ends were advancements of the state's interests in the world and the "hygiene" of its populace

D.N. Weisstub, D.C. Thomasma, S. Gauthier & G.F. Tomossy (eds.), Aging: Decisions at the End of Life,
207-226.
© *2001 Kluwer Academic Publishers. Printed in the Netherlands.*

so that the war effort could be most efficient. As a result, individuals were valued for their contribution to the state. Those who could not so contribute were regarded as "ballast existence" on the state and proper candidates for euthanasia (Thomasma 1994).

To counter both the manipulation of human beings by medicine and the state, and the very power of those forces over individuals (Thomasma, Pellegrino 2000), the requirement of informed consent was born from the Nuremberg trials after the war and subsequent revelations about research in the United States like the Tuskegee Syphilis Study. Recall that informed consent is only a minimal way to protect individuals and honor their personhood. Essentially it defines the borders beyond which we cannot go without the participation of the individual. Without consent we cannot reduce a person to an object for any purpose (Faden, Beauchamp 1986). This is not the same as honoring and promoting individual worth, no matter what the circumstance of their neurological condition (van Leeuwen, Vellinga 2000).

Revelations about the U.S. government and scientific community conducting radiation research without consent on the retarded and pregnant women in the 1950's and 1960's demonstrates that the Nazis did not have a lock on the cavalier treatment of vulnerable populations. Boundaries must be established on the power of the state and on the power of science over individuals in our society. The subsequent Helsinki Accords underscore the importance of the Nuremberg Code's emphasis upon individual rights as a check on the power of modern medicine, as well as the creation of a partnership between patients and physicians in advancing medical knowledge.

Difficulties have arisen over the years with the concept of informed consent, particularly with regard to what counts as sufficient information and what counts as true consent without some form of coercion. These may be called conceptual difficulties. Constant and continued work on clarifying our ideas regarding consent is needed.

For the past 50 years there has also been a set of problems with informed consent we might call procedural problems. These have to do with implementing our theories of informed consent. As a result of problems implementing our conceptions of full and free consent, for example, IRBs have added to the consent form a host of additional sections detailing who will benefit from the research, who will bear the costs of treatment if there are reactions, proximate and remote possible side-effects, economic risks, and the like. An example of a current concern under this heading of procedural problems is whether an open-label extension study would unduly induce potential subjects to sign on to a double-blind study (Micetich 1996).

A third set of problems relate to significant violations, modifications and alterations of the doctrine itself. This is the locus of my argument in this paper.

MODIFICATIONS OF INFORMED CONSENT

The most reprehensible modifications of the doctrine of informed consent have occurred in willful violations of the doctrine. Some of the violations occurred anterior to the explicit formulation of the doctrine into the Nuremberg Code, even though it was a practice to inform subjects when enlisting them in research projects (Wigodsky, Hoppe 1996). Here, in addition to Nazi physicians (Annas, Grodin 1992), one should count Japanese physicians experimenting on Chinese prisoners during the Second World War (Chen 1997), the Tuskegee Syphilis experiments (Roy 1995), and many American researchers doing radiation experiments on subjects without their consent during and after the Nuremberg Code was developed (Advisory Committee on Human Radiation Experiments 1995).

A different form of modification occurs when public proposals are made to modify the doctrine, entailing discussion and subsequent Governmental approval. For example, the FDA has approved controversial exceptions to the rule that requires an individual's informed consent in the U.S. These exceptions govern potentially beneficial research on currently incompetent patients, e.g., patients who have suddenly become incompetent due to trauma head injury or stroke. I will discuss this plan as an example of community consent later. It is instructive to note at this point that in order to proceed this research must be widely publicized in the community from which patients are drawn. A presumption in favor of being put on the protocol is implied by this model, a point I will elaborate shortly.

Resolving problems about informed consent is never more difficult than in cases of presumed incompetence regarding research that may benefit a class of persons similarly affected by the causative agent of the incompetence, e.g., Alzheimer's patients or schizophrenics. For those who have been declared legally incompetent, mechanisms exist whereby the state enters *in loco parentis* and appoints through its legal system a guardian to speak for the patient, or at least designates such guardians or surrogates in the law (Tomossy and Weisstub 1997). For the most part, however, such designations occur with respect to clinical practice in which the incompetent person can be said to directly benefit from a plan of proposed treatment. What of research on such patients?

TYPES OF RESEARCH

The traditional distinction between therapeutic and non-therapeutic research has also been expanded and refined over the past 50 years. Today we speak of "expected benefit research, direct and indirect," where the purpose of the research is therapeutic either directly aimed at the subjects on the protocol, or indirectly aimed at them by focusing upon a class of patients to which the subject belongs by reason of disease, illness, or accident. Examples of direct benefit research would be studies of drugs designed to help prolong the life of AIDS patients, drugs that decrease the risk of an additional heart attack, and so on, when these studies are done on patients with these disorders. Examples of indirect benefit research would be studies not designed to directly benefit the subject, but to benefit future generations of patients with the subject's disease. Examples would be Interleukin II studies on terminally ill cancer patients, or drugs to slow and arrest the process of Alzheimer's Disease. The patients who sign on to these studies often do so in the hopes that they may personally benefit, even though they are warned in the consent form that this will probably not be the case, but their motives also include the possibility of helping others who may contract the disease in the future.

By contrast one could speak of non-benefit research as encompassing research in which the purpose is to study a potential therapy or device by measuring its impact upon the human person. Examples here would be Phase One studies of dosages, experimenting on contrast dyes during routine testing, comparing sleep monitors or new high pressure respirators, and the like.

Many of the most vigorous discussions in IRBs concern projects where either the line is blurred between benefit and non-benefit research, or the potential subject is incapable of fully understanding the intricacies of the research protocol and its side effects. Examples might be: recruiting alcoholics off the street for non-benefit research by offering a place to live and some square meals; doing blood gas research for different respirator setting levels on respirator-dependent respiratory failure patients intended to benefit the class, but most likely will not produce valid scientific results before the current patient expires; or entering incompetent patients by surrogates on protocols not designed for such persons, e.g., the current NIH double-blinded Prostate Cancer Prevention Trial (PCPT) designed to determine if a seven year course of Proscar versus a placebo can significantly reduce the incidence of prostate cancer. Proscar is not an investigational drug and is widely available by prescription for prostate enlargement, but this study represents a potential new usage. Could an adult daughter enter her senile father in such a program?

Two examples of research protocols that have prodded my thinking about community consent/models are the following.

Schizophrenics were sought out by physicians at UCLA for participation in a study to examine the effectiveness of a new drug in controlling their disease. Thus the study had presumed beneficial effects on them and on the class of persons affected by the disease. Nonetheless, as part of the study design there was a deliberate effect of relapse such that the studies are now somewhat infamously called "relapse studies." The patients were taken off their current medications. Some were happy this occurred, and went on a wild manic binge. Patients dislike their current medication anyway since it contains their happiest outbursts and makes them sluggish most of the time. As they came off their medication it is alleged that, besides creating terrible tensions for their loved ones, some even got into criminal trouble by threatening to kill the imposters posing in their loved ones' bodies.[1]

Such patients are legally competent to participate in the research, but are definitely harmed by the study design since the disease is progressive and each relapse sets them back farther when they do go back into remission. More importantly, the families and caregivers of the schizophrenics are also harmed by these studies. Their risks are also increased should their loved one relapse and behave poorly. Indeed, they often are involved to a greater extent in the daily lives of their loved ones than are families of persons with other diseases. They too "have" schizophrenia in the sense that they take on the duties of caring for individuals who suffer from this illness. For these two reasons increased risks to caregivers and long-term duties of caring, families should thus be part of the consent process. As such they constitute the immediate "community" and are affected by the risks and benefits of the research project itself (Shamoo, Keay 1996).

A second example is research on mildly affected Alzheimer's patients in the earlier stages of disease. A public announcement was made by the medical center participating in the project (Loyola University Chicago 1996). A drug is proposed that promises benefit in controlling the effects of their disease in the early stages. This is a Phase One study though, one that evaluates the safety and efficacy of an investigational drug in improving overall performance. Persons already in nursing homes are ruled out.

Persons participating in the study will undergo a year's evaluation of an initial comprehensive examination and seven follow-up visits. Then they are dropped from the study. A key requirement is, besides age and otherwise good health, that volunteers "have a reliable caregiver to accompany them to all physician visits and ensure proper administration of the study drug" (Loyola University Chicago 1996). As in the previous example, then, significant impact will occur upon the caregivers who bring their loved ones in for this research. The research will most certainly have fewer negative effects than the relapse study, although the fact that it is a Phase One trial means that no direct benefit is intended to the patient, but rather to future generations that suffer from this disease.

Both of these studies share a common feature. The subjects are not legally incompetent, but have significant barriers to understanding or consenting. They are especially vulnerable, as are their daily caregivers, because they cannot fully process the consequences of their decisions. Both the vulnerable and their care givers need protection, then, since they are intertwined in the disease process and share the burdens of that process, one suffering them directly, and the caregiver suffering by accepting the duties and responsibilities that emerge from the study. What is needed, then, is a separate model of consent for research involving caregivers in the community of vulnerable elderly persons who suffer some cognitive deficits but have not yet been declared to be incompetent. This is especially needed for studies that are low-risk but might not directly benefit them; instead a future class of patients sharing the bond of the same disease might benefit.

MODELS OF COMMUNITY CONSENT

As noted, the persons I am discussing are not yet legally incompetent. By reason of their disease, they cannot properly process or anticipate the impact of their decision to participate in certain research (but not other research) on their caregivers and loved ones. Thus their autonomy if they do consent to research may impact negatively upon either their own well-being, or upon the well-being of the persons who must take care of them. On the other hand, significant new research is becoming available for these classes of patients. How then can both the advancement of medical science and the proper protection of questionable agents be protected? This is where the community consent model enters.

In the mental health field various consent models already exist. For example, confining a person because she is a "clear and present danger to herself or others" is a therapeutic tool that both helps the patient in distress and protects the community. Consent is obtained if possible, but not obtaining it is not considered a barrier to effective control. Problems emerge especially when family and others wish to confine individuals against their will. This applies especially to minors who become so troublesome to care for that relief by the family and by caregivers is sought by confining them in institutions (Cichon 1997).

Gentler models of guardianship are now appearing that provide for a balance of tribunal, family, and community interests. Within the Australian Federation such laws enable officials to consider the specific and unique narratives of the participants. These laws give recognition to two important features of community consent models: the role of an assistant decision maker designated by the confused but not yet incompetent adult, and the role of informal (if not authorized) decisions made within families and groups of caretakers (Carney, Tait 1997). Similar models are being formed into law, e.g., in British Columbia, where assisted decision-making is seen as an alternative to adult guardianship. These initiatives, called Representation Agreement Acts,[2] provide for designating an assistant decision-maker, or having a court appoint one, and to various degrees, recognizing the assistance of family and friends in the community about specific decisions. While the bulk of these acts are now in force some concerns exist about implementation of these models among indigenous populations and immigrants, especially if their notions of shared decision-making differ from the state's (Gordon 1997).

While these models are targeted to the best therapeutic interests of the participants, and not toward research, I suggest that they are part of a growing recognition of the importance of the *Umwelt* of the patient and could be analogously applied in clinical research for appropriate subjects.

I have already mentioned the current FDA permission for research on emergency interventions. The idea here is that there are a certain class of drugs and devices that may significantly improve the recovery and/or quality of life of those who have experienced sudden neurological damage from either trauma or stroke, although other research arenas are possible too. When the person first arrives in the emergency room there is no one from whom to obtain consent for this research. Note that most research is on therapies that may benefit the victim. Yet some are Phase One trials from which benefit is only coincidental and serious harm may result compared to those not on a study. Furthermore, interventions will be done that may contribute to life prolonged in a vegetative state or other incapacity when, without such interventions, the patient would have died. Harms therefore are real, but potential benefits are also real. The FDA rules require that hospitals publicize their commitment to a specific research project under these rules in the community. In this way, even if there is no real individual consent for research, there is tacit approval for it by the community. Critics note that this is the first time since the Nuremberg Code and Helsinki Accords that research without consent has been approved by a governmental body (Biros et al. 1998).

There is a major difference between this approval and the Nazi use of vulnerable individuals. This difference lies in the purpose of the research to directly or eventually benefit the class of patients in which the potential subject is found, rather than the state as a whole. The duty to protect the vulnerable, especially the neurologically vulnerable, from harm is not abrogated (Thomasma 1990).[3]

THE OBLIGATION TO PARTICIPATE IN RESEARCH

The "virtue of informed consent" is a negative virtue; it sets requirements limiting the power of others over vulnerable individuals. A more positive virtue would be honoring the individual to a greater extent than we honor those who can contribute to society. This honor consists in ensuring that such persons are not treated as objects, but may include their participation in research that may benefit them or persons like them, suffering from a similar disease. In other words, there is no "dishonor" in participating in research and, in fact, there may be honor attached to it.

I will argue here that there is an obligation to participate in research that is stronger than the general obligation to "do something for humanity," but not so strong as to require that we all participate in research. The obligation flows from being a beneficiary of past generations of human beings who have participated in studies that have advanced the state of art of caring for the disease with which one finds oneself, and from the current dedication of caregivers who have devoted their lives to studying and caring for the class of patients in which one finds oneself.

That being acknowledged, however, the only research that ethically can be conducted upon the neurologically impaired is that which either benefits them directly (direct benefit research), or benefits the class of such beings directly (class benefit research, direct type), or themselves only indirectly, but poses either no or only very mild risk (Weisstub, Arboleda-Florez, and Tomossy 1998). Of course, even for direct benefit or class benefit research to be conducted some consent must be obtained. In this regard, a committee of research consent according to the communal model should both provide that consent and monitor the research progress (Keyserlingk et al. 1995). This research consent committee should include but not be limited to the legally authorized surrogates. It should have at least one member of the class of subjects who have the disease upon which the research is being proposed. Preferably this person or these persons would have undergone similar research in the past. Representatives of advocacy groups should also be included (McNeill, 1993).

For high risk but potentially high benefit research, the neurologically impaired cannot be considered suitable subjects, even if down the line they might profit as a class from this research. A general principle should be that for serious disease, such as schizophrenia, a relapse must be also be considered a serious harm, and its potential, even in direct benefit research, would preclude the ability of any committee to subject a person to such harm. With respect to Alzheimer's Research one might imagine a fetal tissue transplant study that has significant risk and is aimed at learning new techniques for implanting the tissue. This too would not be permitted.

Following an intermediate path such as I propose, society can benefit from research and still protect the vulnerable from harm. This doctrine of protecting from harm has served us well, not only by protecting all individuals in medical research, but especially those who might be the most vulnerable to manipulation. Notwithstanding the moderate view I have proposed, it still involves a communitarian rather than libertarian component, namely, permitting a committee representing the community of the person's care givers to replace a seriously ill person's ability to consent to moderate risk research.

A PRIORI CONDITIONS FOR COMMUNITY CONSENT RESEARCH

Thus far, the persons I have discussed have not yet been declared legally incompetent. By reason of their disease they cannot properly process or anticipate the impact of their decision to participate in certain research (but not other research) on their caregivers and loved ones. Presumed autonomy impacts negatively upon either their own well-being, or upon the well-being of the persons who must take care of them. This can be called a state of legal competence but moral incompetence (Thomasma, Weisstub [Forthcoming]). Thus, certain additional protections must be in place before they can be considered to be suitable candidates for research.

However, at this point in the argument we may also add individuals who are not legally competent who have caregivers and community support, such that the model of community consent and community oversight can equally apply. The reason they can be added to the discussion is that the types of research have been delimited to moderate or low risk research. Both the cognitively impaired group and the legally incompetent group of vulnerable elderly should be allowed to participate in research as an opportunity to discharge a duty. Before proceeding further, let us revisit the argument about the obligation to participate in research.

All human beings are finite and suffer from similar patterns of growth and decline. The progress of civilization rests upon the willingness of past individuals to submit to medical experiments so that either they and/or future generations might profit from them. To some extent every medical intervention is an experiment since it applies general, more abstract knowledge, to each individual case. Every time health professionals and patients interact they increase the pool of knowledge about human health and disease. Thus, the obligation to participate in research is a moderately strong obligation stemming at the very least from: 1) a common bond with all persons of existential finitude; 2) a sense of gratitude for the altruism of others; 3) the virtue of justice toward others as beneficiaries of past acts of generosity; 4) easing the burdens of our own disease if possible. I am tempted to add the additional point much emphasized in the debate about embryonic stem cell research that "it might save lives," but this point does not apply to all potential research on the elderly that may be aimed solely at improving quality of life or even at understanding physiological processes.

We are now able to examine certain conditions that would be needed in order to pursue community consent for the impaired elderly.

Moderate Risk/Benefit Ratio

If there is an obligation to participate in research, then, impediments for fulfilling that obligation should be removed. One of those impediments, of course, is too high a risk and too low a concomitant benefit. Since community consent is a type of benign paternalism on behalf of those who cannot adequately participate in the decision, a first condition must be that the research risk be only moderate and the benefit accrue at the very least to the class of persons suffering from the disease possessed by the potential subject. Under this restriction, no cognitively impaired elderly could be entered on a study that has high risk without direct benefit to the subject, or even moderate risk that has little or no benefit to persons like the subject.

Neutrality About the Research

The discussion above, regarding obligations to participate in research, leads to a second condition as well. The special obligation, as I have called it, to participate in research on diseases which fate has dealt the individual can propel the community decision-makers toward entering the subject on a research protocol only if the subject never raised an objection to it. This can only be done if the subject is ascertained to be morally neutral regarding the research, that is to say, has never explicitly rejected it in the past nor does not object to it at present.

We can presume the subject is neutral toward either no risk or moderate risk, direct benefit research. We cannot be certain that the subject would, if able, consent to or deny participation in such research. Thus, we would not violate necessarily their best interests in deciding to enter them in studies from which they might benefit. For this reason, the normal process of relying upon a surrogate decision-maker, most often a parent or spouse, to decide about direct benefit research would be appropriate, especially since this reliance is expanded to include the daily caregivers, if they are different from parents, spouses, or assisted decision-makers, and representatives of the IRB.

With respect to research that holds less direct benefit, but might benefit the class of patients who have the disease of the potential subject, I propose that we rely upon the communal surrogates to decide if the risk is minimal enough. If the risk increases to moderate, however, the communal model should be employed, with the addition of the primary care physician and a representative of the advocacy groups for patients with this disease. I call the latter an objective, disinterested party (disinterested, that is, from the perspective of the drug company or medical research group sponsoring the study). This would ensure that the patient's advocate would critically examine the benefit/risk ratio for this specific person, rather than, as the IRB already does, for all patients to be entered in the study.

If the risk is high, and the disease seriously impairs the person, then the presumption of best interests would deny the possibility of entering the person in the study. No decision making model needs to be employed since by public policy the surrogates could not be approached with a request to participate in this instance. Restricting high-risk research is a thorny point since a number of important discoveries could be made about strokes or sudden heart attacks that might benefit the subjects. Looked at from the point of necessary paternalism (that is, paternalism exercised on behalf of the subject so that he or she can participate in research despite the absence of true consent abilities) seems to require a limit. Otherwise how would we avoid the specter of overweening domination of another human person?

Hence, based upon this benefit presumption *a priori,* the communal model functions in cases of little or no risk, and expands to include potential but intermediate harm between no risk of harm and serious risk of harm. For no risk direct benefit research, communal surrogates decide. For serious risk, public policy denies access to vulnerable subjects. In the middle are research studies that would be decided on the basis of a committee representing the community's care for the individual, composed of those who know and love the individual, and those who are touched by the disease most directly or are advocates for such persons.

The Dominion Condition

Domination is a concern in many different arenas of modern life, from irresponsible disregard of the environment to "playing God" with embryonic stem cells. A corrective is needed for a tendency of scientific civilization to objectify and manipulate human life. This corrective can be called the Principle of Dominion.

The Principle of Dominion would state that for every intervention into a natural process, special care must be taken for any foreseeable consequences. A corollary would be not to objectify or manipulate for our own ends, however noble, the person for whom the intervention is contemplated (Thomasma 2000b). Since in the matter at hand, that individual cannot make judgments for him or herself, analyses of the consequences on his or her life, and on the lives of others, must be an intrinsic part of the decision to intervene.

Under this principle it would seem to be impossible to justify non-therapeutic research on neurologically impaired individuals. Some direct benefit must be contemplated and designed into the protocol. However this need not be an absolute prohibition, especially if the community of caregivers determines that the risk to the subject is low enough to justify placing him or her on the protocol. The principle of dominion only requires that all reasonable foresight be exercised regarding possible consequences to the subject.

A good analogy might come from doing clinical trials in developing countries. Just as with the neurologically impaired, conducting research in developing countries requires that increased care be taken by the ethics review committees because the population is more vulnerable than most to exploitation due to poverty, the range of disease, or a lack of understanding (Shapiro, Merlin, 2001). If one uses similar cautions about entering the neurologically impaired elderly in clinical trials, one has implemented the extra care required by the principle of dominion.

Tracking Variable Competency

A third condition for the communal model comes from the variability of competency (Drane 1985). There is a wide range of persons who lack the capacity to consent, either by reason of age, maturity, or disability. Among those impaired by reason of age are embryos, fetuses, and children until at least the age of reason and beyond. Among those impaired by reason of maturity are those who have suffered a stroke or suffer from sufficient Alzheimer's to limit or impede their capacity to consent. Disability, too, can cover a wide range of lives, from those whose pain overrides their ability to reason things through, to various genetic or organic physiological impairments. For each of these classes there are special considerations. For example, there is a moral difference between those who have never constructed a biography (e.g., the newborn), and those who have (e.g., a grandmother after a severe stroke). In the former case, we must exercise greater efforts to protect from harm precisely because we do not know the value-system of the subject, while in the latter instances, having in hand some statements or life-histories regarding values, we can take those into account in our considerations about research.

Those whose disability is neurological impairment face a particularly difficult process if they are to participate in neurobiological or any other kind of research. The normal standards of informed consent do not seem applicable, particularly since informed consent doctrine requires on the part of the participant at the very least a capacity to process the information and to apply it one's own circumstances. Neurologically impaired individuals may either understand but be unable to foresee the consequences for themselves or, more often, be unable either to process information or apply it, especially to their future capacities.

The class of neurologically impaired might include both those who are temporarily so because of mood swings or personality disorders, and those who are permanently so because of permanent damage to the brain. The ethics of conducting research on Alzheimer's patients, while an analogue of the ethics of research on schizophrenics, does not fit it exactly, since schizophrenia is capable of being controlled and patients may become competent during periods of remission, while Alzheimer's is a progressive and fatal condition. This point need not be belabored. But it does indicate a difference in the way we ought to protect the vulnerable from harm.

In permanent conditions of incapacity, full-scale paternalism requires extreme caution about subjecting individuals to research, even that which might benefit them directly. In temporary conditions, however, greater reliance on individual wishes while competent must occur. Notwithstanding our commitment to self-determination, the cautions for schizophrenics ought to include the communal model already adumbrated. Thus the person's autonomy with respect to medical research is modified to some extent by the consent and oversight committee, regarded as a sort of "committee of the whole community."

Deeper Risks Condition

A fourth concern supporting the communal model comes from a reflection about the deeper risks of relationship alienation. We are all vulnerable when it comes to modern, scientific research within the context of a medical establishment. The normal one-to-one doctor-patient relationship is disrupted in favor of gaining new knowledge. Nonetheless, one must not forget that health professionals' "act of profession," is a declaration of commitment, the act of "consecration" to use Cushing's word, to a way of life that is not ordinary. In that act health professionals promise that they will not place their own interests first (Pellegrino, Thomasma 1993), that they will not exploit the vulnerability of those they serve, and that they will honor the trust illness forces on those who are ill (Kass 1983). This necessity for a higher standard impelled Plato to use medicine as a paradigm for the ethical use of knowledge. Medicine was for him a *tekne*, a craft and art to be sure, but a craft with a very significant difference from all the others.

The normal doctor-patient relationship proceeds through at least three separate steps (Von Gebsattel 1996; Welie 1995). The first is one of compassion or immediacy. This is a personal stage in which both patient and physician respond to the disability, pain, or suffering of the patient on an empathetic level. Becoming a research subject changes this immediate bond between doctor and patient. The second stage is one of alienation, during which the help to be offered by the physician requires an Objectification of the body of the patient, a categorization, a necessary but somewhat Cartesian step that is normally synthesized in a the third step. In a research model of a subject-researcher relationship, the second stage of the doctor-patient relationship becomes the standard. The third stage is a personal synthesis stage, where the physician returns to the particularities of the patient as a person, in the context of his or her own values (Pellegrino, Thomasma 1988). What is most distressing to the neurologically impaired who have been research subjects is a lack of sensitivity to this final, synthetic stage in the research model. Perhaps expecting such concern for values is too much to ask of a research model (Thomasma 1992; 1993; Pellegrino, Thomasma 1997)?

All medical research, then, involves greater risks than those normally detailed in the consent form regarding the protocol itself. There is a risk of alienation, both from oneself and from the physician. If one is incapable, not only of understanding the nature of the research and its consequences, but also of ascertaining, of processing the fundamental changes in character of the physician-researcher and the relationship with that professional, then the risks of the research are too high to permit the neurologically impaired to participate. I submit that this is what happened in the studies about schizophrenia. The subjects and their families found themselves facing the terror of relapse without having fully comprehended its possibility and the damage it would do to relationships within the family, in society, and with the physician-researchers.

That being said, it is important also to note that modern biomedical research advances in very tiny steps. Most often the possible gain for individual subjects is almost minuscule. Rarely are there major leaps in knowledge provided by research protocols. This makes research on all classes of neurologically impaired individuals highly burdensome on those individuals, not to speak of the imbalance in the risk/benefit ratio. Knowing this ahead of time, even when persons might be in remission from their disease, it is important to circumscribe their autonomy with the deliberations of a committee that can assess the dangers and risks to values, and not just to the physiology of the subject. In this way the results of the research, its impact upon family and society, can also be brought to the table. Should consent be offered, this same care committee can monitor the progress of the research very closely and judge its impact upon those same values, even if the subject has no physical side-effects from the drugs or treatment arm. This monitoring is important, too, since not all risks and impacts can be predicted, especially when we consider the broader risks I have suggested in this section. Further a general monitoring of all research on the vulnerable elderly is certainly to be suggested given today's lack of proper due care in medical research, even on healthy subjects. Although instances, like that at Johns Hopkins in 2001 are still infrequent, adequate monitoring of medical research is still a work in progress.

Transcending Autonomy

A fifth condition for using the communal model deals with not only broadening our concerns about risks beyond physiological ones, but also broadening our respect for the neurologically impaired beyond a concern for autonomy itself.

As is well known, the principle of autonomy lies behind the doctrine of informed consent. We are now able to examine its role a little more carefully. Throughout the world there has been an increased emphasis on autonomy, not only in research, but also in clinical practice. The standard for information given a patient is either what the patient requires in order to participate, or what the average, prudent patient, might wish to know, rather than what the medical profession thinks is appropriate. This shift to the patient is based, not only on autonomy, but also on a deeper commitment to respect persons (Gustafson 1982). In this view, the best interests of the patient cannot be formulated without reference to that patient's own desires and preferences.

James Gustafson has said: "A person has a right to determine his or her own destiny...capacity for self-determination is what makes humans distinctive as a species" (1982, 37). It is a testimony to the strength of our commitment to this belief that concern over the treatment of incompetent patients has now gained so much attention (Monagle, Thomasma 1988). If a person has a right to determine his or her own destiny, what must we make of incompetent patients? Are they no longer persons? Have they forfeited the right of self-determination because they are now incapable of sometimes expressing it? What should be our duties toward them?

My response is that autonomy and beneficence are closely linked: one cannot ignore one for the sake of the other. Tom Beauchamp has pointed out that competency and autonomy are quite separate concepts, with quite separate grammars and considerations (Robertson 1991). Customary arguments about making decisions for incompetent patients draw too close a link between the two concepts. As a result, far too much attention is paid to what patients might wish, or formerly wanted, and not enough to medical indications themselves. That is, too much attention is paid to autonomy and consent-related concerns. Quality of life judgments that employ criteria, such as Substituted Judgment or Reasonable Persons, are tinged with a preoccupation about consent or, better, decision-making, as the basis of our commitment to respect the dignity of human beings. Self-determination or autonomy need not be the only basis of personal dignity. There are many other features.

Examples of features of human dignity not necessarily connected with autonomy and consent are first, other forms of freedom and, second, other dimensions of human existence. There are at least five types of freedom, only one of which may properly be called autonomy, and yet none of which could be lacking without serious impairment occurring to human dignity (Beauchamp 1991; Thomasma 1984). One example can suffice. If autonomy or freedom is too closely identified with freedom of choice (decision-making), one neglects therein the primal freedom of creating new choices or of committing oneself to a single choice.

Similarly, human *dignity* should not be totally identified with autonomy (Weisstub, Thomasma, 2001 forthcoming). Loss of autonomy, almost all physicians, ethicists, and lawyers concede, does not mean a person loses status as a person, and no longer deserves respect. Thus Wilfred Gaylin, perhaps acknowledging an earlier 1972 sketch by Joseph Fletcher regarding indicators of humanhood (Fletcher 1972), developed an argument that human dignity should encompass such elements beyond autonomy as responsibility, past achievements, fundamental genetic human nature, a capacity for technology, and freedom from instinctual fixation. Gaylin (1984) notes: "Modern developments in medicine raised questions about individuals whose autonomy was limited, and yet who commanded a sense of special worth – the infant and the child, the senile, comatose, or retarded person. The conflation of dignity into autonomy threatened their position in the moral world, and compromised our treatment of them."

Generally speaking, then, the principle of autonomy can be overridden for reasons of beneficence when doctors and surrogates such as parents believe that a treatment or research protocol would be in the best interests of a minor or offspring who is neurologically impaired. Thus, minors may assent for a procedure but cannot refuse one that the parents and doctor think is necessary. By analogy, I suggest, human dignity may still be preserved when surrogates consent "parentally" to entering a neurologically impaired person on a research study. Yet the parentalism involved in this surrogate consent, as I have said, requires that the research study directly benefit the subject/patient and present little or no risk. If the risk of harms increases from this level, greater protections must be employed.

There is an important objection to this line of reasoning. The objection comes from a sophisticated analysis of autonomy and incompetence provided by Robert Veatch. According to Veatch, the current preferences of even legally and psychologically incompetent patients should generally prevail. The same commitment of respect, according to Veatch, must be made to those who previously expressed their wishes, although the status of those wishes can only be legally confirmed through Living Wills or a mechanism like the durable power of attorney. Only in this instance does the legal doctrine of substituted judgment make much sense, Veatch argues, as it then would be a process of a guardian determining what the expressed wishes of the patient actually entailed.

Veatch includes two other categories of patients: those who were never competent, and have no relatives or other agents to step into the guardian role. The third class is that of incompetent patients who do have guardians or family members able and willing to act as guardians. For Veatch both of these categories include patients who may once have been competent, but failed to make clear their wishes about the situations which eventually befell them. He argues that decisions can be made for such patients without appeal to subjective, substituted judgment criteria, that is, by substituting an earlier judgment they themselves made for a missing one at this time.

He confronts the fact that when patients have never expressed prior wishes, the principle of autonomy has no further merit. It makes no sense to try to respect a person's autonomy (read, "decision-making") when they have left no clues about what they would wish. Instead, and this is the danger in his position, one no longer aims at the best interests of the patient. Veatch holds these as now indeterminable. Instead, as he says, "the goal is not to serve the patient's best interests, but to honor his wishes out of respect for him" (1984, 667) But how can wishes not made be "honored"?

The willingness to abandon a search for the patient's best interests seems to lie in an identification of respecting persons with honoring their wishes, and in a repugnance for substituting one's own judgment about the quality of life for a missing capacity. This is an important, even necessary component to the whole notion of respect for persons. Consequently, there is much to admire in Veatch's position. But respecting wishes is a necessary but not a sufficient way to respect persons, especially in the absence of a decision-making capacity, precisely what is missing in incompetent subjects. Keeping faith with such patients, then, requires some other kind of judgment.

There is a way to keep that faith and not abandon it, as Veatch's reasoning seems to have forced him to do. As noted, the flaw in that reasoning concentrates respect for persons on decision-making capacity. A better approach broadens our concerns for respecting persons to the locus of their life, the context of care in which they thrive best. Persons responsible for maintaining this support are then able to judge best about the impacts of research on that environment. Recall that that environment is only tentatively stable and reached after a long and painful struggle.

Note that quality of life judgments themselves are not the nemesis of care we often take them to be. The real threat is their subjectivity (Gutheil, Appelbaum 1983). One may make quality of life judgments in an objective manner by assessing physiological and social function and comparing this to the subject's perception (if moderately competent) or to other more "normal" categories of function. No matter how many objective criteria are employed, however, at some point the committee will have to make assessments from the standpoint of their own values as well.

Necessary Parentalism

The sixth condition is necessary parentalism. For those who are most vulnerable in a modern, scientific society, greater care must be taken for their vulnerability. In effect this means that no action can be taken that places these beings and persons at greater risk than others of being objectified and manipulated as if they were without subjectivity and personhood. For the majority of us consent is the only way that such "use" of persons can occur.

For those who, since we are able to agree ahead of time to alienate our subjectivity from an objective body to be studied, either in hopes of obtaining some benefit, or even without such hopes. Further we consent in such circumstances to suspend the normal stages of a doctor-patient relationship in favor of altruistic ends both the subject and the researcher presumably employ. Put another way, how can one maintain beneficence in face of uncertain benefit? The usual method is to calculate beneficence on the basis of the wishes of the patient, thus identifying what is best for the patient with his or her previously expressed wishes.

However, in dealing with incompetent patients it is not always the case that one's best interests are served by a respect for prior wishes, or an effort to determine what someone would wish were that person now competent. Even though earlier, under the neutrality condition, I argued that the subject must never have explicitly ruled out research of the type being considered, if there had been a general and not specific advance directive against medical research, it may be possible to ignore that in favor of therapeutic research at present on the presumption that the individual had not anticipated this kind of potential benefit.

Indeed, there is a standard therapeutic assumption that can be applied to the communal model: The physician is justified in treating the disorder that renders the patient incompetent (Thomasma, Pellegrino 1987). By analogy, the committee can place the patient on a study that may remove impediments to competence (e.g., decrease the chances of relapse). In this respect there is nothing essentially sacrosanct about respecting prior wishes of patients who were once competent. Because of the new situation that was perhaps not foreseen, these prior wishes themselves may now be suspect. Morreim (1991) argues that respecting prior wishes of the incompetent may not be the best way to respect autonomy itself. Social pressure to respect prior wishes for utilitarian reasons, given current and future economic pressures, may lead to rapid acceptance of a refusal of therapy (by prior wish). This may come to be the preferred social outcome (Abernathy 1991). If so, either one of two things must give. Either we will no longer seek the best interests of patients (but rather that of society), or we will sacrifice quality of care for inappropriately applied prior wishes.

Two traditional attempts to avoid capricious decision-making in quality of life judgments should now be examined, since the committee would most certainly attempt to employ them in deciding the appropriateness of entering a neurologically impaired person on a study. In general, both often are used to try to decide on treatment plans for incompetent patients.

CONCLUSION

Research on the neurologically impaired and partially competent can be conducted if there is little or no risk to subjects using already existing criteria of surrogate decision-making and IRB requirements and oversight are already in place. Surrogate decision-makers can offer the consent if the subject is currently incapable of offering it. These surrogates can represent the community of caregivers that care for the patient on a daily level and by representatives of the IRB. When appropriate a disinterested, objective human subject should be added to this community consent group. Thus, when individual surrogacy either does not yet exist (since the subject has not been declared legally incompetent) or when it may already exist (a legally appointed guardian) but the research on which the subject is to be entered is not aimed at the neurological problem *per se*, but at other problems the subject may have, then I am suggesting employing additional safeguards that may come from more representation both consenting to and tracking the impact of the research on the subject.

This model expands on methods we already use for assessing the ethics of research, namely, the consensus panel method and combines it with greater oversight by the IRB on the progress of the research. I have sketched in this paper the way that the community of caregivers must become involved in the consent process, because they have an inherent interest in the outcome of the research. As I have detailed this interest, it is more than simply a sense of involvement regarding the possible wishes or values of the subject. It is that indeed. However, I have also suggested that the community of caregivers also stands to either benefit or be harmed by the research and should have a voice in the consent and ongoing monitoring of the research by that very fact.

Thus, the best interests of both the research subject and that subject's caregivers are at stake. Both the subject and the caregiving stakeholders need to be involved in the consent process. The subject's consent, however, is replaced by those very stakeholders. Hence they have a dual duty to the subject and to themselves. The subject's right of non-interference can be overridden for possible research benefit or, with modest risk, for the benefit of the class of those with similar disease, by the community of caregivers. Their judgment about the subject's entering the protocol, however, includes an assessment of possible harms or burdens to themselves as well. This is a legitimate consequence of the community consent model.

David C. Thomasma, Professor and Fr. English Chair of Medical Ethics, Neiswanger Institute for Bioethics and Health Policy, Loyola University of Chicago Medical Center, U.S.A.

NOTES

[1] A trial regarding these allegations is scheduled in Los Angeles County for Oct. 12, 2001.

[2] See, for example, *Representation Agreement Act*, 1996, R.S.B.C., c. 405.

[3] See for example, the State of Maryland, Office of the Attorney General, Research Working Group that has issued a Third Report on a proposed statute to govern participation by decisionally incapacitated individuals, originally issued Aug. 1, 1997. This was again revised and then the initiative was subsequently dropped due to legislative difficulties.

REFERENCES

Abernathy, V. 1991. Judgments about patient competence: Cultural and economic antecedents. In *Competence: A study of informal competency determinations in primary care*, eds. M.A. Cutter, and E. Shelp, 211-26. Dordrecht: Kluwer Academic Publishers.

Advisory Committee on Human Radiation Experiments. 1995. *Final report*. Washington, D.C.: U.S. Government Printing Office.

Annas, G.J., and M.A. Grodin, eds. 1992. *The Nazi doctors and the Nuremburg Code: Human rights in human experimentation*. New York: Oxford University Press.

Beauchamp, T. 1991. Competence. *Competence: A study of informal competency determinations in primary care*, eds. M.A. Cutter, and E. Shelp, 49-77. Dordrecht: Kluwer Academic Publishers.

Biros, M.H., J.W. Runge, R.J. Lewis, and C. Doherty. 1998. Emergency medicine and the development of the Food and Drug Administration's final rule on informed consent and waiver of informed consent in emergency research circumstances. *Academic Emergency Medicine* 5(4): 359-68.

Callahan, D. 1990. *What kind of life: The limits of medical progress*. Washington, D.C.: Georgetown University Press.

Carney, T., and D. Tait. 1997. Caught between two systems: Guardianship and young people with a disability. *International Journal of Law and Psychiatry* 20(1): 141-66.

Chen, Y-F. 1997. Japanese death factories and the American cover-up. *Cambridge Quarterly for Healthcare Ethics* 6: 240-2.

Cichon, D.E. 1997. Third party admission of minors into psychiatric facilities. Paper presented at the XXIInd International Congress on Law & Mental Health, Montreal, Canada, June 19-21.

Derkach, L. 1997. Supported decision-making as an alternative to guardianship: The B.C. model. Paper presented at the XXIInd International Congress on Law & Mental Health, Montreal, Canada, June 19-21.

Drane, J. 1985. The many faces of competency. *Hastings Center Report* 15(2): 17-21.

Faden, R.R., and T.L. Beauchamp. 1986. *A history and theory of informed consent*. New York: Oxford University Press.

Fletcher, J. 1972. Indicators of humanhood. *Hastings Center Report* 2: 1-4.

Gaylin,W. 1984. In defense of the dignity of being human. *Hastings Center Report* 14: 8-22.

Gordon, R.M. 2000. The emergence of assisted (supported) decision-making in the Canadian law of adult guardianship and substitute decision-making, *International Journal of Law and Psychiatry* 23(1): 61-77.

Gustafson, J. 1982. Ain't nobody gonna cut on my head. In *Cases in Bioethics*, eds. R. Levine, and R. Veatch, 37. New York: Hastings Center.

Gutheil, T., and P. Appelbaum. 1983. Substituted judgment: Best interests in disguise. *Hastings Center Report* 13: 8-11.

Kass, L. 1983. Professing Ethically. *JAMA* 249(10): 1305-10.

Keyserlingk, E.W., K. Glass, S. Kogan, and S. Gauthier. 1995. Proposed guidelines for the participation of persons with dementia as research subjects. *Perspectives in Biology & Medicine* 38(2): 319-62.

Leary, W.E. 1996. Subjects in radiation experiment were not informed, panel says. *New York Times* Jan. 31, B6.

Loyola University Chicago. 1996. Patients with Alzheimer's disease needed for Loyola Study. News Release. Sept. 18.

McNeill, P.M. 1993. *The ethics and politics of human experimentation.* Cambridge: Cambridge University Press.

Micetich, K.C. 1996. The ethical problems of the open label extension study. *Cambridge Quarterly of Healthcare Ethics* 5(3): 410-4.

Monagle, J.F., and D.C. Thomasma, eds. 1988. *Medical ethics: A guide for health care professionals.* Rockville, MD: Aspen Publishers.

Morreim, E.H. 1991. Competence at the intersection of law, medicine and philosophy. In *Competence: A study of informal competency determinations in primary care,* eds. M.A. Cutter, and E. Shelp, 93-125. Dordrecht: Kluwer Academic Publishers.

Pellegrino, E.D., and D.C. Thomasma. 1988. *For the patient's good: The restoration of beneficence in health care.* New York: Oxford University Press.

———. 1993. *The virtues in medical practice.* New York: Oxford University Press.

———. 1997. *Helping and healing: Religious commitment in health care.* Washington, D.C.: Georgetown University Press.

Pellegrino, E.D., and D.C. Thomasma. 2000. Dubious premises, evil conclusions: Moral reasoning at the Nuremberg trials. *Cambridge Quarterly of Healthcare Ethics* 9(2): 261-74.

Roy, B. 1995. The Tuskegee syphilis experiment: Medical ethics, constitutionalism, and property in the body. *Harvard Journal of Minority Public Health* 1(1): 11-15.

Shamoo, A.E., and T.J. Keay. 1996. Research with vulnerable subjects: ethical concerns about relapse studies. *Cambridge Quarterly of Healthcare Ethics* 5(3): 373-86.

Shapiro, H.T., and E.M. Merlin. 2001. Ethical issues in the design and conduct of clinical trials in developing countries. *New England Journal of Medicine* 345(2): 139-42.

Thomasma, D.C. 1984. Freedom, dependency, and the care of the very old. *Journal of the American Geriatric Society* 32: 906-14.

———. 1990. The ethics of caring for vulnerable individuals. In *Reflections on ethics,* American Speech-Language-Hearing Assoc., Washington, DC, 39-45. Reprinted in 1992-93, *National Student Speech Language Hearing Association Journal* 20: 122-124.

———. 1992. Models of the doctor-patient relationship and the ethics committee: Part one. Cambridge Quarterly of Healthcare Ethics 1: 11-31.

———. 1993. Models of the doctor-patient relationship and the ethics committee: Part two. *Cambridge Quarterly of Healthcare Ethics* 3: 10-26.

———. 1994. Euthanasia as power and empowerment. In *Medicine unbound: The human body and the limits of medical intervention,* eds. R. Bland, and A. Bonnicksen, 210-27. New York: Columbia University Press.

———. 1996. A communal model for presumed consent for research on the neurologically vulnerable. *Accountability in Research Policies and Quality Assurance* 4: 227-39.

———. 2000a. A model of community substituted consent for research on the vulnerable. *Medicine, Health Care and Philosophy* 3(1): 47-57.

———. 2000b. The principle of dominion. In *The healthcare professional as friend and healer,* eds. D.C. Thomasma, and J.L. Kissell, 133-47. Washington, D.C.: Georgetown University Press.

Thomasma, D.C., and D.N. Weisstub, eds. Forthcoming. *Variables of Moral Capacity,* Dordrecht: Kluwer Academic Publishers.

Thomasma, D.C., and E. Pellegrino. 1987. The role of the family and physicians in decisions for incompetent patients. *Theoretical Medicine* 8: 283-92.

Tomossy, G.F., and D.N. Weisstub. 1997. The reform of adult guardianship laws: The case of non-therapeutic experimentation. *International Journal of Law and Psychiatry* 20: 113-40.

van Leeuwen, E., and A. Vellinga. 2000. Competence, living wills, and natural law. Paper presented at the XXVth Anniversary Congress on Law & Mental Health, July 11.

Veatch, R. 1984. An ethical framework for terminal care decision: A new classification of patients. *Journal of the American Geriatric Society* 3: 665-9.

Von Gebsattel, E. 1996. The meaning of medical practice. Trans. J. Welie. *Theoretical Medicine* 16: 41-72.

Weisstub, D.N., J. Arboleda-Florez, and G.F. Tomossy. 1998. Establishing the boundaries of ethically permissible research with vulnerable populations. In *Research on human subjects: Ethics, law and social policy,* ed. D.N. Weisstub, 355-79. Oxford: Elsevier Science.

Weisstub, D.N., and D.C. Thomasma. 2001 (Forthcoming). Persons, vulnerability, and dignity. In *Personhood in Health Care,* eds. D.C. Thomasma, and D.N. Weisstub., Dordrecht: Kluwer Academic Publishers.

Welie, J. 1995. Viktor Emil von Gebsattel on the doctor-patient relationship. *Theoretical Medicine* 16: 41-72.

Whitehouse, P. 1998. *Genetic testing for Alzheimer disease: Ethical and clinical issues.* Baltimore: Johns Hopkins University Press.

Wigodsky, H., and S.K. Hoppe. 1996. Humans as research subjects. In *Birth to death,* eds. T.K. Kushner, and D.C. Thomasma, 259-69. Cambridge: Cambridge University Press.

GEORGE F. TOMOSSY,
DAVID N. WEISSTUB & SERGE GAUTHIER

REGULATING ETHICAL RESEARCH INVOLVING COGNITIVELY IMPAIRED ELDERLY SUBJECTS

Canada as a Case Study

Canada is experiencing a renaissance in health research. In 1998, the Federal Government responded to the problem of an ailing health research sector[1] by reversing the downward trend of per capita public funding. This was followed in 2000 by the creation of the Canadian Institutes of Health Research (CIHR).[2] This new body, which replaced the Medical Research Council of Canada, has a noble objective:

> to excel, according to international standards of scientific excellence, in the creation of new knowledge and its translation into improved health for Canadians, more effective health services and products and a strengthened Canadian health care system.[3]

Canada has also made important progress towards effective regulation of clinical trials through the implementation of federal regulatory amendments on Sept. 1, 2001.[4] New regulations under the *Food and Drug Act* [hereinafter *Regulations*] oblige sponsors conducting Phase I to III trials to apply under a system now *requiring* (rather than merely encouraging) ethics committee approval and written consent by all subjects.[5] The *Regulations* prescribe: that trials be conducted in accordance with good clinical practice; criteria for composition of research ethics boards, labeling, record-keeping, and monitoring; rules for reporting adverse drug reaction; and inspection powers.

This chapter will address the importance of an effective legal and ethical framework for research involving elderly subjects. First, our discussion of Alzheimer's disease research demonstrates its complexity and the range of ethical issues that are encountered. Second, in our review of international developments, we note the convergence of official policies concerning vulnerable populations and the trend towards stronger regulatory oversight of research activities. Third, we present Canada as a case study of regulatory reform in human experimentation. We identify

D.N. Weisstub, D.C. Thomasma, S. Gauthier & G.F. Tomossy (eds.), Aging: Decisions at the End of Life,
227-254.
© *2001 Kluwer Academic Publishers. Printed in the Netherlands.*

the questionable legality of research involving cognitively impaired elderly subjects in most Canadian provinces, and conclude that further legislative reform should occur across Canada in order to correct this deficiency. Because other jurisdictions are facing similar issues, we submit that the Canadian experience can provide a useful reference point for public policy formulation.

ALZHEIMER'S DISEASE

Thirteen "virtual" institutes were created under the auspices of the CIHR in order to address a wide range of health concerns, which properly includes ethical dimensions.[6] The Institute on Healthy Aging will support research "to promote healthy aging and to address causes, prevention, screening, diagnosis, treatment, support systems, and palliation for a wide range of conditions associated with aging."[7] The creation of an Institute having this special focus recognizes the increasing social relevance of research pertinent to elderly healthcare consumers.

The broad scope of the Institute's terms of reference underscores the importance of adopting a holistic approach to understanding and treating conditions associated with aging, such as Alzheimer's disease (AD).[8] With typical AD, over the span of the expected survival time of eight years (Barclay et al. 1985), a number of clinical milestones have been described, some of which are potentially useful as endpoints for randomized clinical trials (Galasko et al. 1995):[9]

- conversion from mild cognitive impairment to dementia
- loss of selected instrumental activities of daily living
- emergence of psychiatric symptoms
- nursing home placement
- loss of self-care activities of daily living
- death

Throughout the course of the illness, as observed in clinical practice, the patient with AD may exhibit a range of mental states. Patients often show anxiety and depression early in the course of the disease, and neuropsychiatric manifestations will emerge at the intermediate stage, to abate in the late stage where motor signs become prominent. Cognitive and functional decline tend to be more linear, whereas caregiver burden peaks and decreases in parallel to the neuropsychiatric symptoms (Zarit et al. 1986). Consequently, the study of AD must take into consideration a full range of preventative, treatment, palliative and social aspects, including for the patient, family, and caregiver.

The importance of prevention is illustrated in Table 1 below, which demonstrates increasing cost of care as the severity of AD progresses. Cognitive loss – reflected by Mini Mental State Examination (MMSE) scores (Folstein et al, 1975) – associated with each stage of four stages of AD (mild, mild to moderate, moderate, and severe) is correlated with increasing functional losses on the (FAST) Functional Assessment Staging scale (Sclan and Reisberg 1992) and annual costs in Canadian dollars (Hux et al. 1998). From a clinical and societal standpoint, it is desirable to delay progression of moderate and severe stages, using early diagnosis and treatment.[10]

Table 1. Correlation between disease stage, cognitive impairment, functional loss, and costs

Stage of AD	MMSE	FAST	Annual Cost
Mild	21-26	3	$9,451
Mild to moderate	15-20	4	$16,054
Moderate	10-14	5	$25,724
Severe	Less than 10	6,7	$36,794

It is also important to consider the range of ethical issues raised by the diagnosis and treatment of AD. A thorough analysis would exceed the scope of this chapter. To list a few examples, the following issues that raise ethical implications have been addressed in the literature:

- *Genetic screening:* Presymptomatic diagnosis using apoE genotyping has been found to be inappropriate, given the lack of predictability of this test and the lack of effective preventive therapy (NIA AAWG 1996). This issue will need to be revisited when more sensitive and specific biological tests are available (The Ronald and Nancy Reagan Research Institute 1998).

- *Quality of life/Disclosure:* The use of cholinesterase inhibitors (CI) prolongs the time in stages where patients are aware of their disease, which may exacerbate anxiety and depression. This is best resolved by full disclosure of known benefits and risks of treatment, and treatment of depression prior to the use of CI.

- *Clinical Trials/Disclosure:* Stopping symptomatic drug therapy with CI once the severe stage has been reached entails a risk of an abrupt loss of benefit. This is usually dealt with by an honest discussion with the patient's legal representative(s), with assurance that the CI would be restarted if such a decline were observed. This issue will need to be revisited when treatments that slow progression of disease will be available, since an abrupt decline will not be observed.

- *Justice-Access:* Availability of CI and other drugs to follow is not universal, partly because of failure from families and physicians to appreciate the benefit of therapy. This is addressed by public and continuous medical education activities, to a great extent facilitated by Alzheimer's associations.

- *Justice-Access:* The cost of drugs is relatively high, limiting access for many individuals worldwide. This can be addressed by concerted efforts from regulators, manufacturers, and health professionals to reduce costs of drug development, accept reasonable profit, and use the drugs well, including careful assessment of response over time.

• *Clinical Trial Design:* The availability of CI as symptomatic drugs is changing the perspective on the value of placebo-controlled studies in mild to moderate stages of AD, ranging from refusal (Knopman et al, 1998) to conditional support (Karlawish & Whitehouse, 1998; Farlow, 1998). The middle ground will likely be RCT of different designs based on the hypothesis to be tested, including combination studies (vide infra). This issue is revelant to clinical trials in geriatric psychopharmacology at large (Young & Gauthier, 2000).[11]

The above examples were presented to illustrate the complexity of issues permeating research in AD, many of which are generalizable to the study of other illnesses, including those affecting other populations. This list also emphasizes the importance of ethical discussion and awareness while pursuing the equally worthy objective of scientific advancement. Solutions to most ethical dilemmas in research require a range of responses from education to legal sanctions. While not wishing to detract from the importance of professional education in addressing ethical issues, the remainder of this chapter will focus on the regulatory side of this spectrum.

RESEARCH WITH VULNERABLE POPULATIONS

Ethical and legal responses to the issues raised by research involving vulnerable persons have been canvassed in detail elsewhere (Weisstub 1998, 355-530). It is an area where official policies show a "remarkable and stable consensus" (Brody 1998, 199). Research ethics policies consistently provide guidance about permissible levels of risk, which must be minimized; emphasize the importance of risks not outweighing the expected benefits; and reinforce the requirement of informed consent, whether by the subject or legal representative (Weisstub, Arboleda-Flórez, and Tomossy 1998). With regard to elderly subjects, Weisstub, Verdun-Jones, and Walker (1998) propose a number of ethical and legal safeguards, and Keyserlingk et al. (1995) provide a comprehensive set of guidelines for research specifically on persons with dementia.

The objective of this chapter is not to provide a detailed review of these principles, but rather to emphasize the importance of establishing suitable regulatory controls to protect vulnerable subjects – in this case, the cognitively impaired elderly – that are presently lacking throughout most of Canada. It has been argued previously that Canadian jurisdictions[12] should adopt a consistent legislative framework (Verdun-Jones and Weisstub 1996-97; Tomossy and Weisstub 1997). We submit that this remains the appropriate response.

RECENT INTERNATIONAL DEVELOPMENTS

Developments in Canada – and in other jurisdictions as well, for that matter – can be better understood by viewing them in the context of events that have occurred internationally.

Firstly, one can observe a convergence of official policies in support of the stance that research involving vulnerable populations is desirable and should be legally sanctioned within prescribed ethical parameters.[13] In the case of persons unable to provide consent (such as cognitively impaired elderly subjects), these principles can be summarized as: (i) mandatory consent by the subject's legal representative; (ii) limiting research to conditions affecting the class to which the subject belongs; (iii) the impossibility of carrying out the research on "competent" individuals; (iv) restricting research to no more than "minimal" risk. The importance of convergence is that it provides support for the entrenchment of core principles in a legislative regime.

Secondly, an analysis of international developments, particularly in the United States, illustrates how growing concerns about the global transformation of the research enterprise have exposed the inadequacy of existing regulatory models, particularly in dealing with conflicts of interest in the face of financial pressures on researchers and institutions. The international trend in response is towards stronger independent regulation of research activities.

World Medical Association – Declaration of Helsinki

The *Declaration of Helsinki*, revised in October 2000, continues to serve as a useful international consolidation of ethical principles. Although arguably weaker in terms of its commitment to the primacy of subjects' rights (Brennan 1999) and presenting potential barriers to the use of placebos in clinical trials (Koski and Nightingale 2001, 137), the document has improved in several aspects. For example, the nebulous distinction between "therapeutic" and "non-therapeutic" research has now been abandoned (Levine 1999).[14]

The articles concerning research on subjects unable to give lawful consent were also strengthened, and encapsulate the most important safeguards to protect this population:

- Informed consent *must* be obtained from a legally authorized representative;
- The nature of research should be restricted to that which promotes the health of the population represented and which cannot be performed on legally competent persons; and
- The subject's assent *must* be obtained when it is possible to do so.

It is important, however, to emphasize that the *Declaration* itself has no legal force. Guidelines put forward by international organizations, including by the World Medical Association, World Psychiatric Association (1996), and Council for International Organizations of Medical Sciences (1993), rely upon voluntary enforcement of their precepts by funding agencies, professional associations, and publishers of scientific journals. The consistency and frequency with which these codifications of ethical principles are adopted serve only to strengthen their status as *statements* illustrative of internationally convergent ethical principles.

Council of Europe – Convention on Human Rights and Biomedicine

The Council of Europe's *Convention on Human Rights and Biomedicine* is a much more significant instrument. This multilateral treaty came into force on December 1, 1999.[15] It entrenches fundamental rights of subjects, including specifically for persons who are unable to provide consent. For this group of subjects, the *Convention* enshrines the requirements of: (i) consent by a legal representative; (ii) relevance of the research (having the potential to produce real and direct benefits to the subject); and (iii) impossibility of carrying out the research on individuals capable of providing consent. Unlike the *Declaration of Helsinki*, the *Convention* stipulates an absence of the subject's *objection*, rather than a need to obtain *assent*. It is expected that these principles will be expanded in the near future through a Special Protocol to the Convention, as was done in the area of human cloning.

The potential effect of the *Convention* to harmonize ethical and legal standards on research in Europe is great. A signatory must "take in its internal law the necessary measures to give effect to the provisions of this Convention." Although the exact meaning of this article is under some debate, the *Explanatory Report* released prior to promulgation of the *Convention* supports the interpretation that signatories either apply the Convention's provisions directly into domestic law or enact the necessary legislation to give the provisions effect (Steering Committee on Bioethics 1996, para. 20). Because the treaty is not exclusive to European nations, it has the potential to effect global change as well.

However, in the light of the slow pace of ratifications – only ten to date[16] – and the further possibility of reservations, it is difficult to predict the real impact of the *Convention*. It is also worth noting that, by August 29 2001, the treaty had an additional twenty signatories (not followed by ratification), including France and Switzerland, but regrettably not Germany, the United Kingdom, or Russia.[17] Likewise, none of the non-member states that had participated in the elaboration of the treaty, such as Canada, the United States, Australia and Japan, have exercised their right to sign.

It is hoped that the outstanding ratifications will take place in the near future – including by Canada.

Australia

Tackling these issues at a domestic level, Australia provides a good example of a comprehensive set of ethical guidelines promulgated by a national funding agency. First published in 1966, they were most recently revised and improved in 1999.

The principal limitation of the Australian guidelines is that their effect is inherently circumscribed. The guidelines are applicable only to those institutions that receive funding from the National Health and Medical Research Council, which means that they have no influence over research undertaken by industry, unless it is also carried out by an institution that otherwise receives Council funding. Sanctions are limited to loss of access to or withdrawal of funding (NHMRC 1999).[18] This situation is similar to that in Canada.

The Australian guidelines do not contain specific principles pertaining to elderly subjects. Cognitively impaired elderly would fall under the section "persons with intellectual or mental impairment." For this class of subjects, consent must be obtained whenever possible from the prospective subject or, failing that, from the person, authority, or organization authorized by law to give consent on the subject's behalf. Subject refusal must always be respected, and consent cannot be given (nor can research be approved) where the research is "contrary to the best interests of the person with the intellectual or mental impairment."

A deficiency in these principles is that they fail to prevent research involving incompetent persons that could be carried out on other subjects. Secondly, the "best-interests" provision is problematic since non-therapeutic research, by definition, conveys no direct benefit to the subject and hence, according to a strict interpretation, if it exposes the subject to even a minimal level of risk, would be contrary to a subject's best interests. Arguably, research on health issues affecting the subject could hold out a potential long term or indirect benefit, and thus might not breach the rule; but this justification does, however, require ethics committees to make the difficult calculation of balancing relative burdens (risks) and expected benefits, which could lead to inconsistent results.

The "contrary to the best interests of the person" principle should be replaced by language similar to that used in the *Declaration of Helsinki* – namely, that research should be restricted to that which promotes the health of the population represented and which cannot be performed on legally competent persons.

In any event, the decision to enroll an individual with an intellectual or mental impairment in an experiment ultimately must begin with consent provided by either the subject or a legally authorized substitute. In Australia, except where it is explicitly authorized through legislation,[19] it is unlikely, however, that a substitute decision-maker can legally make this decision, or that an individual can provide legally binding advance consent prior to incapacity (Queensland Law Reform Commission 1996; Mendelson 1996).

United Kingdom

A similar problem concerning the validity of third-party consent exists in the United Kingdom. The Law Commission (1995) specifically noted that non-therapeutic research involving cognitively impaired persons cannot be justified under the common law, and recommended a statutory response to fill this gap, coupled with the establishment of a public regulatory body. Regrettably, after a lengthy public consultation process, the UK Government declared that, although it intended to adopt many of the recommended reforms of the law on mental competency, those pertaining to independent supervision of medical and research procedures would "not be taken forward at this time" (Lord Chancellor's Department 1999).

Guidelines by the Medical Research Council (1991) appropriately caution that substitute consent can only be ethically accepted by a research ethics committee if "that participation would place that individual at no more than negligible risk of harm and is therefore not against that individual's interests." Their revised *Guidelines for Good Clinical Practice* (1998) do not provide further clarification on this issue.

Despite the absence of a legislative response, positive changes have occurred in health policy since 1997, which should lead to significant effects on research regulated by the National Health Service (NHS). A Central Office for Research Ethics Committees was established in August 2000 to respond to "the need to re-visit accountability and operating frameworks of Local and Multi-Centre Research Ethics Committees."[20] As part of broader reforms, the Department of Health published a *Research Governance Framework for Health and Social Care* (2000) in the NHS[21] and launched its official website for research governance on February 28, 2001.

United States

Recent trends in the U.S. suggest reforms on the horizon that could lead to increased surveillance of research activities. The inability of Institutional Review Boards to cope with current workloads is well documented (James Bell Associates 1998; Office of Inspector General 1998; 2000a; 2000b). Globalization of the pharmaceutical industry and the conduct of international clinical trials are raising concerns about the equitable distribution of risks and benefits of research, as well as the need for a uniform set of international standards (Dominguez-Urban 1997; NBAC 2001a; Shapiro and Meslin 2001). The impact of industry forces on the research paradigm is fuelling concerns about the impartiality of investigators in the face of not insignificant financial pressures (Office of Inspector General 2000c; 2000d; Angell 2000; Bodenheimer 2000). Finally, scandalous revelations in the media have heightened public awareness about the insufficiency of existing research subject protections. The tragic deaths of Jesse Gelsinger in the course of experimental gene therapy at the University of Pennsylvania (Nelson and Weiss 2000a; 2000b; Thompson 2000) and of Ellen Roche, a healthy volunteer at the John Hopkins University Medical Center (Levine and Weiss 2001; Argetsinger 2001), are but two recent examples.

The need for regulatory reform has also been recognized at official levels. In April 2000 the Office of Inspector General observed that, although there had been a substantial increase in enforcement by the FDA and National Institutes of Health (NIH) of federal subject protections, minimal progress had been made in terms of enacting its earlier recommendations or those of the Advisory Commission. It suggested that federal legislation might lead to quicker reforms than merely revising the "common rule," bypassing the need to obtain the approval of each of the participating federal departments.

On May 23, 2000, the Department of Health and Human Services (DHHS) announced that it would institute important changes in federal research policy (Shalala 2000).[22] The Office for the Protection of Research Risks (and part of the NIH) was elevated to the Office of the Secretary of the DHHS and renamed the Office of Human Research Protections. This was done to provide it with greater effectiveness as a regulatory body, and assign to it the broad responsibility of monitoring institutions' compliance with the rules governing research subjects. A National Human Research Protections Advisory Committee was also created. In addition to expanding education initiatives, the NIH and FDA would issue improved guidelines on informed consent. Most importantly, the DHHS announced that it would pursue legislation to allow the FDA to levy fines of up to $250,000 per clinical investigator and $1 million per research institution for violations of research regulations (HHS 2000).

Then, on June 8, 2000, the *Human Research Subject Protections Act*[23] was introduced to the House of Representatives. Some of the main objectives of the Act were:

- to apply protections to all human research subjects independent of setting and funding source;
- to require research ethics review of all proposals by an accredited institutional review board; and
- to enhance regulatory oversight of human subject research by formally establishing an Office for Protection of Research Subjects within the Office of the Secretary of Health and Human Services.

These objectives are consistent with the recommendations of the National Bioethics Advisory Commission (NBAC 2001b). Although the Bill had been referred to a House Subcommittee, it should be reintroduced in 2001.[24] The impetus for reform is thus clearly present.[25]

It is worth noting that a gap in the existing U.S. regulatory framework is particularly apparent in the case of adults with impaired decision-making ability, which are not covered explicitly by current regulations.[26] General provisions direct institutional review boards to adopt additional safeguards in the case of vulnerable populations, which are defined as including "mentally disabled persons," and to pay special attention to a subject's potential susceptibility to "undue influence or coercion"; but the regulations are otherwise silent with respect to the problem of impaired decision-making capacity (Bonnie 1997). In response to these concerns, the National Bioethics Advisory Commission released its report on research involving persons with mental disorders in December 1998. This report provides a thorough set of ethical guidelines that are consistent with emerging international principles. In terms of regulatory changes, it was suggested that the "common rule" be revised, with the option of adding a new subpart to govern decisionally impaired subjects.

CANADA – THE CASE FOR REGULATORY REFORM

We present Canada as a case study for reform in research regulation with the view that the Canadian experience can provide a useful reference point for public policy formulation in other jurisdictions. It should be noted that arriving at the present stage of federal amendments has been a lengthy process, and that much still needs to be accomplished.[27] Historically, increasing external (legal) regulatory controls has been strongly resisted by the research community.[28] Previous calls for legislative reform were not embraced, such as by the Law Reform Commission of Canada (1989) and the *Enquiry on Research Ethics* for the Province of Ontario in 1995.[29] While ethical guidelines have existed since 1978,[30] research in Canada, with the exception of the Province of Quebec, has been essentially self-regulating. Excluding federal drug regulations, legislation on human experimentation at the provincial level exists only in the Province of Quebec.

We submit that establishing an effective regulatory framework for research involves the acceptance of certain key principles:

(i)   the ethical imperative in the pursuit of "good" science;
(ii) reliance only on official ethical guidelines (those promulgated or formally adopted by a statutory regulatory authority that also enforces them);
(iii) consistent application of ethical and legal standards to *all* research, whether funded by public or private sources;
(iv) independent oversight of research; and
(v)   entrenchment of core ethical principles and oversight mechanisms in legislation and/or regulations.

This final section will identify how each of these principles have been (or should be) recognized in the Canadian context, and will propose further reforms.

The Importance of Ethics

Amidst the flurry of activity surrounding the creation of the CIHR, it is encouraging to note that the importance of ethics has not been overlooked (e.g. CIHR, Interim Governing Council 2000). Indeed, the statutory mandate of the CIHR requires it to: "foster the discussion of ethical issues and the application of ethical principles for health research."[31] This directive is timely considering the gravity of concerns about the current state of research regulation in Canada. In its parting words to the CIHR, the Medical Research Council's Standing Committee on Ethics (2000) commented:

> As it stands now, academic REBs [Research Ethics Boards] in Canada are seriously over-burdened, little-recognized and under-resourced. The real difficulties and challenges they face have been too easily ignored to date, but cannot go unaddressed.
>
> ... true lines of accountability [between funding agencies and academic research] are far from clear and continuing ethical review is virtually non-existent. Lines of accountability and processes for continuing review are even less developed in non-

academic, privately-funded human research. This is a rapidly growing sector about
which very little is known or understood at the present time.

As noted further by colleagues in a subsequent report, "the creation of the CIHR
provides the opportunity of reconsidering the flawed structures now in place for
managing research ethics in Canada" (Sherwin et al. 2000).

The CIHR (2001b) presently requires all funding recipients to adhere to existing
ethical guidelines, and has adopted the 1998 *Tri-Council Policy Statement*
[hereinafter *Statement*].[32] The *Statement*, however, has limitations inherent in ethical
guidelines in general, and is by no means perfect, requiring further modification. It
is therefore appropriate that the CIHR, in partnership with the other two federal
granting agencies,[33] has created a new governance structure to provide advice on the
"development, evolution, interpretation and implementation" of the *Statement* over
the next five years (CIHR 2001a).

Separately, in 2001, Health Canada has undertaken to establish its own Research
Ethics Board (REB).[34] But of greater significance than developments in the CIHR,
the federal *Regulations* (that took effect on Sept. 1, 2001) are testament to the
centrality of ethics in the research approval process. For example, the requirement of
review by an independent research ethics board, which previously was merely
encouraged, is now enshrined under the definition of good clinical practice.[35]

The Limits of Ethical Codes

The shortcomings of ethical guidelines that are not promulgated by a regulatory
body are threefold:

(i) they cannot sanction conduct that might not otherwise be legally
     permissible;
(ii) they might not provide a judicially recognizable standard of professional
     conduct; and
(iii) they depend on voluntary adoption and enforcement.

Hence, adoption of formal guidelines derived from regulations or legislation is
imperative in order to achieve effective research regulation.

To explain: while adherence to ethical guidelines may be consistent with
professional practice, it may be insufficient to protect researchers against liability for
actions that are not otherwise sanctioned by law, such as in the case of non-
therapeutic research with cognitively impaired adults (Tomossy and Weisstub 1997,
127). As we will discuss below, unless such research takes place in Quebec (or in
British Columbia with a subject's valid advance directive), third-party consent could
be meaningless and should not be relied upon.

For example, in the case of *Weiss* v. *Solomon*, the court did not refer to any
"standard of practice" in reaching its decision, including the MRC Guidelines that
were in use during that period.[36] It is therefore appropriate that the *Statement* –
which replaced the MRC Guidelines – uses the proviso, "subject to applicable legal
requirements," as the preface to its recommendations for the ethical enrolment of
persons who are not legally competent to provide consent.

Looking at the legal utility of ethical guidelines from another standpoint: *non-adherence* to ethical codes or guidelines may not be directly legally actionable, except perhaps through the avenue of breach of contract where specific conditions of compliance are prescribed by an institution, funding agency, or sponsor (Dickens 2000). Moreover,

> Whether any purely implied term, in an investigator's research or employment contract, of observance of an ethical code, etc. is enforceable by, for instance, a subject of study or a university or hospital, is unclear. ... [F]ailures to observe a code etc. of ethical conduct may expose investigators to lawful discipline by professional licensing authorities governing them, but when they are not under such governance no sanction for unethical conduct or for breach of an implied term of a contract may be legally imposable by employing universities, hospitals, etc.   (Dickens 2000)

We acknowledge that non-adherence to an ethical code that is formally adopted by a professional body could lead to sanction by that group, and thereby enhance the regulatory effectiveness of "unofficial" guidelines;[37] but in general, we question whether the *Statement* – or any similar set of guidelines – would serve as a reliable source for adjudicating disputes, such as in litigation involving an aggrieved subject.[38]

On a related point, ethical guidelines – that do not derive from a regulatory authority – can suffer from inconsistent application and enforcement. The *Statement*, for example, applies only to research that is funded by agencies, such as the CIHR, that have adopted it. This then presupposes that these agencies actively enforce implementation of the guidelines. Research that is funded by private sources would fall outside this sphere of influence entirely.

Legislation – Providing Consistency

The Law Commission of Canada (2000) concluded that research governance in Canada is complex and "poses major challenges in terms of consistency, transparency and accountability." It commented further that: "Creating a research ethics culture is one of our most important recommendations, but one of the hardest to implement. It is an area that readily invites window-dressing and stalling."

Bioethicists, lawyers, philosophers, scientists (and industry) have debated issues in research ethics for decades, and will likely continue to do so. Although there remains ample space for academic debate on many issues, a review of international policies in research will reveal that international convergence exists in a number of areas, including with respect to vulnerable populations (Brody 1998). While we agree with the Law Commission's recommendation to conduct further study of governance structures, we submit that fundamental principles in research ethics have reached a sufficient degree of convergence internationally to substantiate their formulation in principled legislation.

Legislative entrenchment of fundamental principles would establish the nucleus from which supporting governance structures and expanded principles could evolve. Researchers must currently draw upon multiple sources, including law, ethics, and professional standards, to determine the boundaries of ethical (and legal) conduct, which coupled with the increasingly international nature of research "creates concomitant levels of complexity, and of contradiction if not confusion" (Knoppers 2000).

In our view, despite the inevitable constitutional hurdles to overcome,[39] legislation of core principles (and creation of a statutory authority, which would promulgate official ethical guidelines) remains the best method to ensure consistency in application and enforcement of ethical principles across Canada. The recent *Regulations* are an important step towards achieving this objective.

Independent Oversight of Research Activities

There is a growing consensus that funding agencies should not occupy the dual role of regulator. This view was made clear in recent reports by the Standing Committee on Ethics, Medical Research Council of Canada (2000), Sherwin et al. (2000), and the Law Commission of Canada (2000); and is consistent with opinions expressed in other jurisdictions, such as Australia (Andrews 1998) and the United States (Fletcher 2001). This position holds that a conflict of interest exists between objectives that promote research (and related commercial opportunities) and those that would delimit research freedom, such as the protection of subjects.

Within this debate, the impact of economic or commercial motivations should not be underestimated. In the United Kingdom, policy documents speak to the funding agency's obligation to achieve "value for the money" (Department of Health (UK) 2001). A similar goal – to reap economic returns from publicly-funded research – has underwritten Canadian policy recommendations (Advisory Council on Science and Technology, Canada 1999). A policy of encouraging entrepreneurship and partnerships between researchers with industry is, admittedly, important within the modern research context; however, the potential impact of conflict of interest should not be ignored.

There is also the issue of enforcement. A funding agency, unless specially empowered to go beyond this function, can only influence those institutions or researchers that receive funding from that agency. It lacks the ability to impose sanctions beyond the withdrawal or denial of funding. Privately funded (and conducted) research generally lies beyond a public funding agency's sphere of control.[40] In environments where the bulk of research derives funding from a public agency, external oversight might be considered unnecessary[41] (that is, if one discounts the conflict of interest argument); however, owing to the rapidly rising level of privately funded research over recent years, this rationalization is becoming untenable.

So, although the mandate of the CIHR is considerably broader than that of its predecessor, it remains (and should be treated as) a funding agency – albeit with a potential to influence federal health policy. Its principal roles in ethics should be promotion and education.[42] This is not say that the CIHR should not apply those sanctions that are available to it; indeed, it should pursue these alternatives diligently. Researchers receiving funding from the CIHR should be required to follow the highest standards of good clinical practice and adhere to ethical guidelines. In the case of a breach, the CIHR should withdraw or deny future funding, or even seek damages for breach of contract. It should do so, however, not in the guise of a regulatory authority but as an institution, entrusted with a specific mission, that must maintain public confidence and enhance its reputation in the international community.

Sherwin et al. (2000) recommended the creation of an independent body to oversee research and protect subjects. In an earlier report, the CIHR Interim Governing Council Sub-Committee on Ethics (1999) indicated that this would require broader governmental action by the Department of Health, as it goes beyond the mandate of the CIHR. The assumption of this responsibility by Health Canada is therefore appropriate, provided that adequate resources are assigned to carry out this function:

> Health Canada will inspect clinical trial sites and trial sponsors to ensure that the generally accepted principles of good clinical practice are met. The objectives of the inspection will be to ensure that participants in clinical trials are not subjected to undue risks, to validate the quality of the data generated or to investigate complaints.
>
> The Minister will use the information collected as a result of these inspections to ensure compliance with the regulatory framework and will take enforcement action, when deemed necessary.[43]

Presumably, a national governance regime would include a system of accreditation of research ethics boards.[44]

Entrenching Core Principles – Federal Regulatory Amendments

Prior to the federal regulatory amendments of Sept. 1, 2001, sponsors of clinical trials had to comply with federal drug regulations that have essentially been unchanged since they were developed in the early 1960s. A 60-day default review period was adopted in 1987. Sponsors were required to maintain accurate records, report adverse drug reactions, and monitor investigators' adherence to the protocol.[45]

The *Regulations* are aimed to satisfy two policy objectives: to create a regulatory environment that is competitive with other jurisdictions – namely, a framework that ensures timely decision-making by the regulator in order to encourage research and development of human drugs in Canada;[46] and to ensure that clinical trials are properly designed and undertaken in such a way that subjects are not exposed to undue risk.[47] In response to the first objective, the default review period was reduced to 30 days. Protection of subjects would result from improved regulatory oversight[48] of sponsors for whom the regulations delineate various obligations, such as labeling, record-keeping, reporting adverse drug reactions, and in particular, adherence to good clinical practices, which are defined as:

> generally accepted clinical practices that are designed to ensure the protection of the rights, safety and well-being of clinical trial subjects and other persons, and the good clinical practices referred to in section.

In addition,

Every sponsor shall ensure that a clinical trial[49] is conducted in accordance with good clinical practices[50] and ... shall ensure that

(a) the clinical trial is scientifically sound and clearly described in a protocol;
(b) the clinical trial is conducted, and the drug is used, in accordance with the protocol...;
(c) systems and procedures that assure the quality of every aspect of the clinical trial are implemented;
(d) *for each clinical trial site, the approval of a research ethics board is obtained before the trial begins at the site;*
(e) at each clinical site, there is no more than one qualified investigator;
(f) at each clinical trial site, medical care and medical decisions, in respect of the clinical trial, are under the supervision of the qualified investigator;
(g) each individual involved in the conduct of the clinical trial is qualified by education, training and experience to perform his or her respective tasks;
(h) *written informed consent, given in accordance with the applicable laws governing consent, is obtained from every person before that person participates in the clinical trial* but only after that person has been informed of
  i. the risks and anticipated benefits to his or her health arising from participation in the clinical trial, and
  ii. all other aspects of the clinical trial that are necessary for that person to make the decision to participate in the clinical trial;[1]
(i) the requirements respecting information and records set out in section C.05.012 are met; and
(j) the drug is manufactured, handled and stored with the applicable good manufacturing practices...   [emphasis ours]

The above conditions constitute an excellent consolidation of principles that are consistent with international standards. The emphasis on subject protection is clearly apparent, particularly with regard to the requirements for ethics committee review (d) and informed consent (h).

However, the current stage of reforms in research regulation should be viewed as "work in progress." A number of issues still need to be resolved, such as a system of accreditation for research ethics boards and elaboration (or adoption) of a uniform set of ethical guidelines to guide them.[51] In addition, further legislative intervention will be required – either federally, provincially, or in some combination – in order to address existing deficiencies in Canadian common law relating to informed consent.

The Need for Further Legislation – Closing Gaps in the Common Law

In the course of carrying out what we would expect to be the next logical steps following implementation of the federal amendments – developing an accreditation system for research ethics boards and promulgating (or adopting) official guidelines or standards of practice for boards and researchers – a federal system of oversight under Health Canada should be able to fill most of the gaps in the regulatory framework. However, further legislative efforts will be required to correct the deficiency in the common law concerning research with persons that are unable to provide a valid consent, such as the cognitively impaired elderly and ensure uniform protection of human subjects throughout Canada.

Quebec is currently the only province with legislation on human experimentation. Article 21 of the *Civil Code*,[52] last modified in 1998, provides for research involving adults unable to provide consent. It distinguishes between experiments conducted on individuals and on groups. In the case of a single subject who is incapable of giving consent as an only subject, an experiment must produce a direct benefit to the person's health. If the experiment is being performed on a group, the benefits of the research project must accrue to other persons having the same disease or handicap, and must be approved and monitored by a competent ethics committee formed by the Minister of Health and Social Services.[53] It should also be noted that, in British Columbia, the definition of "health care" includes "participation in a medical research program approved by an ethics committee." As such, individuals can provide advance consent to participate in research while competent, in anticipation of future incapacity.[54]

But for the rest of Canada, as in Australia[55] and the United Kingdom, the common law would most likely render illegal any non-therapeutic research intervention involving a cognitively impaired adult who is unable to provide an informed consent. In brief, the argument is based on the principle that a Canadian court's *parens patriae* powers cannot sanction interventions for incompetent persons that are not in their best interests. Non-therapeutic research, by definition, provides no direct benefit to the subject. The result of this determination would naturally depend on the level of risk associated with the intervention. Negligible or minimal risk might not attract the censure of the courts. At the very least, non-therapeutic research conducted with this class of subjects stands on shaky ground (Tomossy and Weisstub 1997). The practice might even be considered illegal, and expose researchers to civil liability, and possibly criminal sanctions (Law Reform Commission of Canada 1989; Sava, Matlow, and Sole 1994).

In the absence of statutory guidance, the majority of Canadian provinces must therefore rely on the common law or professional guidelines, such as the *Statement*,[56] neither of which provides a legal justification for the practice. We argue that a proactive response by health policy-makers and the research community is required to protect vulnerable subjects and researchers alike. We propose that uniform legislation on human experimentation with vulnerable populations, which specifically sanctions research within defined parameters, should be enacted across Canada. A legally binding set of ethical guidelines or standards of practice should then be promulgated.

For the reasons discussed earlier, the CIHR cannot fulfill this task. The *Regulations* cannot solve this problem either as the requirement for written informed consent must be "given in accordance with the applicable laws governing consent." By deferring to provincial legislation or the common law, the requirement that "written informed consent...obtained from every person" reinforces the view that, with the exception of Quebec (or possibly British Columbia), research involving a cognitively impaired elderly person (who cannot provide written informed consent) is not permissible.

The task therefore, in the light of constitutional considerations related to division of powers, falls upon the federal Department of Health to work with the Provinces to enact uniform legislation across Canada to correct this gap in the common law and to establish more general protections for human subjects throughout Canada – irrespective of whether the research is privately or publicly funded. This could be accomplished, for example, by implementation of a federal legislative regulatory structure that would permit the provinces to develop similar mechanisms on the basis of equivalency (Starkman 1998, 283-4).

CONCLUSION

The need for greater transparency and accountability of review procedures has grown in the light of global changes to the research culture, primarily owing to its increasingly commercial character. There is a corresponding international trend away from reliance on self-regulation by researchers and toward more effective regulatory oversight by independent regulatory authorities (other than the funding agency) with investigative powers and the ability to impose appropriate sanctions. Such a body should promulgate or adopt official guidelines that would establish a uniform standard of professional conduct, and put into practice a system of accreditation for researchers, research ethics boards, and their members. Finally, core ethical principles, such as with regard to the permissibility of research involving vulnerable populations, have reached an acceptable level of international convergence and thereby merit entrenchment into legislation.

The creation of the CIHR and recent federal *Regulations* should provide the impetus to implement much-needed changes in research regulation. It is submitted that current legal structures in Canada are inadequate, which the *Regulations* go partway to solve. Further reforms should be undertaken to provide a uniform legislative regulatory framework to protect human research subjects across Canada, regardless of whether funding derives from public or private sources.

George F. Tomossy, Ross Waite Parsons Scholar, Faculty of Law, University of Sydney, Australia.

David N. Weisstub, Philippe Pinel Professor of Legal Psychiatry and Biomedical Ethics, Faculté de médecine, Université de Montréal, Canada.

Serge Gauthier, Professor of Neurology, Neurosurgery and Psychiatry, Centre for Studies on Aging, Faculty of Medicine, McGill University, Canada.

NOTES

We wish to thank Prof. Terry Carney of the Faculty of Law, University of Sydney, for his comments on drafts of this chapter.

We are also grateful to the following colleagues who provided us with the benefit of their feedback on Canadian developments: Siddika Mithani, Tim Flaherty, and Karen Reynolds from Health Canada; Bernard Starkman from Justice Canada; and Richard Carpentier from the National Council on Ethics in Human Research.

This chapter is a revised and expanded version of an article published by the first author with the permission of the Centre for Human Bioethics at Monash University: "Regulating Ethical Research: Canadian Developments," Monash Bioethics Review, 2001; 20(4).

Any errors and opinions expressed in this chapter are the responsibility of the authors.

[1]   In Canada, between 1994 and 1998, per capita public funding for health research had experienced a sharp downturn relative to the United States, where per capita funding by the National Institutes of Health nearly doubled during the same period (Coalition for Biomedical and Health Research 1998).

[2]   The CIHR was established under federal statute, the *Canadian Institutes of Health Research Act*, S.C. 2000, c. 6. The CIHR officially "opened its doors" on June 7, 2000.

[3]   *Canadian Institutes of Health Research Act*, s. 4.

[4]   See: Regulations Amending the Food and Drug Regulations (1024 – Clinical Trials). Registration SOR/2001-203 (7 June, 2001). *Canada Gazette Part II*, 135(13): 1116-53.

[5]   Sponsors of Phase IV clinical trials are not required under the *Regulations* to make an application prior to commencing the trial, however, they remain bound by the requirements of good clinical practice, which includes the requirement of ethics committee approval.

[6]   Thirteen virtual institutes, including the Institute on Healthy Aging, were announced on July 25, 2000. The concept behind these institutes is that "Scientific Directors will be guided by an advisory group of Canadians and international experts and supported by the best researchers wherever they conduct their work in Canada or abroad." See http://www.cihr.ca/news/press_releases/00/pr-0004_e.shtml. Scientific Directors were appointed for each on December 5, 2000.

[7]   "Ethics issues related to research, care strategies, and access to care" are included within the scope Institute's terms of reference. See: http://www.cihr.ca/institutes/iha/iha_about_institute_e.shtml. Indeed, the application of ethical principles to health research forms part of the statutory mandate of the CIHR. See: *Canadian Institutes of Health Research Act*, S.C. 2000, ss. 4(g).

[8]   Research on the causes and treatment of Alzheimer's disease (AD) has increased significantly in the past 25 years, not only due to growing prevalence of the illness, but because of a better understanding of pathophysiological processes that are currently being targeted for therapy, including cholinergic neuronal loss, excessive inflammation and oxidation, and amyloid deposition (Cummings et al. 1998). Each of these processes is currently being targeted for therapy.

For example, cholinesterase inhibitors (CI), acting as cholinergic enhancers through prolongation of transmitter half-life within the synaptic cleft, have been partially successful in relieving some of the cognitive and non-cognitive symptoms associated with AD in its mild to moderate stages (Gauthier, 1999). Considering the strength of results from randomized clinical trials of three to twelve months duration, CI are now considered as standard therapy for the relief of symptoms of AD, even though their measurable effects above baseline are short-lived, and are reversible when discontinued (Doody et al, 2001). A more realistic perspective for most patients is a slowing of progression symptoms over at least one year (Winblad et al. 2001). Studies are under way to establish the safety and efficacy of CI in mild cognitive impairment (Geda and Petersen. 2001) as well as in more severe stages of AD (Feldman et al. 2001).

Other agents are under evaluation for delaying progression of disease by altering basic processes such as synaptic and neuronal loss. Anti-inflammatory drugs selective for cyclo-oxygenase 2 are being tested to delay conversion from MCI to AD or delay of progression of AD, as well as anti-oxidants such as alpha-tocopherol which has been tested in moderately severe stages of AD (Sano et al. 1997). Drugs modifying the metabolism of amyloid as well as immuno-therapy against amyloid plaques are in early stages of human experimentation.

9   The study of the natural history of AD has been facilitated by clinical research criteria such as the DSM-IV (APA, 1994) and the NINCDS-ADRDA criteria (McKhann et al. 1984).

10   Symptomatic drugs such as cholinesterase inhibitors (see note 8) are currently considered to be cost-neutral or positive, whereby the cost of the drug is offset by the benefit of the treatment (O'Brien et al. 1999). Unfortunately, access by patients who could benefit from such treatments is limited as the drugs are not reimbursable in many jurisdictions. Within Canada, inequities exist for the reimbursement of AD drugs where, for example British Columbia and the Maritime provinces, have so far refused to pay for them.

11   The parallel groups design has been used more widely so far, but there is considerable interest in time to reach clinical milestones design, as demonstrated by the very high clinical impact of Sano et al. study using vitamin E in later stages of AD (Sano et al.1997). From an ethical point of view these designs involve a placebo-arm, with the possibility of switching to active treatment if patients deteriorate below a predetermined point or has reached the target milestone. The staggered-start/withdrawal design has been suggested by Leber (1997) but has proven to be difficult to apply (Kittner.1999), and involves a prolong washout from active treatment, which in itself if ethically objectionable. A short (6 weeks) withdrawal component to a RCT has been useful to demonstrate the reversibility of CI action over six weeks (Doody and Pratt 1999), but it likely that agents slowing disease progression will show a lack of reversibility during such a washout period, limiting its usefulness. Data from open label extensions of RCT have been successful in demonstrating a sustained therapeutic benefit over many months (Rogers and Friedhoff 1998; Raskind et al. 2000), have the ethical value of sustained active treatment under careful observation, but lack a control group. Additive studies where patients on stable doses of a CI will get a new drug or a placebo will be favored in the future, once the safety and minimal efficacy of a new drug has been established in short placebo-controlled RCT.

12   Legislating matters affecting health are complicated in Canada, which as a federation divides this portfolio across the federal and provincial levels.

13   Schafer (2001), in Chapter 10 of this volume, explores the ethical arguments for and against research involving elderly subjects. For the purpose of our present discussion, we wish to promote the view that relevant research – that is, only on health conditions affecting elderly persons – is both necessary and permissible, provided that it is conducted within an appropriate legal and ethical framework.

14   For a detailed discussion of the problem of classifying research as "non-therapeutic" versus "therapeutic," see: Weisstub and Verdun-Jones (1996-97).

15   See: Council of Europe. *Convention for the Protection of Human Rights and Dignity of the Human Being with Regard to the Application of Biology and Medicine: Convention on Human Rights and Biomedicine.* 1997. European Treaty Series, No. 164. Available at: http://conventions.coe.int/treaty/en/Treaties/Html/164.htm

16   As of August 29, 2001, the following states have ratified the *Convention*: Czech Republic, Denmark, Georgia, Greece, Portugal, Romania, San Marino, Slovakia, Slovenia, and Spain.

17   As of August 29, 2001, the following states have signed but not yet ratified the *Convention:* Bulgaria, Croatia, Cyprus, Estonia, Finland, France, Hungary, Iceland, Italy, Latvia, Lithuania, Luxembourg, Moldova, Netherlands, Norway, Poland, Sweden, Switzerland, the former Yugoslav Republic of Macedonia, and Turkey.

[18] It should be noted however, as Chalmers (2001) explains, that there appears to be *de facto* compliance in private institutions with the *National Statement*, and that a "market" in ethics – where sponsors shop for ethics approval – has not arisen.

[19] For example, it should be noted that in New South Wales, under s. 45AA of the *Guardianship Act 1987*, the Guardianship Tribunal may approve clinical trials involving patients who may not be able to provide consent, provided that it is satisfied that:

   (a) the drugs or techniques being tested in the clinical trial are intended to cure or alleviate a particular condition from which the patients suffer, and

   (b) the trial will not involve any known substantial risk to the patients (or, if there are existing treatments for the condition concerned, will not involve material risks greater than the risks associated with those treatments), and

   (c) the development of the drugs or techniques has reached a stage at which safety and ethical considerations make it appropriate that the drugs or techniques be available to patients who suffer from that condition even if those patients are not able to consent to taking part in the trial, and

   (d) having regard to the potential benefits (as well as the potential risks) of participation in the trial, it is in the best interests of patients who suffer from that condition that they take part in the trial, and

   (e) the trial has been approved by a relevant ethics committee and complies with any relevant guidelines issued by the National Health and Medical Research Council.

A similar provision exists in the recent *Guardianship and Administration Act 2000* in Queensland (see s. 72). These changes occurred as a result of the Queensland Law Reform Commission report on substituted decision-making (1996).

[20] See: Letter from Professor Sir John Pattison to MREC & LREC Chairman & Administrators. (31 Aug 2000). Available at: http://www.doh.gov.uk/research/documents/corecletter.pdf.

[21] "The standards in [the UK] framework apply to all research which relates to the responsibilities of the Secretary of State for Health - that is research concerned with the protection and promotion of public health, research undertaken in or by the Department of Health, its non-Departmental Public Bodies and the NHS, and research undertaken by or within social care services that might have an impact on the quality of those services." See Department of Health (UK) (2001, para. 1.2).

[22] See also: Press Release: Secretary Shalala bolsters protections for human research subjects (May 23, 2000). Available at: http://www.hhs.gov/news/press/2000pres/20000523.html; HHS Fact Sheet: Protecting human research subjects (June 6, 2000). Available at: http://www.hhs.gov/news/press/2000pres/20000606a.html.

[23] See: U.S. House. 2000. *Human Research Subject Protections Act.* 106th Cong., 2nd sess., H.R. 4605, s. 2(b).

[24] Representative Diana Degette (2001) indicated in her statement to the National Human Research Advisory Committee on July 31, 2001, that she would be reintroducing the Act this year.

[25] Prompted by the conclusions reached by the Advisory Committee on Human Radiation Experiments (1995), President Clinton laid the foundation for reforms in research regulation when he established the National Bioethics Advisory Commission in October 1995. Since then, the Commission has completed a series of excellent reports: Cloning (1997), Research Involving Persons with Mental Disorders (1998), Research Involving Human Biologic Materials (1999), Human Stem Cell Research (1999), International Research (2001), and Ethical and Policy Issues in Research Involving Human Participants (2001). Each of these reports can be obtained from: http:///www.bioethics.gov.

[26] 45 CFR 46 (18 June 1991); 21 CFR 50 (General Provisions and Informed Consent of Subjects); and 21 CFR 56 (Institutional Review Boards).

[27] Specifically, we refer to the need for a system of accreditation of research ethics boards, developing effective enforcement mechanisms, and further legislative entrenchment of fundamental ethical and legal principles across Canada.

[28] The Medical Research Council of Canada (1987, 10-12), for example, expressed its view strongly that legislation is not an appropriate response for ensuring ethical conduct in research. This bias continued in the revised *Tri-Council Policy Statement* (p. i.8). However, this traditional bias towards self-regulation by the research community appears to be finally falling into disfavor owing to increased public demands for "assurance that regulatory bodies place public interest ahead of professional interests" (Sherwin et al. 2000).

[29] See: Weisstub DN (Chair). *Enquiry on Research Ethics*. Submitted to the Ontario Ministry of Health (August 1995).

[30] Ethical guidelines were first promulgated by the Medical Research Council of Canada in 1978, and revised in 1987. In 1998, these were replaced in 1998 by a *Tri-Council Policy Statement on Ethical Conduct for Research Involving Human*. The *Tri-Council Policy Statement* was prepared by the three federal granting agencies: MRC (Medical Research Council of Canada) – now the CIHR, NSERC (National Sciences and Engineering Research Council of Canada), and SSHRC (Social Sciences and Humanities Research Council of Canada).

[31] *Canadian Institutes of Health Research Act*, S.C. 2000, ss. 4(e), (g).

[32] Nevertheless, it should be noted that although it has been observed that the Tri-Council Statement contains problems that require attention (Sherwin et al. 2000), the articles that would apply to cognitively impaired elderly do incorporate essential ethical safeguards that have evolved internationally to protect this population. The Statement presents the following core principles:

> Article 2.5
> Subject to applicable legal requirements, individuals who are not legally competent shall only be asked to become research subjects when:
> (a) the research question can only be addressed using individuals within the identified group(s); and
> (b) free and informed consent will be sought from their authorized representative(s); and
> (c) the research does not expose them to more than minimal risks without the potential for direct benefits for them.

The guidelines contain further requirements to obtain a subject's *assent* – or to preclude involvement in the case of *dissent* – as well to provide for ongoing monitoring of consent.

Some further elaboration on special concerns for different subpopulations that would fall under the category of "incompetent individuals" – such as the elderly – would be desirable.

[33] See *supra* note 30.

[34] See: Science Advisory Board, Health Canada. "Executive Summary for SAB Meeting on February 13, 2001": http://www.hc-sc.gc.ca/sab-ccs/feb2001_REB_e.html.

[35] See: *Regulations*, s. C.05.010.(h).

[36] Freedman and Glass (1990) and Glass and Freedman (1991) review the decision of *Weiss* v. *Solomon* ([1989] R.J.Q. 731 (S.C.)) and liability implications for researchers. In this case, the heirs of a volunteer who died in an experiment successfully sued the investigator and his university-affiliated hospital. The judgement of the court did not make any reference to a professional standard of practice and found both the investigator and hospital (through its ethics committee) liable.

[37] A unique situation exists in the Province of Alberta. The Alberta College of Physicians and Surgeons created its own research ethics board to provide a mechanism for those who would not otherwise have ready access to a research ethics committee (such as in medical schools). The province's *Health Professions Act* was also amended so as to enable the College to prohibit the use of private, for-profit research ethics boards (Kinsella 2000, 167).

[38] As a regulatory device for ensuring ethical conduct by researchers, specifically with respect to protocols involving cognitively impaired subjects, the Tri-Council's guidelines are of limited utility. It cannot sanction this practice in provinces other than Quebec, and to a limited extent British Columbia.

Legislation and associated regulations (or official guidelines derived from either of these sources), on the other hand, would provide clear legal standards for professional conduct.

[39] Legislative action in Australia would encounter similar difficulties owing to the division of powers within the federal/state structure (Chalmers 2001).

[40] This is a weakness shared by the current Australian and American models. See text accompanying notes 18 and 23.

[41] See note 18.

[42] The National Council on Ethics in Human Research, created in 1989 as a voluntary non-governmental organization, also serves as a national educational and evaluative resource for research ethics boards in Canada.

[43] See: "Regulatory Impact Statement" accompanying the *Regulations* at 1135. These regulatory changes are in response to a request by a House of Commons Standing Committee in consideration of a report by the Auditor General of Canada:

> That Health Canada independently develop an audit system to monitor clinical trials of unlicensed drugs randomly and at regular intervals;
>
> a)   That Health Canada proceed to make the necessary regulatory changes in order to implement this audit system;

See: Reports of the Standing Committee on Public Accounts to the House of Commons. (Wed. 15 Dec., 1999). In *2000 Report of the Auditor General of Canada – December – Appendix C*. Available at: http://www.oag-bvg.gc.ca/domino/reports.nsf/html/00ac_e.html; and Report of the Auditor General of Canada – April 1999. Other Audit Observations. Available at: http://www.oag-bvg.gc.ca/domino/reports.nsf/html/9900ce.html.

[44] The issue of "Establishing a Canada-wide policy for accreditation of research ethics boards" was canvassed at a meeting of the Science Advisory Board, Health Canada. See: Science Advisory Board, Health Canada. "Agenda, Science Advisory Board, February 13-14, 2001": http://www.hc-sc.gc.ca/sab-ccs/feb2001_agenda_e.html.

The National Council on Ethics in Human Research is expected to release a report on accreditation of ethics boards in the near future.

[45] See: "Regulatory Impact Statement" accompanying the *Regulations* at 1131.

[46] The *Regulations* achieve the objective of creating a rapid approval framework by providing for only a 30-day waiting period. After this period the sponsor may, provided that conditions such as research ethics board approval are satisfied, proceed with the study unless the Minister has given to the contrary (s. C.05.006).

[47] See: "Regulatory Impact Statement" accompanying the *Regulations* at 1130.

[48] The Minister has investigative powers and the discretion to terminate a trial or refuse a sponsor's application where there is a concern in respect of the scientific validity of the protocol or safety of the subjects involved. The conditions where the Minister may refuse an application to commence a clinical trial include:

> (a)   the use of the drug for the purposes of the clinical trial endangers the health of a clinical trial subject or other person;
> (b)   the clinical trial is contrary to the best interests of a clinical trial subject; or
> (c)   the objectives of the clinical trial will not be achieved.

See: *Regulations*, s. C.05.006.

[49] The *Regulations* (s. C.05.001) define "clinical trial" as:

> an investigation in respect of a drug for use in humans that involves human subjects and that is intended to discover or verify the clinical, pharmacological or pharmacodynamic effects of the drug, identify any adverse events in respect of the drug, study the

absorption, distribution, metabolism and excretion of the drug, or ascertain the safety or efficacy of the drug.

50   The *Regulations* (s. C.05.001) define "good clinical practices" as:

> generally accepted clinical practices that are designed to ensure the protection of the rights, safety and well-being of clinical trial subjects and other persons, and the good clinical practices referred to in section.

51   The Therapeutic Products Directorate has produced some specific guidance documents, including on clinical trials in the geriatric population. "Guidance for Clinical Trial Sponsors" (in draft form at the time of writing – http://www.hc-sc.gc.ca/hpb-dgps/therapeut/htmleng/draft_guide_industry.html) and other documents are available http://www.hc-sc.gc.ca/hpb-dgps/therapeut

52   Article 21 of the Civil Code of Québec states:

> A minor or a person of full age who is incapable of giving consent may not be submitted to an experiment if the experiment involves serious risk to his health or, where he understands the nature and consequences of experiment, if he objects.
>
> Moreover, a minor or a person of full age who is incapable of giving consent may be submitted to an experiment only if, where the person is the only subject of the experiment, it has the potential to produce benefit to the person's health or only if, in the case of an experiment on a group, it has the potential to produce results capable of conferring benefit to other persons in the same age category or having the same disease or handicap. Such an experiment must be part of a research project approved and monitored by an ethics committee. The competent ethics committees are formed by the Minister of Health and Social Services or designated by that Minister among existing research ethics committees; the composition and operating conditions of the committees are determined by the Minister and published in the Gazette officielle du Québec.
>
> Consent to experimentation may be given, in the case of a minor, by the person having parental authority or the tutor and, in the case of a person of full age incapable of giving consent, by the mandatary, tutor or curator. Where a person of full age suddenly becomes incapable of consent and the experiment, insofar as it must be undertaken promptly after the appearance of the condition giving rise to it, does not permit, for lack of time, the designation of a legal representative, consent may be given by the person authorized to consent to any care the person requires; it is incumbent upon the competent ethics committee to determine, when examining the research project, whether the experiment meets that condition.
>
> Care considered by the ethics committee to be innovative care required by the state of health of the person concerned does not constitute an experiment.

53   Further regulatory reforms may be forthcoming, including with respect to extending protections to subjects in privately funded research (Ministère de la Santé et des Services sociaux (Québec), *Plan d'action* 1998). However, the Auditor General of Quebec noted that, after three years since the *Plan d'action* was released, the majority of items proposed have yet to be acted on. Moreover, ethical oversight of research was found to be deficient; most hospitals reviewed had failed to adopt appropriate oversight mechanisms for human subjects research conducted at their sites (Le vérificateur général du Québec 2001, §4.4, 4.6).

54   See: Health Care (Consent) and Care Facility (Admission) Act. R.S.B.C. 1996, c. 181; Representation Agreement Act. R.S.B.C. 1996, c. 405.

55   See note 19 and accompanying text.

[56] See: Medical Research Council of Canada, Natural Sciences and Engineering Research Council of Canada, Social Sciences and Humanities Research Council of Canada. *Tri-Council policy statement: ethical conduct for research involving humans.* (August 1998) Available at: http://www.nserc.ca/programs/ethics/english/policy.htm.

REFERENCES

Advisory Committee on Human Radiation Experiments. 1995. *Final report.* Washington, D.C.: U.S. Government Printing Office.

APA (American Psychiatric Association). 1994. *Diagnosis and Statistical Manual of Mental Disorders,* 4th ed. Washington, DC: APA.

Argetsinger, A. 2001. Panel blames Hopkins in research death. *The Washington Post,* 30 Aug., p. B03.

Advisory Council on Science and Technology, Canada. Expert Panel on the Commercialization of University Research. 1999. Public investments in university research: reaping the benefits. (May 4). Available at: http://acst-ccst.gc.ca/acst/comm/rpaper/finalreport4.pdf.

Andrews, K. 1998. Regulating human experimentation. In: *Experimenting with humans: dilemmas for medical research and practice,* ed. J. Martins, 53-70. Hobart, Australia: Catholic Chaplaincy and Faculty of Health Science, University of Tasmania.

Angell, M. 2000. Is academic medicine for sale? *New England Journal of Medicine* 342(20): 1516-18.

Arboleda-Flórez, J., and D.N. Weisstub. 1998. Ethical research with vulnerable populations: The mentally disordered. In *Research on human subjects: Ethics, law and social policy,* ed. D.N. Weisstub, 433-50. Oxford: Elsevier Science.

Barclay, L.L., A. Zemcov, J.P. Blass, and J. Sansone. 1985. Survival in Alzheimer's disease and vascular dementia. *Neurology* 35: 834-40.

Bodenheimer, T. 2000. Uneasy alliance – Clinical investigators and the pharmaceutical industry. *New England Journal of Medicine* 342(20): 1539-44.

Bonnie, R.J. 1997. Research with cognitively impaired subjects: Unfinished business in the regulation of human research. *Archives of General Psychiatry* 54(2): 105-11.

Brennan, T.A. 1999. Proposed revisions to the Declaration of Helsinki - Will they weaken the ethical principles underlying human research? *New England Journal of Medicine* 341(7): 527-31.

Brody, B.A. 1998. *The ethics of biomedical research: An international perspective.* New York: Oxford University Press.

Chalmers, D. 2001. Research ethics in Australia. In: *Ethical and policy issues in research involving human participants. Volume II: Commissioned papers and staff analysis,* A1-66. Bethesda, MD: NBAC. Available at: http://www.bioethics.gov.

CIHR (Canadian Institutes of Health Research). 2001a. Governance structure for the *Tri-Council Policy Statement: Ethical Conduct for Research Involving Humans.* Terms of Reference (2001-2006). Available at: http://www.nserc.ca/programs/ethics/context_e.htm.

———. 2001b. *2001-2002 Canadian Institutes of Health Research general information guide.* Available from : http://www.cihr.ca/funding_opportunities/guide_to_applicants/gag0001/intro/ggamp_e.shtml.

CIHR, Interim Governing Council. 2000. *Final report: where health research meets the future.* (June). Available at: http://www.cihr.ca/about_cihr/interim_gc_report/final_report_english.pdf.

CIHR, Interim Governing Council Sub-Committee on Ethics. 1999. Working Paper – The ethics mandate of the Canadian Institutes of Health Research: implementing a transformative vision. (November 10). Available at: http://www.cihr.ca/about_cihr/interim_gc_report/ethicsenglishfinal.pdf.

CIOMS (Council for International Organizations of Medical Science), in collaboration with the World Health Organization. 1993. *International ethical guidelines for biomedical research involving human subjects.* Geneva: CIOMS.

Coalition for Biomedical and Health Research. 1998. *A crisis in health research. A report based on a survey of Canadian academic health centres.* (January). Available at: http://www.cbhr.ca/briefs/presur.htm.

Cummings, J.L., H.V. Vinters, G.M. Cole, and Z.S. Khachaturian. 1998. Alzheimer's disease. Etiologies, pathophysiology, cognitive reserve, and treatment opportunities *Neurology* 51(Suppl 1): S2-S17.

Degette, D. 2001. Statement by Representative Diana Degette before the National Human Research Protections Advisory Committee. July 31, Rockville, MD. Available at: http://ohrp.osophs.dhhs.gov/nhrpac/mtg07-01/di0731.pdf.

Department of Health (UK). 2001. *Research Governance Framework for Health and Social Care.* (March). Available at: http://www.nhsetrent.gov.uk/trentrd/resgov/govhome.htm.

Dickens, B. 2000. Governance relations in biomedical research. In: Law Reform Commission of Canada. The governance of health research involving human subjects (HRIHS). (May). Available at: http://www.lcc.gc.ca/en/papers/macdonald/macdonald.pdf (section C-1).

Dominguez-Urban, I. 1997. Harmonization in the regulation of pharmaceutical research and human rights: The need to think globally. *Cornell International Law Journal* 30: 245-86.

Doody, R.S., and R.D. Pratt. 1999. Clinical benefits of donepezil: Results from a long-term phase III extension trial. *Neurology* 52(suppl 2): A174.

Doody, R.S., J.C. Stevens, C. Beck, R.M. Dubinski, J.A. Kaye, L. Gwyther, R.C. Mohs, L.J. Thal, P.J. Whitehouse, S.T. DeKosky, and J.L. Cummings. 2001. Practice parameter: Management of dementia (an evidence-based review). Report of the Quality Standards Subcommittee of the American Academy of Neurology. *Neurology* 56: 1154-66.

Farlow, M.R. 1998. New treatments in Alzheimer Disease and the continued need for placebo-controlled trials. *Archives of Neurology* 55: 1396-8.

Feldman, H., S. Gauthier, and J. Hecker. 2001. A 24-week, randomized, double-blind study of donepezil in moderate to severe Alzheimer's disease. *Neurology* [in press].

Fletcher, J.C. 2001. Location of the Office for Protection from Research Risks within the National Institutes of Health: Problems of status and independent authority. In: *Ethical and policy issues in research involving human participants. Volume II: Commissioned papers and staff analysis,* B1-21. Bethesda, MD: NBAC. Available at: http://www.bioethics.gov.

Folstein, M.F., S.E. Folstein, and P.R. McHugh. 1975. Mini Mental State: a practical method for grading the cognitive state of patients for the clinician. *Journal of Psychiatric Research* 12: 189-98.

Freedman, B., and K.C. Glass. 1990. Weiss v. Solomon: A case study in institutional responsibility for clinical research. *Law, Medicine & Health Care* 18: 395-403.

Galasko, D., S.D. Edland, J.C. Morris, C. Clark, R. Mohs, et al. 1995. The Consortium to establish a Registry for Alzheimer's Disease (CERAD). Part XI. Clinical milestones in patients with Alzheimer's disease followed over three years. *Neurology* 45: 1451-5.

Gauthier, S. 1999. Acetylcholinesterase inhibitors in the treatment of Alzheimer's disease. *Expert Opinion on Investigational Drugs* 8: 1511-20.

Geda, Y.E., and R.C. Petersen. 2001. Clinical trials in mild cognitive impairment. In *Alzheimer's disease and related disorders annual 2001,* eds. S. Gauthier, and J. Cummings, 69-83. London: Martin Dunitz.

Glass, K.C., and B. Freedman. 1991. Legal liability for injury to research subjects. *Clinical Investigation and Medicin* 14(2): 176-80.

HHS announces steps to strengthen protection of research subjects. 2000. *News Along the Pike* [Newsletter for the Center for Drug Evaluation and Research, FDA (US)] 6(6): 6-7. Available at: http://www.fda.gov/cder/pike/jun2000.pdf.

Hux, M.J., B.J. O'Brien, M. Iskedjian, R. Goeree, M. Gagnon, et al. 1998. Relation between severity of Alzheimer's disease and costs of caring. *Canadian Medical association Journal* 159: 457-65.

James Bell Associates. 1998. *Final Report: Evaluation of NIH implementation of section 491 of the Public Health Service Act, mandating a program of protection for research subjects.* (15 June). Available at: http://ohrp.osophs.dhhs.gov/hsp_report/hsp_final_rpt.pdf

Karlawish, J.H.T., and P. Whitehouse 1998. Is the placebo control obsolete in a world after donepezil and vitamin E? *Archives of Neurology* 55: 1420-4.

Keyserlingk, E.W., K. Glass, S. Kogan, and S. Gauthier. 1995. Proposed guidelines for the participation of persons with dementia as research subjects. *Perspectives in Biology and Medicine* 38(2) : 319-62.

Kinsella, T.D. 2000. Research involving humans: Current regulatory status of the Canadian medical profession. In: Law Reform Commission of Canada. The governance of health research involving human subjects (HRIHS). (May). Available at: http://www.lcc.gc.ca/en/papers/macdonald/macdonald.pdf (section D-3).

Kittner, B. (for the European – Canadian Propentofylline Study Group). 1999. Clinical trials of propentofylline in vascular dementia. *Alzheimer Disease and Associated Disorders* 13(Suppl 3): S166-71.

Knopman, D., J. Kahn, and S. Miles. 1998. Clinical research designs for emerging treatments to Alzheimer's diseaase: moving beyond placebo-controlled trials. *Archives of Neurology* 55: 1425-9.

Knoppers, B.M. 2000. Ethics and human research: Complexity or confusion? In: *The governance of health research involving human subjects (HRIHS)*. Law Reform Commission of Canada. (May). Available at: http://www.lcc.gc.ca/en/papers/macdonald/macdonald.pdf (section c-2).

Koski, G., and S.L. Nightingale. 2001. Research involving human subjects in developing countries. *New England Journal of Medicine* 345(2): 136-8.

Law Commission of Canada. 2000. The governance of health research involving human subjects (HRIHS). (May). Available at: http://www.lcc.gc.ca/en/papers/macdonald/macdonald.pdf

Law Reform Commission of Canada. 1989. *Working paper no. 61: Biomedical experimentation involving human subjects*. Ottawa: Law Reform Commission.

Leber P. 1997. Slowing the progression of Alzheimer's disease: methodological issues. *Alzheimer's Disease and Associated Disorders* 11(suppl 5): S10-21.

Le vérificateur général du Québec. 2001. *Rapport à l'Assemblée nationale pour l'année 2000-2001. Tome 1*. (Québec : Bibliothèque nationale du Québec). Available at : http://www.vgq.gouv.qc.ca/rappann/rapp_2001_1/Rapport/Index.html

Levine, R.J. 1999. The need to revise the Declaration of Helsinki. *New England Journal of Medicine* 341(7): 531-34.

Levine, S., and R. Weiss. Hopkins told to halt trials funded by U.S. – Death of medical volunteer prompted federal directive. *The Washington Post*. 20 July, p. A01.

Lord Chancellor's Department (UK). 1999. *Making Decisions. The Government's proposals for making decisions on behalf of mentally incapacitated adults*. A Report issued in the light of responses to the consultation paper *Who Decides?* (Cm 4465) October. Available at: http://www.open.gov.uk/lcd/family/mdecisions/indbod.htm.

McKhann, G., D. Drachman, M. Folstein, R. Katzman, D. Price, et al. 1984. Clinical diagnosis of Alzheimer's disease: report of the NINCDS-ADRDA workgroup. *Neurology* 34: 939-44.

Medical Research Council (Canada). 1987. *Guidelines on Research Involving Human Subjects* (Ottawa: Supply & Services Canada).

Medical Research Council (UK). 1998. *Guidelines for good clinical practice*. Available at: http://www.mrc.ac.uk/Clinical_trials/ctg.pdf.

Medical Research Council (UK), Working Party on Research on the Mentally Incapacitated. 1991. The ethical conduct of research on the Mentally Incapacitated. (December). Available at: http://www.mrc.ac.uk/ethics_c.html.

Medical Research Council of Canada, Standing Committee on Ethics. 2000. A proposed ethics agenda for CIHR: Challenges and opportunities – recommendations of the standing committee on ethics to council as it prepares its parting advice to CIHR governing council. (March 2000). Available at: http://www.cihr.ca/governing_council/working_groups/mrc_standing_committee_ethics_e.pdf.

Mendelson, D. 1996. Statutory regimes relating to third party consent by patients with Alzheimer's type dementia. *Psychiatry, Psychology and Law* 3(1): 11-23.

Ministère de la Santé et des Services sociaux (Québec). *Plan d'action ministériel en éthique de la recherche et en intégrité scientifique*. Québec : Gouvernement du Québec, 1998. Available at: http://206.167.52.1/fr/document/publication.nsf/933f276880164d6685256809007069af/4dbb7899d93 b164c85256753004be0a9/$FILE/98_759.pdf

NBAC (National Bioethics Advisory Commission) (US). 1998. *Research involving persons with mental disorders that may affect decisionmaking capacity*. (Volume I: Report and recommendations of the National Bioethics Advisory Commission). Available at: http://bioethics.gov

———. 2001a. *Ethical and policy issues in international research: Clinical trials in developing countries*. (April 30). Available at: http://bioethics.gov

———. 2001b. *Ethical and policy issues in research involving human participants*. (August 20). Available at: http://bioethics.gov/human/overvol1.pdf.

NHMRC (National Health and Medical Research Council) (Australia). 1999. *National statement on ethical conduct in research involving humans.* Canberra: National Health and Medical Research Council, 1999. Available at: http://www.health.gov.au/nhmrc/publicat/pdf/e35.pdf.

National Institute on Aging – Alzheimer Association Working Group. 1996. Apolipoprotein E genotyping in Alzheimer's disease. *Lancet* 347: 1091-5.

Nelson, D., and R. Weiss. 2000a. Penn researchers sued in gene therapy. *The Washington Post,* 19 Sept., p. A03. Available at: http://www.washingtonpost.com/wp-dyn/articles/A30464-2000Sep18.html

———. 2000b. Penn settles gene therapy suit. *The Washington Post,* 9 Nov., p. A04.

O'Brien, B.J., R. Goeree, M. Hux, M. Iskedjian, G. Blackhouse, et al. 1999. Economic evaluation of donepezil for the treatment of Alzheimer's disease in Canada. *Journal of the American Gerontological Society* 47: 570-8.

Office of Inspector General, Department of Health and Human Services (US). 1998. Institutional review boards: a time for reform. (June). Available at: http://oig.hhs.gov/oei/summaries/b276.pdf

———. 2000a. Protecting human research subjects – status of recommendations. April 2000. Available at: http://www.hhs.gov/oig/oei/reports/a447.pdf.

———. 2000b. FDA oversight of clinical investigators. (June). Available at: http://www.hhs.gov/oig/oei/reports/a457.pdf.

———. 2000c. Recruiting human subjects – pressures in industry-sponsored clinical research. (June). Available at: http://www.hhs.gov/oig/oei/summaries/b459.pdf.

———. 2000d. Recruiting human subjects – sample guidelines for practice. (June). Available at: http://www.hhs.gov/oig/oei/reports/a458.pdf.

Queensland Law Reform Commission. 1996. *Assisted and substituted decisions: decision-making by and for people with a decision-making disability.* Vol. 1. Brisbane: Queensland Law Reform Commission.

Raskind, M.A., E.R. Peskind, T. Wessel, and W. Yuan. 2000. Galantamine in AD: A 6-month randomized, placebo-controlled trial with a 6-month extension. *Neurology* 54: 2261-8.

Rogers, S.L., and L.T. Friedhoff. 1998. Long-term efficacy and safety of donepezil in the treatment of Alzheimer's disease: an interim analysis of the results of a U.S.multicentre open label extension study. *European Neuropsychopharmacology* 8: 67-75.

Sano, M., C. Ernesto, R.G. Thomas, M.R. Klauber, K. Schafer, et al., for the members of the Alzheimer's Disease Cooperative Study. 1997. A controlled trial of selegiline, alpha-tocopherol, or both as treatment for Alzheimer's disease. *New England Journal of Medicine* 326: 1245-7.

Sava, H., Matlow, P.T., and M.J. Sole. 1994. Legal liability of physicians in medical research. *Clinical & Investigative Medicine* 17: 148-184.

Schafer, A. 2001. Research on elderly subjects: Striking the right balance. In *Aging: Decisions at the end of life,* eds. D.N. Weisstub, D.C. Thomasma, S. Gauthier, and G.F. Tomossy, 171-205. Dordrecht: Kluwer.

Sclan, S.G., and B. Reisberg. 1992. Functional Assessent Staging (FAST) in Alzheimer's disease: reliability, validity, and ordinality. *International Psychogeriatrics* 4(suppl 1): 55-69.

Shalala, D. 2000. Protecting research subjects – what must be done. *New England Journal of Medicine* 343(11): 808-10.

Shapiro, H.T., and E. Meslin. 2001. Ethical issues in the design and conduct of clinical trials in developing countries. *New England Journal of Medicine* 345(2): 139-42.

Sherwin S., *et al.* 2000. *Integrating bioethics and health law in the Canadian Institutes of Health Research.* Report to the CIHR.

Starkman, B. 1998. Models for regulating research: The Council of Europe and international trends. In *Research on human subjects: Ethics, law and social policy,* ed. D.N. Weisstub, 264-85. Oxford: Elsevier Science.

Steering Committee on Bioethics, Council of Europe. 1996. Convention for the protection of human rights and dignity of the human being with regard to the application of biology and medicine: Convention on Human Rights and Biomedicine – Explanatory Report, para. 20 17 December 1996. Available at: http://conventions.coe.int.

The Law Commission (UK). 1995. *Mental incapacity (Report 231).* London: HMSO. Available at: http://www.lawcom.gov.uk/library/lc231/summary.htm.

The Ronald and Nancy Reagan Research Institute of the Alzheimer's Association and the National Institute on Aging Working Group. 1998. Consensus report of the Working Group on molecular and biochemical markers of Alzheimer's disease. *Neurobiology of Aging* 19: 109-16.

Thompson, L. 2000. Human gene therapy harsh lessons, high hopes. *FDA Consumer Magazine* 34(5). Available at: http://www.fda.gov/fdac/features/2000/500_gene.html.

Tomossy, G.F., and D.N. Weisstub. 1997. The reform of adult guardianship laws: The case of non-therapeutic experimentation. *International Journal of Law and Psychiatry* 20: 113-39.

Verdun-Jones, S.N., and D.N. Weisstub. 1996-97. The regulation of biomedical research experimentation in Canada: Developing an effective apparatus for the implementation of ethical principles in a scientific milieu. *University of Ottawa Law Review* 28: 297-341.

Weisstub, D.N., ed. 1998. *Research on human subjects: Ethics, law and social policy.* Oxford: Elsevier Science.

Weisstub, D.N., and J. Arboleda-Flórez. 1998. Ethical research with vulnerable populations: The developmentally disabled. In *Research on human subjects: Ethics, law and social policy,* ed. D.N. Weisstub, 479-494. Oxford: Elsevier Science.

Weisstub, D.N., J. Arboleda-Flórez, and G.F. Tomossy. 1998. Establishing the boundaries of ethically permissible research with vulnerable populations. In *Research on human subjects: Ethics, law and social policy,* ed. D.N. Weisstub, 355-79. Oxford: Elsevier Science.

Weisstub, D.N, and S.N. Verdun-Jones. 1996-97. Pour une distinction entre l'expérimentation thérapeutique et l'expérimentation non thérapeutique. *Revue de droit de l'Université de Sherbrooke* 27: 49-87.

Weisstub, D.N., S.N. Verdun-Jones, and J. Walker. 1998. Biomedical experimentation involving elderly subjects: The need to balance limited, benevolent protection with recognition of a long history of autonomous decision-making. In *Research on human subjects: Ethics, law and social policy,* ed. D.N. Weisstub, 405-32. Oxford: Elsevier Science.

Winblad, B., H. Brodaty, S. Gauthier, J.C. Morris, J.M. Orgogozo, K. Rockwood, L. Schneider, M. Takeda, P. Tariot, and D. Wilkinson. 2001. Pharmacotherapy of Alzheimer's disease: Is there a need to redefine treatment success? *International Journal of Geriatric Psychiatry* 16(7): 653-66.

WMA (World Medical Association). 2000. *Recommendations guiding physicians in biomedical research involving human subjects (Declaration of Helsinki),* as amended by the 52nd World Medical Assembly, held in Edinburgh, Scotland, October 3-7. Available at: http://www.wma.net/e/policy/17-c_e.html.

WPA (World Psychiatric Association). 1996. *Madrid declaration on ethical standards for psychiatric practice.* (August). Available at: http://www.wpanet.org/generalinfo/ethic1.html.

Young, S.N., and S. Gauthier. 2000. Is placebo control in geriatric psychopharmacology clinical trials ethical? *International Journal of Geriatric Psychopharmacology* 2: 113-8.

Zarit, S.H., P.A. Todd, and J.M. Zarit. 1986. Subjective burden of husbands and wives as caregivers: a longitudinal study. *The Gerontologist* 26: 260-6.

INDEX

International Library of Ethics, Law, and the New Medicine

KLUWER ACADEMIC PUBLISHERS – DORDRECHT / BOSTON / LONDON

Printed in the United States
46477LVS00001B/112